Christian Physicalism?

Christian Physicalism?

Philosophical Theological Criticisms

Edited by
R. Keith Loftin and Joshua R. Farris

Foreword by
Thomas H. McCall

LEXINGTON BOOKS
Lanham • Boulder • New York • London

Published by Lexington Books
An imprint of The Rowman & Littlefield Publishing Group, Inc.
4501 Forbes Boulevard, Suite 200, Lanham, Maryland 20706
www.rowman.com

Unit A, Whitacre Mews, 26-34 Stannary Street, London SE11 4AB

British Library Cataloguing in Publication Information Available

Library of Congress Cataloging-in-Publication Data

Names: Loftin, R. Keith, 1981- editor.
Title: Christian physicalism? : philosophical theological criticisms / edited by R. Keith Loftin and Joshua R. Farris.
Description: Lanham : Lexington Books, 2017. | Includes bibliographical references and index.
Identifiers: LCCN 2017049612 (print) | LCCN 2017048252 (ebook) | ISBN 9781498549240 (electronic) | ISBN 9781498549233 (cloth : alk. paper)
Subjects: LCSH: Human body—Religious aspects—Christianity. | Theological anthropology—Christianity. | Materialism—Religious aspects—Christianity. | Matter—Religious aspects—Christianity. | Substance (Philosophy)
Classification: LCC BT741.3 (print) | LCC BT741.3 .C47 2017 (ebook) | DDC 233/.5—dc23
LC record available at https://lccn.loc.gov/2017049612

♾™ The paper used in this publication meets the minimum requirements of American National Standard for Information Sciences—Permanence of Paper for Printed Library Materials, ANSI/NISO Z39.48-1992.

Printed in the United States of America.

Keith: For Julie, my love.

Joshua: To John Cooper, for his spirited theological defenses of the soul and the intermediate state.

Contents

Foreword

The past few decades have witnessed some truly remarkable shifts in Christian views of the human person. Among the most important changes is the large-scale movement from various versions of mind-body dualism to the adoption and defense of various forms of physicalism. For centuries, Christian theologians—patristic, medieval, and modern, and Orthodox, Roman Catholic, and Protestant—have understood the mind or soul as essential to humanity. There have, of course, been significant disagreements among these theologians; some have thought that a human person *is* a soul that has a body (or at least usually or ideally has a body), others have thought that the human person *has* a soul. Various accounts of the mind-body relation have been proffered, and it would be a mistake to lump them all together—clearly, Origen's theory, for instance, is not to be confused with that of Thomas Aquinas. Despite the real and important differences, however, there is significant continuity within the broad Christian tradition (especially when seen in comparison to the recent developments).

Matters have changed significantly. Many Christian theologians and philosophers, often in dialogue with such widely diverse fields as philosophy of mind, biblical studies, cognitive science, and neuroscience, have begun to adopt and defend distinctly physicalist accounts of the human person. Arguments in favor of physicalism come from various angles; some make a case that the presumption in favor of dualism was really grounded in Greek philosophy and was always really unworthy of Christian theology, some argue that the case for dualism was based upon poor exegesis of Scripture, some go further and argue that Scripture actually supports physicalism, and some make the case that physicalism best accounts for the Christian hope of the resurrection *of the body*. Appeals are made to exegesis, science, and metaphysics; concerns are also raised about the ethical implications (either perceived or real) of the various versions of dualism. And some systematic and philosophical theologians are now taking the further steps of applying physicalist accounts of anthropology to other areas of Christian doctrine. Indeed, we have now reached the point where it is sometimes taken for granted that physicalism is the most fitting Christian view—or even the *only* truly Christian position.

But questions abound; questions about the alleged weaknesses of various forms of dualism, questions about the supposed strengths of physicalism, questions about the various levels and kinds of arguments for and against physicalism, and questions about the theological fecundity (or aridity) of physicalism. Does physicalism really offer practical advantages in moral theology and pastoral care (as is sometimes claimed)? Does physicalism offer a better interpretation of Scripture? Indeed, can it even offer an adequate understanding of biblical passages that seem to presuppose or teach dualism—how might the physicalist provide an appropriate account of Paul's statement about those who are "absent from the body but present with the Lord" (2 Cor 5:8)? Is physicalism any better off with respect to the problem of "interaction," or does any Christian physicalist who believes in divine agency not encounter this problem elsewhere? Can physicalism cohere with classical Christian commitments to the resurrection of the body (and the intermediate state)? What is the relation of physicalism to such important doctrines as original sin and sanctification? Is physicalism adequate for Christology, or does it run aground on the shoals of heterodox views? Clearly, much work remains—there is much to be done not only at the exegetical, historical-theological and theology-and-science levels but also with respect to the ongoing metaphysical and constructive-systematic theological arguments and issues. This volume takes up these challenges with sophistication, rigor, and a deep sense that these things really matter. The essays in this book push back against physicalism, and they push hard. Taken together, they present a significant challenge to the physicalist option, and the book offers a careful and spirited defense of dualism. All who take it seriously will be better off for having engaged with it. *Tolle lege.*

Thomas H. McCall
Trinity Evangelical Divinity School

Acknowledgments

A number of people have helped this project along in various ways, whether by providing feedback on chapters, encouragement, or helpful discussion. Our thanks to Paul Gould, Ross Inman, Travis Dickinson, S. Mark Hamilton, Jon Loose, Bruce Gordon, Jerry L. Walls, and Oliver Crisp for their help, advice, or encouragement concerning the present project. A special thanks goes to Charles Taliaferro for his advice and his stepping in to fill a need for the volume. Thanks to all the Christian philosophers who have devoted their lives to advancing substantive critiques of physicalism and defending the need for the soul. We have in mind several examples including John Cooper, Charles Taliaferro, Richard Swinburne, Howard Robinson, J. P. Moreland, Alvin Plantinga, Stewart Goetz, Daniel Robinson, Keith Yandell, and Dean Zimmerman, among others.

Many thanks to Sarah Craig for her editorial guidance from start to finish. We want to thank Julia Torres for her excellent service by offering careful proofs of the whole manuscript. Special thanks to the contributors who have given us many reasons for doubting the viability of Christian physicalism. The success of this volume goes to them.

Christian Physicalism?

An Introduction

Joshua R. Farris and R. Keith Loftin

What does it mean to be a human? What are humans? Are they souls, souls and bodies, or merely bodies and brains? These and other questions still confront us today. And these questions connect us to a longstanding tradition of reflection on what it means to be human. However, the dialectic has changed quite significantly from the belief that we are or have souls to the belief that we are wholly physical in nature. Such a change has impacted our perception not only on what it means to be human, but, to what extent we are connected to the animal kingdom and our profound connection to robots as seen in the growing transhumanist literature. With all that has changed in the developing portrait of humanity, there is something often missing in the anthropological discussions only theology can satisfy. Based on what follows, we are convinced that physicalism has very little support and that Christians should resist the trend to mold Christianity into its frame.

We take this one step further. We liken the recent attempts to bring Christianity and physicalism together to reflect the larger recalcitrant narrative characterizing the physicalism literature.[1] As children sometimes develop blind spots to the truth due to stubborn refusal to heed the parent's instruction, so it is with the physicalist. Characteristic of the physicalist literature is a stubborn refusal to carefully attend to the reality that there is more to the world than meets the physical eye. So it is with Christian Physicalism.

A BRIEF STATE OF THE ART: WHY PHYSICALISM FAILS?

Having advanced throughout the twentieth and into the twenty-first century, wholly physicalist accounts of human persons today are dominant within contemporary philosophy of mind. From Gilbert Ryle to Jaegwon Kim, physicalists agree that no explanations of mental phenomena require appeal to any nonphysical substance. Beyond this general agreement though a bevy of views are on offer, with versions of the constitution view and animalism gaining in popularity today. Within this context, especially within the past twenty-five years, there has appeared a growing body of literature arguing for the confluence of wholly materialist ontologies of human persons and Christian theological commitments.[2]

While the influence of Christian physicalism has increased in recent years, it does not at present dominate as the assumed ontology in biblical studies or theology. In some circles of biblical scholarship, especially in the literature on critical biblical studies, there is a common belief that some form of monism (often conflated with materialism) is the teaching of Scripture, or, at a minimum, the teaching that the collective voice of the Scriptural authors yields. We are seeing theologians who are not tightly bound to the catholic, Nicene tradition move in the direction of monism, if not materialism.[3] However, this is out of sync with the Church's recognition of the soul's persistence between somatic death and somatic resurrection. The dogmatic teaching of the Catholic Nicene tradition reflects the collective agreement of theological authorities that affirm with one voice the doctrine of the soul (that is, as an immaterial substance, or something near it) that has the possibility of persisting disembodied during the interim state.[4] Buttressed by the belief that Christ persisted as a human on Holy Saturday, it is arguable that he models what will occur for humans who die somatically prior to the somatic resurrection. While uncharacteristic of the Catholic Nicene tradition, many Roman Catholic, as well as reformed Catholics or Catholic reformed Christians, are moving in the direction of affirming some sort of monistic anthropology and rejecting dualistic anthropologies, especially as theological interpreters denounce the doctrine of the interim state between somatic death and somatic resurrection.[5]

In fact, it has been argued quite vigorously that the rejection of the interim disembodied state is a rejection of what is the natural and/or common interpretation of Scripture, evident in the Catholic Church. John Cooper points this out as a summary of his research:

> The historic position seems to be the most natural interpretation of the biblical text in its historical context. Scripture has been understood that way in ecumenical Christianity since the early church. It not only affirms the biblical emphasis on the unity of human life, but also accounts for its two-stage eschatology—personal existence between death and resurrection. It takes the biblical perspective as the framework for philosophical and scientific reflection on the human constitution. In addition, this anthropology shares with most of the world's religions the belief that embodiment is not necessary for the soul or consciousness—Islam, Hinduism, Buddhism, animism, and popular deism.[6]

Such a view, what Cooper calls two-stage eschatology, wherein the soul persists between death and resurrection, is a fixture in historic Catholic Christianity (found in all three traditions) and reflects the belief in a soul, or consciousness, as distinct from materiality, in all the major religions of the world.

Reflecting the tendency to downplay the soul, substantial dualisms, and the other-worldly message of the Bible, it is not uncommon to read biblical scholars elevating the this-wordly message over the other-worldly message with its attending reticence to affirm the doctrine of a soul. Famous and regarded biblical scholar N. T. Wright reflects this sort of trend to hold off commitment to the doctrine of the soul. While Wright affirms the two-stage eschatology countenanced so rigorously by John Cooper, Matthew Levering, and others, Wright is also reluctant to affirm the logical entailment that the doctrine of the intermediate state yields some kind of substantive dualism (where the soul is, at a minimum, separable as a substance weakly construed).[7] It seems the reticence is motivated by the desire to avoid any association with ancient dualisms or the nasty denigration of the body so commonly perceived in Cartesian dualism rather than a positive reason to affirm its denial.[8]

However, one need not affirm the two-stage eschatology to affirm the need for the soul and reject the doctrine of materialism. As seen in the pages that follow, one could affirm the need for a soul as that part that unifies the disparate material parts and binds them together in such a way as to allow for consciousness. Rather than affirm two-stage eschatology, one could affirm an immediate resurrection view, all the while still holding firm to the need for an immaterial part. This is seen in J. T. Turner's chapter in the pages that follow. While immediate resurrection has been commonplace for Christian materialists, and there are obvious motivations for a materialist to affirm immediate resurrection, Turner makes it clear that one need not be a materialist to affirm immediate resurrection.

A part of the trend away from variations of dualism toward materialism and/or monism has to do with the preposterous cluster of beliefs in the success of the physical sciences coupled with the belief that the scientific community assumes a physicalist worldview—whereby the world can be explained by biological and/or physical causes and effects. Theologies dominated by the biological sciences or brain sciences often reflect what so often dominates the scientific community, namely, the commitment to naturalism. Yet naturalism has no firm situation in a worldview presupposing supernatural agents and events (as in Christianity).[9]

Some theologians have bought into some, or many, of the trends and assumptions typically characteristic of the scientific community.[10] Taking her cues from the supposed scientific consensus on human nature, theologian Susan A. Ross constructively takes up Philip Clayton's emergentism as a way to avoid the challenges to theological views of humans. She says,

"Philosophical and especially theological understandings, to the extent that they are grounded in some nonphysical basis for the human, such as rationality or the soul, no longer have any credibility, at least for some scientists, in an era when we are increasingly able to take the measure of the whole person with observable data."[11] In the context of discussing neuroscience and its compatibility with religion and the soul, Aku Visala rejects a strictly naturalistic worldview and shows the compatibility neuroscience has with the soul and its relation to God. He states,

> Most theological anthropologies have maintained that humans are naturally and essentially open to non-natural realities, revelation, or the experience of God. Traditionally, John Calvin and Thomas Aquinas, for instance, maintained that humans have something like a basic, natural ability to know God. Paul Tillich, Karl Rahner and Wolfhart Pannenberg attempted to flesh out the implication of this with tools they often derived from European philosophies of their time. Similarly, late nineteenth century and early twentieth century theories in religious studies and sociology of religion emphasized universal religious experience and the fundamental social nature of religion, respectively. Some contemporary theologians, such as Robert Jenson and J. Wentzel van Huyssteen, have emphasized the close links between the emergence of religion, humanity, and ritual behavior. All these converging threads seem to point in the direction of religion and religiousness being deeply ingrained into human nature and human beings.[12]

Similarly, philosopher of psychology Daniel N. Robinson pushes against any physicalist attempt to naturalize religion in human nature. He states,

> Moreover, the biographical facts gleaned from the lives of persons of faith scarcely support the generalization that religion yields a calm and comforting perspective on oneself and one's world. Even expressing such a qualification grants too much explanatory power to what are finally neurochemical and neuroelectrical events in brain tissue. The brain has no motives and seeks no solace. That actual persons—possessed of brains and other anatomical structures—are, indeed, motivated and do, indeed, strive to find deeper meaning in an otherwise indifferent cosmos is beyond dispute. That such motives and longings are somehow enabled by the brain should be readily granted but not as a fact that would give the motives and longings to the brain or locate them in the brain. Such inferences might well trigger activity in the anterior cingulate cortex in any creature expecting propositions to be meaningful.[13]

The sciences, however, are not beholden to a physicalist ontology as the following chapters make clear. In fact, the sciences are inadequately assessed without self-awareness, something we suggest is rooted in an immaterial substance. The commitment to science does not necessitate a commitment to naturalism or the influence of naturalism and its effects on theology.

One common motivation in the biblical and theological literature toward a belief in materialism as the assumed ontological stance of the interpreter is the belief that the Scriptures yield a portrait of humans as unified, functionally integrated agents. Pushing against the Hellenistic influence of the Patristic and Medieval interpreters of Scripture, there has been a sense that ancient dualisms imposed an illicit or inappropriate dualism of parts that fractured the image of humanity rather than providing support for the unified picture found in the Bible.[14] Not uncommon to Old Testament scholarship is the assumption that humans are functionally integrated agents representing their creator. In fact, some Old Testament interpreters have taken this to mean that the Old Testament authors assume a monistic anthropology.[15] There are two important items worth noting about how this reasoning has gone wrong. First, it is important to note that while there is a common assumption that monism is synonymous with materialism, this is just simply not true. There are other variations of monism that fall outside of the materialist camp.[16] Second, it is worth noting that the Old Testament does not clearly yield a teaching of monism but rather functional holism. It is important to make this distinction because functional holism is a term representing the integrity of the human (contra some ancient dualisms) in practice as image bearers of God, but it does not necessitate ontological holism (i.e., monism). John Cooper has rigorously argued in several places against the monist assumption so often smuggled into biblical studies, and by default into theological studies.[17]

To date, the most substantive contributions to the Christian physicalist literature have come from philosophers, although key figures also include biblical and theological scholars. Nancey Murphy, herself a nonreductive physicalist, has taken perhaps the most overtly theological approach to Christian physicalism. In her "Human Nature: Historical, Scientific, and Religious Issues," Murphy argues that the history of theological anthropology has brought us to a point of decision between nonreductive physicalism and holistic dualism, substance dualism (which Murphy labels "radical dualism") being dismissed as not "compatible with Christian teaching."[18] Murphy's 2006 work, *Bodies and Souls, or Spirited Bodies?*, is a sustained argument that physicalism—over against dualism—provides the best fit with Christian commitments.[19] Warren Brown, though a neuroscientist and not a theologian, takes an approach similar to that of Murphy. Though not strictly speaking a work of *Christian* physicalism, Murphy and Brown's cowritten *Did My Neurons Make Me Do It? Philosophical and Neurobiological Perspectives on Moral Responsibility and Free Will* makes a case for how physicalists might preserve the notions of rationality, meaning, moral responsibility, and free will.[20] In his 2012 book, *The Physical Nature of Christian Life: Neuroscience, Psychology, and the Church* (cowritten with Brad Strawn), Brown rejects dualism (which, it is maintained, leads to Gnosticism) and reconsiders such theological matters as spiritual formation and the mission of the Church in physicalist terms.[21] Joel Green (a professor of New Testament studies) stands out among biblical scholars for his attempts to defend physicalist readings of Scripture, most notably in his *Body, Soul, and Human Life: The Nature of Humanity in the Bible*, by claiming that the exegetical task must be carried out with the neurosciences fully in view.[22] Green's "neuro-hermeneutic" is deployed to significant theological consequence in his work, notably in undercutting "the presumption of the centrality to biblical eschatology of a disembodied intermediate state."[23]

Among philosophers, the contributions of Lynne Rudder Baker have outstripped those of her fellow Christian physicalists. Beginning with her 1995 article "Need a Christian be a Mind/Body Dualist," Baker has argued "that what we now know about nature renders untenable the idea of a human person as consisting, even in part, of an immaterial soul."[24] Defending instead a version of the constitution view, Baker argues in *Persons and Bodies: A Constitution View* that one is a human in virtue of being constituted by a human body and one is a person in virtue (essentially) of having the capacity for the first-person perspective.[25] Christian physicalists tend to view the Christian doctrine of resurrection as a test case. Baker applies her constitution view to that doctrine in her 2001 article "Material Persons and the Doctrine of Resurrection"[26] as well as her 2007 article "Persons and the Metaphysics of Resurrection," wherein she argues the persistence conditions of the constitution view, according to which "sameness of pre- and postmortem person is sameness of first-person perspective," make constitution the most attractive option for Christians.[27] Kevin Corcoran also advocates the constitution view in his *Rethinking Human Nature: A Christian Materialist Alternative to the Soul*,[28] and his edited 2001 volume, *Soul, Body, and Survival: Essays on the Metaphysics of Human Persons*, remains among the most frequently cited works in the literature.[29]

Peter van Inwagen and Trenton Merricks have made notable contributions to the Christian physicalist literature, each endorsing the animalist view of human persons. As early as 1978, van Inwagen (characteristically) claimed to "have no idea" what sort of object is a "body," arguing that upon death God may well replace one's corpse (or at least the "core person") with a simulacrum in order to preserve one for future resurrection.[30] Later, in his "Dualism and Materialism: Jerusalem and Athens?," van Inwagen fortifies this position.[31] Offering physicalist-friendly readings of key biblical and creedal passages, van Inwagen seeks to decouple Christians from (their traditional) commitment to dualism. Doubtlessly van Inwagen's most momentous contribution, though, is his 2007 article "A Materialist Ontology of the Human Person," in which he rejects property dualism.[32] Trenton Merricks is well-known for denying there are any criteria for identity over time,[33] yet he insists on the resurrection of one's numerically identical body: "if you are not numerically identical with a person who exists in Heaven in the distant future, then you do not have immortality—so *bodily* identity is crucial to resurrection."[34] By his lights, physicalism can make the best sense of this, for "*life after death* and *resurrection* are, for physical organisms like us, *one and the same thing*."[35] Merricks's "How to Live Forever without Saving Your Soul: Physicalism and Immortality" focuses on the challenge(s) presented to his view by the fact that upon death human persons cease to exist and there is a temporal gap before their resurrection.[36] Any plausible personal ontology for Christian theology must also be coherently worked out with the essentials of Christian doctrine. Christian physicalists must attend to the full scope of doctrine in order to motivate a plausible defense for their view (e.g., sin, original sin, knowledge of God, salvation, sanctification). In what follows, the authors have attempted to explore some of these topics with physicalism in mind. The results are less than positive.

Despite reports to the contrary, the success of physicalism is overstated. The preponderance of evidence from Christian sources, rather than favoring physicalism, support some version of dualism with a view that an immaterial part is necessary to ground central Christian teachings.

CONCLUSION

What you have in your hands is a set of critiques against physicalism, generally, and the *supposed* compatibility of Christianity with physicalism, specifically. The integration of Christianity with physicalism has gained some prominence in recent years, as seen above. It is most evident in the philosophical literature, which includes the Christian philosophy of religion, but the impact of physicalism reverberates in biblical studies and theology, in some circles more than others. This is due in part to the overwhelming success of the sciences and the attending belief that science sits firmly in a physicalist worldview, where events are explicable by their underlying physical causes and effects or that organisms find their explanation in biological evolution. Another practical consideration is the common belief that dualism's effect on theology tends to bifurcate the person and denigrate the body. This is something we are loath to do, as reflected in virtually all contemporary Christian dualist defenses of human nature.

This trend toward physicalism is unhelpful and unmotivated, in our assessment. As shown here and in the chapters that follow, we are convinced that the motivations to affirm physicalism are actually quite thin and baseless. Christians who are committed to Nicene Catholic Christianity are nearly compelled to believe in the doctrine of the soul, however one may work that out (for example, as a hylomorphist, a Thomist, a Cartesian, a Berkeleyan). Short of calling Christian materialism a heresy, it is a deviation from the received wisdom of ecumenical Christianity. The Church has made plain the near universal agreement that some doctrine of immateriality is central to our confession of the *anthropos*.

Some Christians may not feel the compulsion to stand so close to the received ecumenical tradition of Christianity reflected in its three expressions (e.g., EO, RC, and Protestant Christianity). Let's assume one's theological method basically reduces to biblical exegesis and philosophical reasoning with a nod to the tradition. Even still, the present resource has something to contribute to these Christians. There remain good biblical reasons for rejecting physicalism, for instance, the need to account for the self-same person identified throughout salvation history and enduring from somatic death into immediate somatic resurrection. Even more, there remain overwhelming philosophical reasons for rejecting physicalism on the basis that it cannot account for the unity of consciousness (see J. P. Moreland's chapter), the enduring self, or the epistemology of religious experience (see Angus Menuge's chapter), among other theological considerations (such as the fact of Holy Saturday and Christ's persistence with human nature intact).

Here is a question for the reader: does even science favor physicalism? We have suggested some reasons above that it does not. Furthermore, if one of the hallmarks of physicalism is its underlying basis in physics, then, according to Bruce Gordon, physicalists don't even have physics on their side. If he is right, then physicalists lack any ground on which to stand. In the end, one may conclude that there are no good reasons to affirm physicalism. This is our conclusion. What is yours?

NOTES

1. J. P. Moreland helpfully uses this term to describe the physicalism movement in his, *The Recalcitrant Imago Dei: Human Persons and the Failure of Naturalism* (London: SCM Press, 2009).

2. For our purposes here, we use the terms "physicalism" and "materialism" interchangeably. This usage is not uncommon in the contemporary philosophy of mind literature. We recognize that some might want to distinguish the two terms. For example, if one is a Berkeleyan idealist, then one might refer to the physical as phenomenological products of the mind, yet use the term materialism to reference physical substantial ontology.

3. Michael Welker, *The Depth of the Human Person: A Multidisciplinary Approach* (Grand Rapids, MI: Eerdmans, 2014). In this multiauthored and multidisciplinary work, the authors collectively represent some of the trends toward some variant of monism, or the move to what might be considered a pluralist ontology. However, the authors, collectively, are quite disparaging of both reductive materialism or substantial dualisms. Otherwise, an excellent collection of essays.

4. See Matthew Levering, *Jesus and the Demise of Death: Resurrection, Afterlife, and the Fate of the Christian* (Waco, TX: Baylor University Press, 2012), 15–27. In fact, the separation of body and soul has been the dominant theological interpretation throughout Church history. This is reflected, for example, in the *Catechism of the Catholic Church* (Mahwah, NJ: Paulist Press, 1994). The argument may be that this is simply reflective of the Roman Catholic teaching, but it is important to note that EOs do not depart from this, as shown by Levering, and Reformed Christians did not often depart from RC on these standard interpretations of Scripture.

5. See Stephen Yates, *Between Death and Resurrection: A Critical Response to Recent Catholic Debate Concerning the Intermediate State* (New York: Bloomsbury, 2017). In this impressive study in analytic theology, Stephen Yates critically assesses the recent moves away from what is considered a dogmatic teaching of Roman Catholicism in the disembodied intermediate state to the affirmation of "immediate resurrection" view of the afterlife. He notes the correlation between the assumption of materialism and/or monism and the assumption of the immediate resurrection position and the resultant rejection of both disembodied intermediate state view with the rejection of the doctrine of the soul. He finds these moves problematic on dogmatic, theological, and biblical grounds and seeks to remotivate a case for the traditional dogmatic view of the soul's persistence disembodied during the intermediate state upon somatic death to somatic resurrection. While Yates is focused on contemporary Roman Catholic theology, his findings are relevant both to Eastern Orthodoxy and Reformed Christianity because of the emphasis on traditional theological teaching, which is collectively agreed upon in all three traditions. He also notes some of the trends reflected in Reformed Christianity toward monism and away from dualism.

6. See John Cooper, "Scripture and Philosophy on the Unity of Body and Soul," in *The Ashgate Research Companion to Theological Anthropology*, eds. Joshua R. Farris and Charles Taliaferro (Farnham, UK: Ashgate, 2015), 39.

7. See N. T. Wright, "Mind, Spirit, Soul and Body: All for One and One for All, Reflections on Paul's Anthropology in his Complex Contexts," http://ntwrightpage.com/2016/07/12/mind-spirit-soul-and-body/ [accessed on August 31, 2017].

8. For a positive defense and construction of substance dualism generally and Cartesianism, specifically, see Joshua R. Farris, *The Soul of Theological Anthropology: A Cartesian Exploration* (New York: Routledge, 2017). Farris advances the first constructive theological account of Cartesianism in the literature that is motivated by Scripture, dogma, and philosophical considerations.

9. "Scripture and Philosophy on the Unity of Body and Soul," 37. John Cooper carefully pushes this point.

10. See for example Malcolm Jeeves, ed., *From Cells to Souls and Beyond: Changing Portraits of Human Nature* (Grand Rapids, MI: Eerdmans, 2004). Taking their cues from what is considered the scientific consensus about the overwhelming shift toward a neurological and biological basis for human nature, the authors affirm either nonreductive physicalism or dual-aspectism. The doctrine of the soul or some version of substantial dualism is largely rejected (see page xii).

11. Susan A. Ross, *Anthropology, Engaging Theology: Catholic Perspectives* (Collegeville, MN: Liturgical Press, 2012), 145. Also see Hans Schwarz, *The Human Being: A Theological Anthropology* (Grand Rapids, MI: Eerdmans, 2013), 127. Theologian Hans Schwarz recognizes the common views of the scientific community and

how that has influenced theological construction. However, he does not see this as yielding materialism or monism. He does not succumb to the pressures of the scientific community, but, instead, is open to some variant of dualism or pluralism. Many have, rather than succumb to the pressure, turned to highlight a relational (albeit passive) ontology as the way to make sense of Scripture in light of the sciences. For one such example, see Ingolf U. Dalferth's excellent work, *Creatures of Possibility: The Theological Basis of Freedom* (Grand Rapids, MI: Baker Academic, 2016), 15, 20–21, 52. This move, we believe, misses the necessity of the substantial ground for relations.

12. See Aku Visla, "Theological Anthropology and the Cognitive Sciences," in *The Ashgate Research Companion to Theological Anthropology*, 70.

13. See Daniel N. Robinson, "Theological Anthropology and the Brain Sciences," in *The Ashgate Research Companion to Theological Anthropology*, 79. Robinson, himself, affirms some form of substance dualism, with its attending Cartesian intuitions. However, here, he is pushing against a common stance often held in the scientific community concerning the brain sciences, and arguing for the need of something that is not housed in the brain itself.

14. See Gerd Theissen, "Sarx, Soma, and the Transformative Pneuma: Personal Identity Endangered and Regained in Pauline Anthropology," in *The Depth of the Human Person*, 166–167. He explicitly points out holistic anthropology in Paul's anthropological view and rejects dualism as out of bounds in Paul's texts.

15. See Bruce Waltke, *Old Testament Theology* (Grand Rapids, MI: Eerdmans, 2007), 209–232. Waltke represents a move wherein some recognize the emphasis of holism in the Old Testament, but in the New Testament some interpreters recognize the need for a soul in Paul's anthropology (see 2 Corinthians 5). There are different ways of making this move. One could understand an actual change that occurs in the Scriptures, according to progressive revelation. Wherein the Old Testament view is just different from the New Testament view. Alternatively, one might seek to harmonize the views by way of highlighting not monism but holism as the integrative motif. Also, one could argue that the Old Testament does not yield a clear teaching on anthropology, at least not definitively. Joel Green makes a distinct, but important argument, that the Bible in no way gives us a need to affirm the soul. See Joel Green, "Why the Imago Dei Should Not Be Identified with the Soul," in *The Ashgate Research Companion to Theological Anthropology.*

16. We have in mind panpsychism generally speaking. We also have in mind a more specified version of panpsychism, namely, Russellian monism, advanced by the famous atheist philosopher Bertrand Russell.

17. See John Cooper, *Body, Soul and Life Everlasting: Biblical Anthropology and the Monism-Dualism Debate* (Grand Rapids, MI: Eerdmans, 2000), xxi–xxviii.

18. Nancey Murphy, "Human Nature: Historical, Scientific, and Religious Issues," in *Whatever Happened to the Soul? Scientific and Theological Portraits of Human Nature*, eds. Warren S. Brown, Nancey Murphy, and H. Newton Malony (Minneapolis, MN: Fortress Press, 1998), 25.

19. Nancey Murphy, *Bodies and Souls, or Spirited Bodies?* (New York: Cambridge University Press, 2006).

20. Nancey Murphy and Warren. S. Brown, *Did My Neurons Make Me Do It? Philosophical and Neurobiological Perspectives on Moral Responsibility and Free Will* (New York: Oxford University Press, 2007).

21. Warren S. Brown and Brad D. Strawn, *The Physical Nature of Christian Life: Neuroscience, Psychology, and the Church* (New York: Cambridge University Press, 2012).

22. Joel B. Green, *Body, Soul, and Human Life: The Nature of Humanity in the Bible* (Grand Rapids, MI: Baker Academic, 2008). See also his "'Bodies—That Is, Human Lives': A Re-Examination of Human Nature in the Bible," in *Whatever Happened to the Soul?*

23. Joel B. Green, "Eschatology and the Nature of Humans: A Reconsideration of Pertinent Biblical Evidence," *Science & Christian Belief* 14, no. 1 (2002): 50. See also Green's framing of the debate over the ontology of human persons in his "Body and Soul, Mind and Brain: Critical Issues," in *In Search of the Soul: Four Views of the Mind-Body Problem*, eds. Joel B. Green and Stuart L. Palmer (Downers Grove, IL: InterVarsity Press, 2005).

24. Lynne Rudder Baker, "Need a Christian be a Mind/Body Dualist?" *Faith and Philosophy* 12, no. 4 (1995): 502.

25. Lynne Rudder Baker, *Persons and Bodies: A Constitution View* (New York: Cambridge University Press, 2000). Baker summarizes her view and addresses some objections in "Materialism with a Human Face," in *Soul, Body, and Survival*, ed. Kevin Corcoran (Ithaca, NY: Cornell University Press) as well as in "Christian Materialism in a Scientific Age," *International Journal for Philosophy of Religion* 70, no. 1 (2011): 47–59.

26. Lynne Rudder Baker, "Material Persons and the Doctrine of Resurrection," *Faith and Philosophy* 18, no. 2 (2001): 151–167.

27. Lynne Rudder Baker, "Persons and the Metaphysics of Resurrection," *Religious Studies* 43, no. 3 (2007): 333–348. Baker argues strenuously that the first-person perspective cannot be accounted for by naturalism in *Naturalism and the First-Person Perspective* (New York: Oxford University Press, 2013).

28. Kevin Corcoran, *Rethinking Human Nature: A Christian Materialist Alternative to the Soul* (Grand Rapids, MI: Baker Books, 2006).

29. Kevin Corcoran, ed., *Soul, Body, and Survival: Essays on the Metaphysics of Human Persons* (Ithaca, NY: Cornell University Press).

30. Peter van Inwagen, "The Possibility of Resurrection," *International Journal for Philosophy of Religion* 9, no. 2 (1978): 114–121.

31. Peter van Inwagen, "Dualism and Materialism: Jerusalem and Athens?" *Faith and Philosophy* 12, no. 4 (1995): 475–488.

32. Peter van Inwagen, "A Materialist Ontology of the Human Person," in *Persons: Human and Divine*, eds. Peter van Inwagen and Dean Zimmerman (Oxford: Clarendon Press, 2007), 213–215.

33. Trenton Merricks, "There are no Criteria for Identity over Time," *Noûs* 32 (1998): 106–124.

34. Trenton Merricks, "The Resurrection of the Body and the Life Everlasting," in *Reason for the Hope Within*, ed. Michael J. Murray (Grand Rapids, MI: Eerdmans 1999), 268.

35. Merricks, "The Resurrection of the Body and the Life Everlasting," 283.

36. Trenton Merricks, "How to Live Forever without Saving Your Soul: Physicalism and Immortality," in *Soul, Body, and Survival*, ed. Corcoran.

Chapter 1

The Incorporeality of the Soul in Patristic Thought

Paul L. Gavrilyuk

Most premodern thinkers, both philosophers and theologians, could wonder what a soul was and how it was different from a body, but they did not doubt the soul's existence. In contrast, it is increasingly common in contemporary western scholarship to deny the existence of the soul and to assume that substance dualism has been definitively refuted. The thinkers of materialist persuasion commonly present caricatures of premodern views of the soul as convenient foils. Consider, for example, the following provocative statement by Daniel Dennett:

> One widespread tradition has it that we human beings are responsible agents, captains of our fate, *because* we really are *souls*, immaterial and immortal clumps of Godstuff that inhabit and control our material bodies rather like spectral puppeteers. It is our souls that are the source of meaning, and the locus of all our suffering, our joy, our glory and shame. But this idea of immaterial souls, capable of defying the laws of physics, has outlived its credibility thanks to the advance of the natural sciences.[1]

Dennett's "spectral puppeteer" is a variation on Gilbert Ryle's description of "Descartes' dream" as the "ghost-in-the-machine."[2] Such caricatures produce a false impression that all dualisms—Platonic, Gnostic, Cartesian, early Christian, medieval, and so on—are equally and obviously inadequate.

Some Christian philosophers have met this challenge by accommodating the metaphysics of personhood to a physicalist claim that "we are our bodies—there is no additional metaphysical element such as a mind or soul or spirit."[3] Such accommodation may take different forms, including "nonreductive physicalism," which denies dualism and asserts that consciousness and mental states are causally connected to matter, while trying to avoid the atheist and deterministic implications of materialism.[4]

In addition to philosophers, some leading twentieth-century theologians and biblical scholars have come to contrast the biblical view of humanity with Greek philosophical anthropology. This contrast has been drawn in stark terms by such an eminent authority as Karl Barth: "The Greek conception of the soul as a second and higher 'part,' as an imperishable, if possible preexistent, and in any case immortal spiritual substance of human reality, contrasted with the body as its lower and mortal part—the conception of the soul as a captive in the prison of its body, is quite unbiblical."[5] The Swiss theologian protests most strongly against the "abstract dualistic conception" of the "Greek picture of man" and issues the following verdict:

> In general, the character and result of this anthropology are marked by a separation of the soul over the body, a humiliation of the body under the soul, in which both really become not merely abstractions but in fact two "co-existing" figments—a picture in which probably no real man ever recognized himself, and with which one cannot possibly do justice to the biblical view and concept of man. It was disastrous that this picture of man could assert and maintain itself for so long as the Christian picture. We must earnestly protest that this is not the Christian picture.[6]

Barth's categorical judgment was generally shared by his contemporaries, including such an influential biblical scholar as Oscar Cullman. According to Cullman, "the Jewish and Christian interpretation of creation excludes the whole Greek dualism of body and soul."[7] This sharp dichotomy between the "Christian interpretation of creation" and "Greek dualism" is commonly based on the alleged contrast between the pagan Greek belief in the immortality of the soul and the Christian teaching concerning the resurrection of the dead. While Cullman's view has been nuanced in various ways, the general trajectory of his thought remains influential in contemporary biblical scholarship.[8]

This chapter offers an account of the Late Antique and early Christian anthropologies that will challenge a facile dichotomy between the "Christian interpretation of creation" and "Greek dualism." The immediate problem with using the expression "Greek dualism" is a failure to differentiate with sufficient clarity between ontological dualism postulating two independent sources of good and evil, and anthropological dualism, which takes human beings to be soul-body composites. Those scholars who intend only the anthropological application of "Greek dualism" often unintentionally evoke the ontological dualism, too. This is a serious problem, as most Christian authors who accepted soul-body dichotomy at the same time rejected ontological dualism. Moreover, later Platonists, who were anthropological dualists, were also quite critical of ontological dualism of the kind propounded by the Gnostics. Hence, the presumption that both dualisms usually appear in tandem is a serious confusion of terms that clouds the study of late antiquity.

I will further problematize the concept of anthropological dualism by demonstrating that there was no consensus among ancient Greek and Hellenistic philosophical schools on the metaphysics of personhood. I will subsequently show that a widespread view that early Christian theologians uncritically accepted Platonic anthropological dualism misrepresents the tradition. In fact, the nature of the soul remained a contested issue among Christian theologians well into the fifth century and beyond. As I will illustrate by considering the work of Origen of Alexandria, Gregory of Nyssa, and Augustine of Hippo, early Christian thinkers had general philosophical as well as particular theological reasons for holding that the soul was irreducibly immaterial.

THE SOUL IN GREEK PHILOSOPHICAL THOUGHT

Ancient Greek thought exhibits a broad range of opinion regarding the nature and powers of the soul (ψυχή). In Homer, the *psyche* is associated with breath, which leaves the body at death.[9] Upon separating from the body, the *psyche* continues to lead a shadowlike existence in the underworld (Hades) with some of its powers diminished. Since the living body moves and the dead body is motionless, the *psyche* is also responsible for movement. Some ancient Greeks went so far as to attribute the *psyche* to everything that moves or is movable, including stars, rivers, and stones.[10] The view that all things are living or ensouled (ἔμψυχος) is commonly designated as "vitalism," or more precisely, "panpsychism." Prephilosophical vitalism assumed that the soul was a body of a special kind and had conceptual difficulties with articulating incorporeality (ἀσωματία, τὸ ἀσωματὸν).

In the Greek culture, the archaic and prephilosophical views of *psyche* continued to circulate side by side with the later philosophical attempts to define the nature and powers of the *psyche* more precisely. In *On the Soul*, Aristotle critically sifts through the opinions of his philosophical predecessors before carving out his own position. The vast majority of the pre-Platonic thinkers conceived of the soul in physical terms: Hippo associated the *psyche* with water and seed, Critias with blood, and Diogenes with air.[11] According to Aristotle, the interesting thing about air is that it "seems to be ἀσώματοσ," yet on closer inspection it is not. Still, air is "thinner and more incorporeal than water," which is to say that it is made of a less dense substance.[12] In non-Christian Greek sources, the term ἀσώματοσ is at times used loosely as a relative concept denoting a substance that is more refined than the corporeal substance to which it is compared. As discussed later in this chapter, the idea of relative incorporeality of the soul in relation to the body is not foreign to patristic thought, beginning with Irenaeus of Lyons.

Aristotle also noted that the *psyche* was often associated with such functions as movement, sensation, affectivity, imagination, and thinking.[13] Pythagoras ascribed the *psyche* to humans and animals, viewing it as a center of sensations, desires, and emotions, but not of thoughts. He located the *psyche* in the heart, and taught that it migrated from body to body, and in this sense was deathless (ἀθάνατοσ). While professing the *psyche*'s immortality, Pythagoras associated the soul with the "particles in the air," which would seem to entail that he agreed with the other pre-Socratics that the soul was corporeal.[14]

In contrast to Pythagoras, Democritus identified the *psyche* with the mind (νοῦσ), holding that it was made of fire or heat and consisted of atoms. Having discussed Democritus's atomism and its limitations, Aristotle makes the following foundational observation:

> As to the nature and number of the first principles opinions differ. The difference is greatest between those who regard them as corporeal and those who regard them as incorporeal (οἱ σωματικὰς ποιοῦντες τοῖς ἀσωμάτους), and from both dissent those who make a blend and draw their principles from both sources. The number of principles is also in dispute; some admit one only, others assert several. There is a consequent diversity in their several accounts of soul.[15]

Aristotle postulates a correspondence between the first principles of a given philosophical system and the nature of the soul, that is, between ontology and psychology. For example, it was natural for an atomist, such as Democritus, to hold that the soul consists of atoms. The clearest statement that the soul is corporeal comes from Epicurus:

> The *psyche* is a fine-structured body diffused through the whole aggregate, most strongly resembling wind with a certain blending of heat, and resembling wind in some respects but heat in others. But there is that part which differs greatly also from wind and heat themselves in its fitness of structure, a fact which makes it the more liable to co-affection with the rest of the aggregate.[16]

According to Epicurus, the soul does not survive the dissolution of the body. Epicurean materialism is the closest ancient Greek analogy to contemporary reductive physicalism, that is, the view that only bodies exist and that all mental properties are reducible to underlying physical properties.

Similar to the Epicureans, the Stoics understood the *psyche* to be corporeal. This view followed from their understanding of God as a both corporeal and intelligent spirit (πνεῦμα) that suffuses and animates the material universe.

However, Stoic pantheistic corporealism was not as straightforwardly reductionist as Epicurean atomism. According to A. A. Long, "the Stoic God is not a body primarily and a mind secondarily. It is his nature to be an intelligent body."[17] Just what this meant the Stoics did not always make clear, for they denied the existence of incorporeal substances. On their account, the soul, being corporeal, does not survive for long after the dissolution of the body. Corporealism was also a broadly shared assumption of ancient Greek and Hellenistic medicine.

It seems that prior to Plato, the concept of incorporeality (τὸ ἀσωματὸν) remained foreign to Greek philosophers.[18] In *Sophist* 246a-b, Plato announces a break with the preceding philosophical tradition, especially with the ancient proponents of reductive physicalism:

> Some of them drag down everything from heaven and the invisible to earth, actually grasping rocks and trees with their hands; for they lay their hands on all such things and maintain stoutly that that alone exists which can be touched and handled; for they define existence and body, or matter, as identical, and if anyone says that anything else, which has no body, exists (μὴ σῶμα ἔχον εἶναι), they despise him utterly, and will not listen to any other theory than their own. . . .
>
> Therefore those who contend against them defend themselves very cautiously with weapons derived from the invisible world above, maintaining forcibly that real existence consists of certain ideas which are only conceived by the mind and have no body.[19]

It is possible that the term ἀσώματοσ was of Plato's coinage.[20] Plato's conjoining of intelligible entities (νοητὰ) with incorporeal entities (ἀσώματα) was a conceptual and metaphysical breakthrough. If prior to Plato the difference between soul and body seems to have been that of rarer versus denser corporeality, Plato understands the difference to be qualitative: the mental power of the soul is incorporeal.[21] The incorporeal soul is simple and self-moving and, therefore, immortal by nature.[22] These teachings were defended by later Platonists and their followers.

Even a brief overview of the Greek understandings of the nature of the soul demonstrates with a great degree of historical probability that a significant number of philosophers favored a position that the soul was in some sense corporeal, although they differed widely with regard to how the soul's nature was to be understood and its powers described. In the philosophical market of ideas, Platonic dualism of the incorporeal soul joined to the body represented a departure from this widespread view, not a standard position. Therefore, it is quite inaccurate and misleading to imagine, as Oscar Cullman does, early Christian theologians as confronting "Greek dualism" *tout court*, since no such monolithic philosophical view was on offer in late antiquity. In fact, a prevailing position was not dualism of any sort, but rather different versions of corporealism and materialism, from lingering traditional folk beliefs about soul-breath to more refined philosophical views, especially those associated with Hellenistic and Roman Stoicism. While we do not know the particulars of the Apostle Paul's debate with the Epicurean and Stoic philosophers in the Areopagus, the resurrection of the dead would have been an entirely plausible target of philosophical "scoffing" and questioning (Acts 17:18, 32).

Early patristic authors were aware of the wide discrepancy of philosophical opinions on the matter of the soul's nature and its powers.[23] To represent early Christian theologians as being mindlessly swayed by "Greek dualism" is to ignore much of the available evidence. In fact, as will be shown later, a view that the soul possesses some kind of corporeality continued to have supporters from the second century to the fifth century. The teaching that the soul was incorporeal, which gradually gained broad acceptance in the church, was as much a result of internal debates among Christian theologians as a product of selective repurposing of valid Platonic insights with considerable metaphysical modifications.

PATRISTIC AUTHORS BEFORE ORIGEN

Origen of Alexandria considers the question of the incorporeality of God and the soul repeatedly in his extant writings. In the preface to his magnum opus *On First Principles*, he registers the fact that in the first part of the third century the matter was far from settled: "The term ἀσώματον, that is, incorporeal, is unused and unknown, not only in many other writings but also in our scriptures."[24] Having set aside an imprecise use of ἀσώματον in the sense of being "thin like air," Origen states,

> We shall inquire whether the actual thing which Greek philosophers call ἀσώματον or incorporeal is found in the holy scriptures under another name. We must also seek to discover how God himself is to be conceived, whether as corporeal and fashioned in some shape, or as being of a different nature from bodies, a point which is not clearly set forth in the teaching. The same inquiry must be made in regard to Christ and the Holy Spirit, and indeed in regard to every soul and every rational nature also.[25]

The Alexandrian theologian acknowledges that in the absence of an explicit scriptural teaching, the matter of God's and the soul's (in)corporeality is fraught with difficulties and is bound to cause controversy. If we are to follow Aristotle's dictum that the account of the nature of the *psyche* must follow the account of the investigation of the first principles, we could treat the conjunction of propositions about the (in)corporeality of God and of the soul as two parts of one metaphysical problem. Such a conjunction generates four logically possible conclusions:

1. God is corporeal, the soul is incorporeal.
2. Both God and the soul are corporeal.
3. God is incorporeal, the soul is corporeal.
4. Both God and the soul are incorporeal.

We may set aside the first proposition as something no one was prepared to bet any metaphysical money on. The second proposition was held by some of Origen's contemporaries and, in a more philosophically rigorous way, was defended by Tertullian. As we shall see, the third proposition held a certain attraction for Origen and other early patristic authors. In the end, the fourth proposition came to be well-established in patristic theology for reasons that we shall explore more closely by considering the thought of Gregory of Nyssa and Augustine.

The second proposition, that God and the soul are corporeal, found supporters among the uneducated as well as the philosophically trained. In Origen's time, there were simple believers in Alexandria, who took the biblical references to divine eyes, hands, feet, bowels, and so on, literally, reaching the conclusion that God has a body. Such a conclusion was unsurprising in a polytheistic milieu in which gods were expected to be anthropomorphic.

Setting aside anthropomorphic conceptions of God, Tertullian offers a more philosophically refined defense of the second proposition. In order to combat the Docetic rejection of the reality of Christ's flesh, Tertullian leans on Stoic corporealism, which postulates that all beings, including God and the soul, are in some sense corporeal. He states his fundamental metaphysical presupposition as follows: "Everything that exists is a body of some kind or another. Nothing is incorporeal except what does not exist."[26]

Tertullian treats the nature of the soul more extensively in his *On the Soul*, the first systematic Christian presentation of the subject. After noting disagreement among the philosophers on the issue of the soul's corporeality and criticizing the Platonists with considerable rhetorical flair, Tertullian draws on the following three arguments of the Stoics in support of the soul's corporeality: (1) the spirit or soul that leaves the body at death must be corporeal in order to make a difference to the body's corporeal state; (2) the fact that children inherit not only bodily traits but also character traits from their parents indicates that this resemblance is transmitted from body to body and from soul to soul in a corporeal fashion; (3) the soul-body interaction or "sympathy" is only possible on the assumption that the soul has a nature akin to the body, but there is nothing in common between the corporeal and incorporeal nature.[27] Tertullian subsequently proceeds to dismantle the arguments of the Platonists for the soul's incorporeality. What is important to emphasize is that Tertullian does not see himself as falling into or being bound by "Greek dualism," because such a view was by no means a default option either in philosophical circles or in a wider culture. On the contrary, he knows that the majority opinion favors corporealism and supports it by rehearsing select Stoic arguments.

In addition, Tertullian adduces scriptural evidence in support of his view. He observes that the description of the eschatological state in the story of the rich man and Lazarus in Luke 19:16–31 includes physical torments, such as fire and thirst, implying embodiment.[28] Tertullian's central proof-text is Gen 2:7: "Then the Lord God formed man from the dust of the ground, and breathed into his nostrils the breath of life (LXX: πνοὴν ζωῆς) and the man became a living being (LXX: ἐγένετο ὁ ἄνθρωπος εἰς ψυχὴν ζῶσαν)." The African theologian interprets this passage to mean that the soul was formed "by the breathing of God and not out of matter," and that the soul has an origin: it is made by God.[29] Tertullian contrasts his view with the Platonic tenet that the soul is unborn or unmade. According to Tertullian, God is a unique uncreated corporeal spirit who forms the corporeal soul by spiration, out of his breath. The soul is a simple corporeal spirit that is different in kind from God's corporeal spirit, just as it is different in kind from the bodies that are not souls. Whatever one makes of Tertullian's metaphysical explorations, it is not an obvious ally of contemporary physicalism, whether overtly materialist and atheist, or emergentist. This is primarily because Tertullian shares the fundamental assumption of the second-century apologists that God is the creator of everything, including matter, and therefore transcends matter.

It was the ontological distinction between the uncreated God and creation that drove the early Christian understanding of God as immaterial. For example, the second-century apologist Tatian wrote, "God is spirit. He does not extend through matter, but is the author of material spirits and of the figures in matter. He is invisible and intangible."[30] Addressing an imaginary audience of Greek philosophers more directly, Tatian adds, "One of you asserts that God is body, but I assert that he is without body."[31] According to Tatian, God is an uncreated incorporeal spirit, who is not limited by space or matter. God is the creator of everything in the world, including matter and material spirits. Tatian's position appears to be close to the third view, namely that while God is incorporeal, the soul and spirit are corporeal.

This seems to have been the view of Justin Martyr, as well. In his *Dialogue with Trypho*, Justin narrates his spiritual pilgrimage from one philosophical school to another, culminating in a period of adherence to Platonism before his eventual conversion to Christianity. Justin understands the later Platonists to be teaching that the soul is unbegotten, immortal, and immaterial because it is a part of the divine mind.[32] In contrast to Platonism, Christianity teaches that God is the creator of everything, including the soul. It follows that the soul is created, has a beginning in time, depends on God, and has life by participation in the divine life, not by nature.[33] This foundational difference between Christian and Platonist accounts of the soul, first recorded by Justin, would become commonplace in patristic literature.

More peculiar to Justin is his understanding, possibly influenced by Stoicism, that immaterial entities are devoid of sensation. If the soul were immaterial, reasons Justin, it would not be susceptible to divine punishment after death. Such a conclusion goes against the Christian teaching concerning the last things. While Justin does not deny explicitly that the soul is immaterial, the logic of his argument would seem to favor the soul's corporeality over incorporeality.

Irenaeus of Lyons is also somewhat ambiguous on the subject of the soul's (in)corporeality. In *Against Heresies*, he claims that "souls are incorporeal when put in comparison with mortal bodies (*incorporales animae, quantum ad comparationem mortalium corporum*); for God breathed into the face of man the breath of life, and man became a living soul. Now the breath of life is an incorporeal thing. And certainly they cannot maintain that the very breath of life is mortal."[34] Irenaeus could be interpreted as claiming that both the divine breath of Genesis 2:7 and the soul are incorporeal, with the difference that the soul's existence and all of its properties, including immortality and incorporeality, depend on God. Alternatively, he could be interpreted as claiming that the soul is incorporeal relative to the "mortal body," which would not exclude a possibility that the soul is a body of a different, more refined sort. In other words, Irenaeus could follow earlier thinkers in understanding corporeality and incorporeality not as two discrete states, but as properties that admit of degrees. As we are about to see, such an understanding of incorporeality was present in Origen's thought as well.

ORIGEN OF ALEXANDRIA

"We have learned from the holy Scripture that the human being is a composite," observes Origen, musing over 1 Thessalonians 5:23. "May your spirit and soul and body be kept sound and blameless at the coming of our Lord Jesus Christ."[35] Origen views human nature as a hierarchical system, in which the spirit corresponds to the highest aspect, the soul corresponds to the intermediate aspect, and the flesh corresponds to the lowest aspect of the self. As an intermediate entity, the soul is capable of aligning itself either with the higher or with the lower aspects of the self. Origen conceives of the soul as a dynamic entity, which becomes mind when it draws closer to God, and becomes dull-witted and entangled with the body when its desire for God cools.[36] Under the influence of Middle Platonism, Origen hypothesized that the souls preexisted their embodied state and were assigned to bodies according to the degree of their voluntary separation from God; he even entertained the possibility of transmigration. He emphasized the speculative and exploratory character of his views and did not expect them to be accepted as authoritative Christian teaching. More generally, he conceded that "the subject of the soul is a wide one, and hard to be unraveled, and it has to be picked out of scattered expressions of Scripture."[37]

Concerning the nature of the soul and its relation to the body, Origen maintains: "All souls and all rational natures, whether holy or wicked, were made or created. All these are incorporeal with respect of their proper nature, but though incorporeal they were nevertheless made."[38] By insisting that the soul is created, Origen clearly differentiates his view concerning the nature of the soul from that of the Platonists, who taught that the soul's incorporeality implied it was by nature eternal and divine. As far as the soul's mode of existence is concerned, Origen further specified that "the soul, which in its own nature is incorporeal and invisible, is in any material place, it requires a body suited to the nature of that environment."[39] While immaterial by nature, the soul owes its spatial location and the visible manifestation of its agency to bodies of different kinds.

More problematically, Origen also repeatedly asserts that incorporeality in the strictest sense belongs to God alone. For example, in a chapter preceding the earlier assertion about the incorporeality of the soul's "proper nature," he writes, "We believe that to exist without material substance and apart from any association with a bodily element is a thing that belongs only to the nature of God, that is, of the Father, the Son and the Holy Spirit." He later repeats that "life without a body is found in the Trinity alone."[40] For Origen, this assertion served as a safeguard against both the anthropomorphism of simple believers and the philosophical corporealism of the Stoics, which influenced such early Christian theologians as Tertullian.[41]

Origen's exposition is marked by a tension between the assertion that all rational beings are incorporeal in their "proper nature," and the assertion that incorporeality is an exclusively divine property. One possible way of dealing with this tension is to question the reliability of Rufinus's translation. While Rufinus's tampering with *On First Principles* cannot be ruled out, in this particular case his theological motivation for reserving incorporeality for God alone would remain inexplicable in the context of the late fourth century. If we accept Rufinus's translation as a more or less faithful rendering of Origen, then Origen might be interpreted as claiming that while God is incorporeal in the sense of not being limited by either body or anything else, the soul is incorporeal, but its existence is marked by its association with the body.

As Brian Daley explains, Origen "shared the Platonic notion that every finite intellectual being needs to make use of some kind of body as the 'vehicle' (ὄχημα) or instrument of its motion and activity, even though its own nature is incorporeal."[42] This interpretation has the distinct advantage of relieving the tension between the claim that incorporeality uniquely distinguishes God and at the same time characterizes the soul's proper nature.

While Origen does incorporate some Platonic elements into his system, he also departs from Platonism when he denies that the soul is divine and asserts, following an already established tradition, the soul's created nature. For him, the soul is a substance "intermediate between God and the flesh"[43] and a "medium between the weak flesh and the willing spirit."[44] When the soul's cognitive powers are directed to embodied things, its association with the body becomes stronger; when its cognitive powers are directed toward the divine reality, its rational and incorporeal nature becomes more fully manifest. When the soul's love for God cools, it draws away from God and as a result becomes less spiritual and more preoccupied with the body. Conversely, when the soul draws near to God, it becomes progressively more spiritual and rational.[45]

Origen is the first Christian theologian to offer both generally philosophical and specifically theological reasons for the soul's incorporeal element. General philosophical reasons are marshaled in the form of rhetorical questions in the first chapter of *On First Principles*, indicating the overall importance of the subject to Origen's theological system:

> If there are any who consider the mind itself and the soul to be a body, I should like them to tell me how it can take in reasons and arguments relating to questions of great importance, full of difficulty and subtlety. Whence comes it that the power of memory, the contemplation of invisible things, yes, and the perception of incorporeal things reside in a body? How does a bodily nature investigate the teachings of the arts and the meanings and reasons of things? And divine doctrines, which are obviously incorporeal, how can it discern and understand them?[46]

According to Origen, the powers of the mind and the rational powers of the soul cannot be corporeal because they operate on incorporeal entities, such as memories, abstract concepts, causal explanations, theological truths, and so on. That which is capable of handling incorporeal entities cannot be purely corporeal.

Origen subsequently argues that just as the bodily senses operate on the properties of material things, such as color, shape, size, and smell, so the mind operates on immaterial objects, such as abstract concepts, reasons, and ideas about God. "To see and to be seen," says Origen, "is a property of bodies; to know and to be known is an attribute of intellectual existence."[47] It is not sufficient for the mind to be an accident or epiphenomenon of the body, because the mind has its distinct incorporeal sphere of operation.

For Origen, this general philosophical argument also has a more specific theological implication. Since God is incorporeal, "there is a certain affinity between the mind and God, of whom the mind is an intellectual image, and that by reason of this fact the mind, especially if it is purified and separated from bodily matter, is able to have some perception of the divine nature."[48] Unlike the Platonists, Origen did not hold that the mind was divine by nature. His point is rather that the mind or rational soul must be sufficiently like God in order to be able to have some understanding of God, for like is known by like.

Origen speculates that since God is immaterial, "in creation, therefore, the human being first created was the one *in the image* (Gen. 1:26) in whom is nothing material. For what is made *in the image* is not made from matter."[49] According to Origen, the first two chapters of Genesis teach about two stages in the creation of humanity: Genesis 1:26 refers to the creation of the incorporeal "inner man," capable of communicating with the incorporeal God; Genesis 2:7 refers to the formation of the "outer man," who is embodied, illustrated by God's "taking dust from the ground." For Origen, only the Son of God, the eternal Logos, is the "image of the invisible God" (Col. 1:15), whereas humans are made "in the image" (Gen. 1:26), with the preposition "in" marking their ontological difference from God.[50] The aspect of the human being that is made in the image of God is the rational soul in its original purity and alignment with God's Spirit:

> [T]hat which is made in the image of God is to be understood of the inward man, as we call it, which is renewed and has the power to be formed in the image of the Creator, when a man becomes perfect as his heavenly Father is perfect, and when he hears "Be holy because I the Lord your God am holy," and when he learns the saying "Become imitators of God" and assumes into his own virtuous soul the characteristics of God. Then also the body of the man who has assumed the characteristics of God, in that part which is made in the image of God, is a temple, since he possesses a soul of this character and has God in his soul because of that which is in His image.[51]

Origen understands the spiritual condition of being "in the image" as dynamic and contingent on the soul's obedience to the divine commandments. When the soul is voluntarily aligned with the spirit of God, its body becomes instrumental in the process. When, on the contrary, the soul chooses to draw away from God, it no longer fully functions as that which is made "in the image," a condition that may even result in death.

In *Dialogue with Heraclides*, when challenged to answer the question of whether the soul is immortal, Origen replies by distinguishing three kinds of death of the soul: "death to sin" (Rom. 6:10), "death to God" (cf. Ezek. 18:4), and physical death as a separation from the body. Since souls survive physical death, they are "immortal" in the sense of enduring after the decomposition of the body and being subject to divine judgment. The soul can be subject to the spiritual death to God only to the extent that it sins. As for the "death to sin," it is a state in which the soul "becomes confirmed in blessedness so that it is inaccessible to death, in possessing eternal life it is no longer mortal but has become, according to this meaning too, immortal."[52] Although God "alone has immortality" (1 Tim. 6:16) by nature, humans may become free participants in God's eternal life. Origen consistently holds that the soul's Godlike properties, such as incorporeality and immortality, depend upon its orientation toward God. In the resurrected state, souls are likely to be joined to ethereal bodies of "heavenly purity and clearness," although it is also possible, speculates Origen, that souls will exist in a bodiless state.[53]

In conclusion, it is clear that Origen does not espouse "Greek dualism" *tout court.* He rejects Stoic corporealism for philosophical reasons and the Platonic view of the soul as divine for theological reasons. His partial and critical adoption of certain elements of Platonist psychology is controlled by the fundamental distinction of early Christian theology between the uncreated and incorporeal God and everything else in creation. Origen's position is closer to the fourth conjunction, that both God and soul (in its proper nature or highest state) are incorporeal, than to the third conjunction, that God is incorporeal while the soul is not. It must be emphasized that the Alexandrian theologian understands the soul as a dynamic medium between the invisible, intelligible, and incorporeal realm of the spirit and the visible, sensible, and corporeal realm of the flesh. The individual soul's share of incorporeality depends upon its closeness to God.

GREGORY OF NYSSA

Origen's teachings concerning the nature of God and the soul were widely circulated and debated in later centuries. To be fair, Origen did not insist that his anthropological speculations should be hardened into a dogma, but intended them as exploratory. Among the theologians who continued these explorations, an eminent place belongs to the fourth-century Cappadocian bishop Gregory of Nyssa. Two works of Gregory, *On the Making of Man* and *On the Soul and the Resurrection*, exemplify a discriminating reception of Origen's speculations, as well as a rigorous engagement of philosophical sources, especially Plato and the later Platonists, in light of scripture. Gregory argues for the fourth conjunction that both God and the soul are incorporeal.

On the Soul and the Resurrection begins with an observation that the materialist argument against the existence of the soul—that the soul is not available to external observation and therefore does not exist—also applies to God. Human beliefs in the existence of God and the soul are closely related (a point that was not lost either on René Descartes, or more recently, on Alvin Plantinga[54]). In response to the materialist challenge, Gregory formulates a version of an argument from design: the order, beauty, attunement, and harmony of an otherwise mindless universe seem to point beyond themselves. These features could be taken as evidence of the existence of a single divine power that accounts for the motion and order of various parts. The fact that God is not available to sense-perception does not make him any less real in the universe. Gregory then observes that each human being is a microcosm, whose soul functions in the body in a way similar to God's agency in the world. Like God, the soul is incorporeal (ἀσώματος).[55] One argument for the soul's immateriality is its ability to reason from known features (design in the world) to unknown entities (the existence of a divine designer), which themselves are not objects of sense-perception.

The soul resembles God in being "intellectual, incorporeal, unconnected with any notion of weight, and in eluding any measurement of its dimensions."[56] For this reason, it would be an exercise in futility to try to locate the soul in a particular part of the body, whether heart, brain, or liver. According to Gregory, the soul "is not restricted to any part of the body, but is equally in touch with the whole, producing its motion according to the nature of the part which is under its influence."[57] The soul is not spatially contained in the body and vice versa.[58] The soul communicates its "vivifying energy" (ζωτικὴ ἐνέργεια)[59] to the body, but the precise nature of soul-body interaction is beyond human understanding. The soul is omnipresent and invisible in the body in a manner similar to God's omnipresence and invisibility in creation. The unity of the soul is not broken by the multiplicity of its powers; similarly, God's simplicity is not undermined by the plurality of his operations in creation.

Another important feature that connects the rational soul or mind to God is free will.[60] The mind, "as being in the image of the most beautiful, itself also remains in beauty and goodness so long as it partakes as far as possible in its likeness to the archetype; but if it were at all to depart from this it is deprived of that beauty which it was."[61] Similar to Origen's view, human likeness to God is a dynamic rather than permanent feature of human existence, crucially depending upon the measure of voluntary participation in God. Rational creatures are capable of losing divine likeness if they choose against God.

The soul is similar, but not identical, to God. The soul is created, whereas God is the creator of all things. The creation of all things out of nothing presents the following difficulty: how can an immaterial God be the creator of material things? Gregory asks, are material things in some sense contained in

God? He answers that material things are "in" God not in any spatial sense, for this would make God himself material, but in the sense that God has the power to create them. Gregory suggests that the creation of any particular body is a convergence of different properties, such as color, shape, and so on. Since these properties before instantiation are not bodies, it follows that the creation of bodies is a convergence of intelligible or nonmaterial properties. The precise mechanism of this idealist scheme of creation remains somewhat mysterious.[62]

The soul does not create its own body. *Pace* non-Christian Platonists and Origen, Gregory denies both the preexistence and the transmigration of souls.[63] The soul and body are created at the same time.[64] Gregory offers the following working definition of the soul: "The soul is an essence which has a beginning; it is a living and intellectual essence which by itself gives to the organic and sensory body the power of life and reception of sense-impressions as long as the nature which can receive these remains in existence."[65] Gregory does not object to the Aristotelian taxonomy of the vegetative soul of plants, sensitive soul of animals, and rational soul of humans, and even finds this classification partially justifiable on scriptural grounds.[66] Nevertheless, only humans possess "the true and perfect soul," whereas plants and animals have "vital energy" rather than the soul proper.[67] Such a narrowing of the definition, for which there were already some precedents in the Greek philosophical tradition, had the advantage of making the soul something distinctly human and blocking possible speculation about the exchange of souls between humans and animals. At the general resurrection, each human soul reconstitutes its body by attracting the atoms scattered after the body's dissolution.[68]

It is clear that Gregory of Nyssa was not an uncritical recipient of Origenist and Plotinian thought. Crucially, he departs from Origen and late Platonism in rejecting the preexistence and the fall of the soul. Like Origen, he insists on the created and changeable nature of the soul. The soul's incorporeality is important in order to safeguard its freedom and its ability to operate with nonphysical entities, such as abstract concepts and the idea of God. The soul's presence in the body parallels the omnipresence of God in creation. If Gregory accepts this particular Plotinian trope, it is only because it was useful for conveying the biblical understanding of the immanence and transcendence of God.

AUGUSTINE

Augustine is often portrayed as the main culprit who infected biblical anthropology with "Greek dualism" (with attendant phobias concerning the body and sexuality), which needs to be exposed and rooted out of Christian theology once and for all.[69] In our discussion of Tertullian, Origen, and Gregory of Nyssa, we have already undertaken to correct this caricature by pointing out that Greek philosophical thought was not monolithically dualist and that corporealism, whether popular or philosophical, was a broadly held view. More generally, "[t]he corporeality and incorporeality of God indeed seems to occupy a central place in the structure of Late Antique thought; it delineates a fundamental demarcation among basic religious attitudes, as well as among major philosophical schools."[70] Christian theologians were aware of the popularity of corporealism and rejected it both on general philosophical and specifically theological grounds. Augustine's intellectual evolution also involved overcoming corporealism and building the case for the incorporeality of God and the soul.

As Augustine stated in his *Soliloquies*, the main aim of his work was "to know God and the soul," and when pressed to tell if he wished to know anything else, he responded, "Absolutely nothing."[71] His views concerning the divine nature underwent an evolution from naïve anthropomorphism, to philosophical corporealism of a Stoic kind, to skepticism stimulated by his engagement of academic arguments, to the view that God transcends physical reality, espoused by Ambrose of Milan.[72] Concurrently, Augustine's anthropological views underwent an even more complex evolution from the Manichean idea of two souls in one human being, to the Platonic teaching that the soul was a part of the intelligible divine realm and preexisted its embodiment, and finally to a Christian understanding, influenced by Ambrose, that the soul is incorporeal yet created, although precisely how it is created and comes to be united to the body remained an especially difficult issue in light of Augustine's doctrine of original sin. Augustine's writings also reflect a polemic against those in the church who took exception to the incorporeality of the soul, while accepting the unique incorporeality of God (the third conjunction in our taxonomy). Our brief survey will focus on the period when Augustine came to believe that both God and the soul were incorporeal (the fourth conjunction in our taxonomy).

In *Confessions* VII, Augustine discusses his early struggles with understanding the nature of God. While it was relatively easy for him to banish the anthropomorphic depictions of the divine nature, it was far more difficult to imagine God as anything other than a body. Augustine was persuaded by the corporealist postulate that "everything from which space was abstracted was non-existent" and on those grounds held that God was "something physical occupying space diffused either in the world or even through infinite space outside the world."[73] Similar to the Stoics, Augustine conceived of God as a refined physical substance permeating everything and extending in infinite space. Augustine's early corporealism shows an enduring influence of Tertullian and Stoic corporealism in Latin theology. According to François Masai, corporealism was the dominant view in much of western theology until the time of Augustine. While Masai may have exaggerated his case, Augustine certainly took the intuitive appeal of corporealism and the attendant difficulty of conceiving the incorporeal substance with utter seriousness, even if he eventually found corporealism untenable.[74]

Augustine also encountered a peculiar version of corporealism in Manichaeism. This religious teaching was a form of ontological dualism, which postulated an opposition between the divine sphere of light and the sphere of darkness, which rebelled against the light. The Manicheans explained the division within the human self between good and evil desires in terms of the anthropological dualism of two natures or two souls, both of which were conceived in corporeal terms. While such a teaching could provide a convenient excuse for Augustine's waywardness during his Manichean period, later in life he no longer found this view to be philosophically satisfactory. The postulate of two souls could not explain how the decision to do good or evil could belong to a single decision maker. Besides, in any decision making there were often more than just two options involved. To allow a separate soul for each option was to open doors not just to the duality, but to the indefinite plurality of souls (or minds or natures) in one human agent, which was absurd. While Augustine took seriously the reality of inner conflict and divided will, he rejected Manichean corporealism and the attendant ontological and anthropological dualisms.[75] The problem of inner conflict could be resolved through the integration of the self rather than through its division into opposing substances.

In his letter 166 to Jerome, Augustine carefully defines bodily substances in order to allow for the possibility of incorporeal substances:

> If every substance, or essence, or—if that which exists somehow in itself is more suitably called anything else—is a body, then the soul is a body. So too if one prefers to call only that nature incorporeal which is immutable in the highest degree and whole everywhere, the soul is a body, because it is not such a thing. But if only that is a body which is at rest or in motion through space with length, breadth and height so that it occupies a larger place with a larger part of itself and a smaller place with a smaller part and is smaller in a part than in the whole, then the soul is not a body.[76]

Augustine's definition of the body is designed to prevent confusion over the corporealist postulate that everything that exists is a body of some kind. He also seeks to avoid an imprecise use of the term "incorporeal" in the sense of "rarified" or "airy," which was still current in his time.[77] Augustine defines the soul as "a certain substance partaking in reason and suited to rule the body" and ascribes it to all living beings, although his main focus is the rational soul. According to Roland Teske, Augustine "uses the Latin *anima* for soul in general, while reserving *animus* or *mens* for the rational soul."[78] Augustine accepts the threefold division of the human being into spirit, soul, and body on the authority of scripture (1 Thess. 5:23), but recognizes that the twofold division, which identifies the spirit with the rational aspect of the soul, is equally acceptable.[79]

According to Augustine, one reason why the soul is not a body is because it is present in its entirety in each part of the body.[80] This exploratory point of Aristotelian psychology became commonplace in later Platonism, especially as it came to differentiate itself from Stoic corporealism.[81] Although Augustine rejects corporealism, he builds on the Stoic concept of *intentio* in order to account for the soul-body relation: "The soul is stretched out through the entire body that it animates, not by a local diffusion, but by a certain vital intention."[82] As he elaborates elsewhere, "The soul is not of a bodily nature, nor does it fill the body as its local space, like water filling a bottle or a sponge, but in wonderful ways it is mixed into the body it animates, and with its incorporeal nod (*nutu*), so to say, it powers or steers the body with a kind of concentration (*intentio*), not with any material engine."[83] *Intentio* marks the soul's capacity to focus its energies on different bodily functions, perhaps somewhat similar to Gregory of Nyssa's "vivifying energy," discussed earlier. As Kevin Corrigan explains, "*Intentio* is also connected with the will, intention, or activity of the subject, and this is important since we may obviously be curious about what in the [soul-body] compound it is that actually does the willing, seeing, imagining, and so on, and this is primarily for Augustine the rational soul or *mens*."[84] The presence of the soul in the body renders possible simultaneous awareness of different bodily experiences as experiences of one and the same subject.

The second reason the soul is incorporeal is because it is capable of generating and storing images of corporeal things that are far greater than its body. If the soul were purely corporeal, its size would be limited by its body, with the result that there would not be enough space to store the spatially extended images of corporeal things. However, the soul's powers of memory and imagination are quite capable of storing and retrieving such images, which implies that such images are apprehended in a manner excluding spatial extension and, therefore, by definition, incorporeal.[85]

The third reason the soul, or to be more precise, the rational soul or the mind (*mens*), is incorporeal is a combination of the first two reasons. The mind has the capacity to be present to itself, to make itself the object of thought and attention in its entirety. In so doing, the mind has a tendency to confuse itself with the physical things that it perceives through the senses and thinks about, but the mind is none of those things. It is precisely because the mind is capable of holding the images of material things in itself by means of the power of memory after those things are no longer directly perceived that its activity is not reducible to anything corporeal.[86]

The fourth and final reason why the soul is incorporeal is theological. In *On the Soul and its Origin*, Augustine argues against a young theologian who claims that God is incorporeal, while the soul is not (corresponding to the third conjunction in our classification). Augustine regards such a view as an inconsistent halfway option between Tertullian's corporealism, which postulates that both God and the soul are bodies sui generis, and the view that both God and the soul are incorporeal. Augustine takes care to differentiate the soul from divine nature. God is unchangeable and omnipotent, while the soul is changeable and limited in power.[87] Augustine argues that if the soul were corporeal, it would not be capable of receiving the image of the incorporeal God. It is precisely as incorporeal that the rational soul is made in the image of God, more specifically, in the image of the trinity.[88]

Augustine's doctrine of original sin presented certain problems for theorizing about the soul's origin. Augustine considered different hypotheses, including traducianism, creationism, and the fallen-soul view.[89] Traducianism had the advantage of accounting for the transmission of original sin from parents to children, but the disadvantage of having materialist implications and the association with Tertullian, whose corporealism Augustine dismissed as "madness."[90] Creationism, which was a view that the individual souls were specially created by God, had broad support in the church, but did not have an obvious explanation for original sin in a freshly created soul. The fallen-soul view had the support of Origen and his followers, but came with the baggage of the theory of transmigration, which Augustine rejected. Augustine was aware of the flaws of each hypothesis and for this reason was reluctant to settle on any one in particular. He believed that any hypothesis about the soul's origin had to be compatible with the claims that the soul was incorporeal and that the human race had fallen "in Adam" with all the attendant difficulties of interpreting Romans 5:12. Augustine was also careful not to associate original sin exclusively with the body and located the problem, or rather the symptoms of the fallen human condition, in the inability of the soul to fully control the body and in the disorder (*concupiscentia*) of human desires. He was also concerned with making his protological and anthropological views consistent with eschatology by emphasizing the integration of the soul with the transfigured body in the resurrection. Like Origen, Augustine was careful not to ascribe to his speculations on the soul's origin and on the soul-body relation undue finality and dogmatic significance. When all was said and done, Augustine acknowledged that these matters remained a profound mystery. To understand themselves properly, humans must relearn to see themselves in God.[91]

Whatever the undeniable difficulties of Augustine's system, the caricature that he injected into biblical anthropology a fatal dose of "Greek dualism" needs to be set aside. Augustine did not succumb to an intellectual monolith called "Greek dualism," because such a monolith is a product of historical fancy. Instead, he initially interiorized the corporealism of the Stoic type, then for a short time toyed with the Manichean brand of dualism, which was also a corporealism of sorts, then became persuaded on philosophical grounds by the claim that both God and the soul were incorporeal, which claim he faithfully integrated with the biblical teaching of humanity's creation in the image of God. It cannot be doubted that he was strongly influenced by later Platonism, especially by Plotinus; it is equally certain that he discriminated between the Platonist claims that were consistent with and those that were contrary to the revealed truth.

CONCLUSION

The Niceno-Constantinopolitan creed draws on Colossians 1:16, describing God as the "maker of all things, visible and invisible." The Origenist tradition construed "invisible" as "incorporeal." The Chalcedonian Definition (451) speaks of Christ's humanity as consisting of "a rational soul and body." While these authoritative conciliar documents do not have recourse to the concept of incorporeality, they are consistent with the fourth conjunction that both God and the soul are incorporeal. Although scriptural authors were not explicit on the matter, early Christian theologians had sound philosophical and theological reasons for defending incorporeality. Origen, Gregory of Nyssa, and Augustine were far from ignoring the scriptural account of human nature in favor of "Greek dualism." In fact, a monolith that goes under the title of "Greek dualism" is a profoundly misleading scholarly construct, which needs to be very carefully qualified or, better still, set aside in future historical studies of early Christian anthropology. We have seen that dualisms came in different shapes and sizes; that the varieties of ontological dualism should not be confused with the varieties of anthropological dualism; that both prephilosophical and Stoic corporealism remained attractive for some in the church, including such prominent figures as Tertullian and early Augustine; that the Late Antique market of ideas was no less complex than our own; that many Christian theologians were aware of this complexity and had to navigate this difficult terrain by making intellectual choices that are no less challenging than those presented by modern science; that ultimately the Church Fathers settled on an account of human nature that included an incorporeal soul, bearing God's image and made immortal by God's grace. The defense of the soul's incorporeality was motivated by the need to provide a suitable metaphysical framework for human freedom, interiority, awareness of bodily experiences, personal identity through memory and self-awareness, the self's continued existence after the dissolution of the earthly body, and, most importantly, human ability to think about and communicate with the incorporeal God.

NOTES

1. Daniel C. Dennett, *Freedom Evolves* (New York: Viking, 2003), 1.

2. Gilbert Ryle, *The Concept of Mind*, 3rd ed. (London: Penguin Books, 1973), 13–25.

3. Nancey Murphy, *Bodies and Souls, or Spirited Bodies?* (Cambridge, UK: Cambridge University Press, 2006), ix.

4. See Nancey Murphy, "Nonreductive Physicalism," in *In Search of the Soul: Four Views of the Mind-Body Problem*, ed. Joel B. Green and Stuart L. Palmer (Downers Grove, IL: InterVarsity Press, 2005), 116.

5. Karl Barth, *Church Dogmatics* 3.2, eds. G. W. Bromiley and T. F. Torrance (Edinburgh: T&T Clark, 1960), 378.

6. Ibid., 382.

7. Oscar Cullman, *Immortality of the Soul: or, Resurrection of the Dead? The Witness of the New Testament* (New York: Macmillan, 1958), 30. Cf. Krister Stendahl: "The world that comes to us through the Bible, OT and NT, is not interested in the immortality of the soul. And if you think it is, it is because you have read this into the material" ("Immortality is Too Much and Too Little," in *Meanings: The Bible as Document and as Guide* [Philadelphia: Fortress, 1984], 196).

8. "Biblical scholarship has established quite conclusively that there is no dichotomous concept of man in the Bible, such as is found in Greek and Hindu thought. The biblical view of man is holistic, not dualistic. The notion of the soul as an immortal entity which enters the body at birth and leaves it at death is quite foreign to the biblical view of man," Lynn de Silva, *The Problem of Self in Buddhism and Christianity* (London: Macmillan, 1979), 75.

9. The standard etymology of ψυχή connects it to the verb ψύχω, to breathe. See Carl Huffman, "The Pythagorean Conception of the Soul from Pythagoras to Philolaus," in *Body and Soul in Ancient Philosophy*, ed. Dorothea Frede and Burkhard Reis (Berlin: De Gruyter, 2009), 21–43. The term ἄψυχος is attested as early as Archilochus (seventh c. BCE), fr. 193 W; see R. Renehan, "On the Greek Origins of the Concepts Incorporeality and Immateriality," *Greek, Roman and Byzantine Studies* 21.2 (Summer 1980): 125.

10. Aristotle, *De anima*, 1.2.403b29.

11. Aristotle, *De anima*, 1.2.405b.

12. Aristotle, *Physica*, 212a12 and 215b5, trans. R. P. Hardie and R. K. Gaye, in Jonathan Barnes, ed., *Complete Works of Aristotle, Volume 1: The Revised Oxford Translation* (Princeton, NJ: Princeton University Press, 1984), 366. See Renehan, "On the Greek Origins of the Concepts Incorporeality and Immateriality," 112.

13. Aristotle, *De anima*, 1.2.403b–404b.

14. Aristotle, *De anima*, 1.2.404a16. See Huffman, "The Pythagorean Conception of the Soul," 29.

15. Aristotle, *De anima*, 1.2.404b31–405a4; trans. J. S. Smith, in Jonathan Barnes, ed., *Complete Works of Aristotle, Volume 1: The Revised Oxford Translation* (Princeton, NJ: Princeton University Press, 1984), 645.

16. Epicurus, *Epistula ad Herodotum* 63, quoted in Stephen Everson, "Epicurean Psychology," in Keimpe Algra *et al.*, eds., *The Cambridge History of Hellenistic Philosophy* (Cambridge, UK: Cambridge University Press, 1999), 543.

17. A. A. Long, "Stoic Psychology," in *The Cambridge History of Hellenistic Philosophy*, 561.

18. A possible exception is Anaxagoras, although I agree with Renehan (115) that the claim is an Aristotelian interpretation.

19. Plato, *Sophist*, 246a–b; trans. Harold North Fowler, *Theaetetus, Sophist* (Cambridge, MA: Harvard University Press, 1921), 373.

20. Renehan, "On the Greek Origins of the Concepts Incorporeality and Immateriality," 129–130.

21. In *Phaedo* 85e, Plato compares the soul to the "invisible and incorporeal" harmony of the lyre. Cf. *Timaeus* 47c–d.

22. Plato, *Phaedrus* 245c.

23. Tertullian uses this common skeptical trope in *De anima*, 3, to great dramatic effect. Having branded Greek philosophers "the patriarchs of heretics," he observes: "Some of them deny the immortality of the soul; others affirm that it is immortal, and something more. Some raise disputes about its substance; others about its form; others, again, respecting each of its several faculties" (trans. Peter Holmes, in *The Ante-Nicene Fathers*, ed. Alexander Roberts and James Donaldson, vol. 3 [New York: Scribners, 1903], 184. Cf. Tatian, *Ad Graecos*, 25).

24. Origen, *De principiis*, 1. Praef. 8, trans. G. W. Butterworth, *Origen: On First Principles* (Gloucester, MA: Peter Smith, 1973), 5, after the Latin translation of Rufinus.

25. Origen, *De principiis*, 1. Praef. 9, trans. Butterworth, 6.

26. Tertullian, *De carne Christi*, 11.4: "*Omne quod est, corpus est sui generis, nihil incorporale, nisi quod non est*"; see Petr Kitzler, "Tertullian's Concept of the Soul and His Corporealistic Ontology," in J. Lagouanere and S. Fialon, eds., *Tertullianus Afer: Tertullien et la littérature chrétienne d'Afrique*, *Instrumenta Patristica et Mediaevalia* 70 (Turnhout: Brepols, 2015), 43–63. Cf. Tertullian, *De anima*, 7: "*Nihil enim nisi corpus.*" In *Adversus Praxeam*, 7.8, Tertullian asks rhetorically: "For who will deny that God is body, although God is a spirit? For spirit is body of its own kind, in its own form" (trans. Ernest Evans, *Tertullian's Treatise against Praxeas* [London: SPCK, 1948], 138). How the divine body was to be differentiated from the body of the world in order to avoid pantheistic implications remained a somewhat moot point. See Carl W. Griffin and David L. Paulsen, "Augustine and the Corporeality of God," *Harvard Theological Review* 95, no. 1 (2002): 97–118, esp. 101.

27. Tertullian, *De anima*, 5. For the discussion of Tertullian's Stoic sources, see Kitzler, "Tertullian's Concept of the Soul," 49. Tertullian's version of this argument is clearly Stoic, not Epicurean, as in Lucretius, *De rerum natura*, III.160–169.

28. Tertullian, *De anima*, 7.
29. Tertullian, *De anima*, 3, 4.
30. Tatian, *Ad Graecos*, 4.2.
31. Tatian, *Ad Graecos*, 25, in *The Ante-Nicene Fathers*, vol. 2, trans. Alexander Roberts, 76.
32. Justin, *Dialogus cum Tryphone*, 1.6. There was no consensus among the Middle Platonists on the issue of the soul's divinity. See John Dillon, *The Middle Platonists, 80 B.C. to A.D. 220* (Ithaca, NY: Cornell University Press, 1977), 100, 292.
33. Justin, *Dialogus cum Tryphone*, 6.1.
34. Irenaeus, *Adversus Haereses*, 5.7.1; see J. Behr, *Asceticism and Anthropology in Irenaeus and Clement* (New York: Oxford University Press, 2000), 91.
35. Origen, *Dialogus cum Heraclide* 6.23–24, trans. Robert J. Daly, *Origen: Treatise on the Passover and Dialogue of Origen with Heraclides and his Fellow Bishops on the Father, the Son, and the Soul* (New York: Paulist Press, 1992), 62.
36. Origen, *De principiis*, 2.8.3–4.
37. Origen, *Commentary on the Gospel of John*, 6. 7.
38. Origen, *De principiis*, 1.7.1, trans. Butterworth, 59.
39. Origen, *Contra Celsum*, 7.32, trans. Henry Chadwick (Cambridge, UK: Cambridge University Press, 1965), 420.
40. Origen, *De principiis*, 1.6.4; cf. 2.2.2, trans. Butterworth, 58, 81.
41. "God is not a material substance. We would not fall into the absurd ideas held by the philosophers who follow the doctrines of Zeno and Chrysippus" (*Contra Celsum*, 8.49, trans. Chadwick, 488).
42. Brian E. Daley, "Incorporeality and 'Divine Sensibility': The Importance of *De Principiis* 4.4 for Origen's Theology," *Studia Patristica* 41 (2006): 143, n. 24. See also Henri Crouzel, "Le theme platonicien de 'véhicule de l'âme' chez Origène," *Didaskalia* 7 (1977): 225–237.
43. Origen, *De principiis*, 2.6.3; cf. Plato, *Timaeus*, 35a 1–3.
44. Origen, *De principiis*, 2.8.4.
45. Origen, *De principiis*, 2.8.5.
46. Origen, *De principiis*, 1.1.7, trans. Butterworth, 12.
47. Origen, *De principiis*, 1.1.8, trans. Butterworth, 13.
48. Origen, *De principiis*, 1.1.7.
49. Origen, *Dialogue with Heraclides* 15.30–34, trans. Daly, 69.
50. See Arne J. Hobbel, "The *Imago Dei* in the Writings of Origen," *Studia Patristica* 21 (1989): 301–307.
51. Origen, *Contra Celsum*, 6.63; trans. Chadwick, 378–379. Cf. *De principiis*, 2.10.7; 3.1.13.
52. Origen, *Dialogus cum Heraclide*, 26.30–27.6, trans. Daly, 77.
53. Origen, *De principiis*, 1.6.4; 2.3.7. The fact that Origen's investigations into the nature of the resurrected state were intentionally tentative was very often lost on his later critics.
54. See Plantinga, *God and Other Minds: A Study of the Rational Justification of Belief in God* (Ithaca, NY: Cornell University Press, 1967).

55. Gregory of Nyssa, *De anima et resurrectione*, in *Gregorii Nysseni De anima et resurrectione, Opera dogmatica minora*, part III, eds., Andreas Spira and Ekkehardus Mühlenberg (Leiden: Brill, 2014), 14.15

56. Gregory of Nyssa, *De anima et resurrectione*, in *Nicene and Post-Nicene Fathers*, vol. 5, trans. H. A. Wilson, 26.12–13.

57. Gregory of Nyssa, *De hominis opificio,* in *Nicene and Post-Nicene Fathers*, vol. 5, trans. H. A. Wilson, 14.1; cf. 15.3.

58. Gregory of Nyssa, *De hominis opificio*, 15.3.

59. Gregory of Nyssa, *De anima et resurrectione*, 28.4.

60. Gregory of Nyssa, *De hominis opificio*, 4.1.

61. Gregory of Nyssa, *De hominis opificio*, 12.9, trans. Wilson, 399.

62. Gregory of Nyssa, *De hominis opificio*, 24.2.

63. Gregory of Nyssa, *De hominis opificio*, 28.1. He mentioned that both issues were a subject of ongoing discussion in the church.

64. Gregory of Nyssa, *De hominis opificio*, 29.3.

65. Gregory of Nyssa, *De anima et resurrectione*, 15.6–9.

66. Gregory of Nyssa, *De hominis opificio*. 8.4.

67. Gregory of Nyssa, *De hominis opificio*, 15.2.

68. Gregory of Nyssa, *De anima et resurrectione*, 55.17–56.14.

69. For an alternative interpretation of Augustine's role, see J. Patout Burns, "Variations on a Dualist Theme: Augustine on the Body and Soul," in *Interpreting Tradition: The Art of Theological Reflection*, ed. Jane Kopas (Chico, CA: Scholars Press, 1983), 13.

70. Gedaliahu Stroumsa, "The Incorporeality of God: Context and Implications of Origen's Position," *Religion* 13 (1983): 353.

71. Augustine, *Soliloquia*, 2.2.7.

72. Augustine recollects that as a young student in Tagaste, under the influence of Manichean teaching, "when I wanted to think of my God, I knew of no way of doing so except as a physical mass. Nor did I think anything existed which is not material. That was the principal and almost sole cause of my inevitable error" (*Confessiones*, 5.10.19, cf. 5.14.25).

73. Augustine, *Confessiones*, 7.1.1 (cf. 7.1.2; 7.5.7; 7.14.20).

74. François Masai, "Les conversions de Saint Augustin et les débuts du spiritualisme en Occident," *Le Moyen Âge* 67(1961): 1–40.

75. Augustine, *Confessiones*, 8.10.23–24.

76. Quoted in R. J. Teske, "Saint Augustine on the Incorporeality of the Soul in Letter 166," *The Modern Schoolman* 60, no. 3 (1983): 175. Cf. Augustine, *De trinitate*, 10.7.9. This definition draws on Plotinus, *Enneads*, 4.2.

77. Augustine, *De anima et ejus origine*, 4.12.18.

78. Roland Teske, "Augustine's Theory of Soul," in *The Cambridge Companion to Augustine*, eds. Eleonore Stump and Norman Kretzmann (Cambridge, UK: Cambridge University Press, 2001), 116.

79. Augustine, *De fide et symbolo*, 10.23.

80. Augustine, *De immortalitate animae*, 16.25: "The soul, however, is present at the same time and entire not only in the entire mass of its body, but also in each of its

individual parts. For, it is the entire soul that feels the pain of a part of the body, yet it does not feel it in the entire body" (trans. Ludwig Schopp, *The Immortality of the Soul; The Magnitude of the Soul; On Music; The Advantage of Believing; On Faith in Things Unseen*, Fathers of the Church, vol. 4 [Washington, DC: Catholic University of America Press, 1947], 46).

81. Aristotle, *De anima*, 1.5 411b19–28; Plotinus, *Enneads*, 6.4.4; 6.5.12.

82. Teske, "Saint Augustine on the Incorporeality of the Soul in Letter 166," 176; see also John M. Rist, *Augustine* (Cambridge, UK: Cambridge University Press, 1994), 108 n. 51.

83. *De genesi ad litteram*, 8.21.42; trans. Edmund Hill, *On Genesis, The Works of Saint Augustine: A Translation for the 21st Century*, part I, vol. 13 (Hyde Park, NY: New City Press, 2002), 370.

84. Corrigan, "The Soul-Body Relation in and before Augustine," 72.

85. Augustine, *De anima et ejus origine*, 4.17.25; *Confessiones*, 7.1.2; cf. Plotinus, *Enneads*, 4.2.1.

86. Augustine, *De trinitate*, 10.5.7–10.14. For a brief summary of these three arguments, see R. J. Teske, "Soul," in *Augustine through the Ages: An Encyclopedia*, eds. Allan D. Fitzerald *et al.*, (Grand Rapids, MI: Eerdmans, 1999), 808.

87. Augustine, *De anima et ejus origine*, 4.12.18.

88. Augustine, *De anima et ejus origine*, 4.14.20; *De trinitate*, 14.2.4. Augustine associates that which is made "in the image" with the soul so as to avoid the Manichean objection that if the human body is made in God's image then God, like the human body, is anthropomorphic.

89. The last two designations in this classification are from Robert J. O'Connell, *The Origin of the Soul in St. Augustine's Later Works* (New York: Fordham University Press, 1987).

90. Augustine, *Letter* 190.4.14–15; discussed in Rist, *Augustine*, 318.

91. Augustine, *Confessiones*, 10.8.15, 10.17.26, 13.31.46; *De Genesi ad litteram*, 4.32.50; see Rist, *Augustine*, 146.

SUGGESTED FURTHER READING

Aristotle. *De anima*. Trans. J. S. Smith. In *Complete Works of Aristotle*, vol. 1*: The Revised Oxford Translation*, edited by Jonathan Barnes. Princeton, NJ: Princeton University Press, 1984.

Burns, Patout J. "Variations on a Dualist Theme: Augustine on the Body and Soul." In *Interpreting Tradition: The Art of Theological Reflection*, edited by Jane Kopas. Chico, CA: Scholars Press, 1983.

Cary, Phillip. *Augustine's Invention of the Inner Self: The Legacy of a Christian Platonist*. New York: Oxford University Press, 2000.

Crabbe, James C., ed. *From Soul to Self*. London: Routledge, 1999.

Cullman, Oscar. *Immortality of the Soul: or, Resurrection of the Dead? The Witness of the New Testament*. New York: Macmillan, 1958.

Daley, Brian E. "Incorporeality and 'Divine Sensibility': The Importance of *De Principiis* 4.4 for Origen's Theology." *Studia Patristica* 41 (2006): 139–144.

Gregory of Nyssa. *De anima et resurrectione*. In *Opera dogmatica minora*, part III, edited by Andreas Spira and Ekkehardus Mühlenberg. Leiden, Netherlands: Brill, 2014.

Long, A. A. "Stoic Psychology." In *The Cambridge History of Hellenistic Philosophy*, edited by Keimpe Algra, et al. Cambridge, UK: Cambridge University Press, 1999.

Origen. *De principiis*. In *Origen: On First Principles*. Translated by G. W. Butterworth. Gloucester, MA: Peter Smith, 1973.

Origen. *Dialogus cum Heraclide*. In *Origen: Treatise on the Passover and Dialogue of Origen with Heraclides and his Fellow Bishops on the Father, the Son, and the Soul*. Translated by Robert J. Daly. New York: Paulist Press, 1992.

Stroumsa, Gedaliahu. "The Incorporeality of God: Context and Implications of Origen's Position." *Religion* 13 (1983): 345–358.

Teske, Roland J. *To Know God and the Soul: Essays on the Thought of St. Augustine.* Washington, DC: Catholic University of America Press, 2008.

Chapter 2

Christian Physicalism

Against the Medieval Divines

Thomas Atkinson

Christian physicalism has not had a time-honored history. No ecumenical council, denominational synod, or inquisitorial office, no Pope or archbishop or reformer, has, to my knowledge, ever endorsed physicalism.[1] In fact, one may note that the doctrinal statements of a large variety of Christian denominations have been taken to be inconsistent with physicalism.[2] This, one may think, is to the advantage of the Christian antiphysicalist. Departure from the Church's teaching for nineteen and a half centuries is no insignificant matter. While one may think it fallacious to reason that because the Church has taught something from time immemorial it must, therefore, be true, it is not fallacious to take the Church's creeds and other statements to hold an authority higher than one's own reason (or even, perhaps, the collective reason of a small group of thinkers).

The historical chapters in this volume together attempt to make clear how it is that Christian physicalism departs from the teaching of the Church. In an earlier chapter, Paul Gavrilyuk considered the early Church Fathers. It is the purpose of this chapter to argue that Christian physicalism is inconsistent with some theses held by the medieval divines. In this respect, Christian physicalism marks a departure from the Church's teaching. This will probably not come as a surprise to anyone. What makes this project especially interesting, however, is that if there were to be a time in Church history (before the nineteenth century) wherein physicalism would have been taken most seriously by Christians, it would likely have been the medieval period.

In this chapter, therefore, I will do two things. First, I will give an overview of some of the reasons why philosophers and theologians understand the medieval divines (particularly Thomas Aquinas) to take seriously a position closer to the contemporary physicalist view of the human person, than the dominant view of the human person in the Church (namely, substance dualism). I will do this by highlighting several points of agreement between Christian physicalists and the views of the medieval divines against the views of Christian substance dualists. Second, however, I will also highlight several points of conflict between the views of medieval divines and Christian physicalists. In the last analysis, I shall conclude that although physicalism is, in some respects, consistent with the thought of the medieval divines in a way that substance dualism is not, the central thesis of physicalism is still inconsistent with the thought of the medieval divines.[3]

CHRISTIAN PHYSICALISM

Given that this chapter aims to display the consistencies and inconsistencies between Christian physicalism and the views of the medieval divines, I must define "Christian physicalism." Moreover, given that this chapter aims to display the inconsistencies between Christian substance dualism and the views of the medieval divines, I should also define "Christian substance dualism."

First, consider physicalism. Physicalism is the view that

> (P) every instantiated property, *F*, is necessitated by, and not metaphysically distinct from, some physical property *G*.[4]

Here are some examples of instantiated properties. *Being made of aluminium* is a property instantiated by my computer. *Being an uncle* is a property that I instantiate. *What it is like to taste ice cream* is a property instantiated by a human being eating an ice cream. This last kind of property is known as a "mental" or "psychological" property. Consider another mental property: *what it is like to see red.* Consider a physical property (or collection of properties) that is plausibly instantiated at the same time as the mental property just mentioned; let's say, the collection of physical properties "oscillations in V4" (an area of the visual cortex). According to (P) an instantiated mental property like the property *what it is like to see red* is necessitated by a physical property (or collection of physical properties) like certain oscillations in V4. This just means that it is impossible for the mental property—*what it is like to see red*—to be instantiated and no physical property be instantiated at the same time.

Not only this, but according to (P) every instantiated mental property is "not metaphysically distinct" from some physical property. Different kinds of physicalism will spell out this part of (P) in different ways.[5] For our purposes, it will do to say that two properties are *metaphysically distinct* when they belong to *different metaphysical kinds*. The two metaphysical kinds that we are interested in (and which are perhaps the most dissimilar of all metaphysical kinds) are "physical" and "non-physical."

This, of course, raises the question "what is a physical property?"[6] This is a vexed question.[7] Daniel Stoljar understands the concept "physical property" to be a cluster concept. He thinks that a "physical property" is, for example, (a) a distinctive property of intuitively physical objects, (b) expressed by a predicate of physics, (c) objective, (d) knowable through scientific investigation, and (e) not a distinctive property of substances such as Cartesian souls and ectoplasm, etc.[8]

Christian physicalists cannot affirm this kind of physicalism. This is because Christian physicalists do not think that *every* instantiated property is metaphysically indistinct from a physical property. Christian physicalists believe in things such as God, angels, demons, the Holy Spirit, etc., and they believe these things instantiate properties that are metaphysically distinct from physical properties. The Christian physicalist believes that God, for example, instantiates properties such as the property of *loving the Son*—a property that is, I take it, uncontroversially nonphysical.[9] In consequence, Christian physicalists must be (and are) only physicalists with regards to *human persons*. Call this *local* physicalism:

> (LP) every property, *F*, instantiated by a human person is necessitated by, and not metaphysically distinct from, some physical property, *G*, instantiated by that human person.[10]

Some mental properties are properties that are instantiated by human persons and, according to the physicalist, are necessitated by, and are not metaphysically distinct from, some physical property of that human person.[11]

I frequently am asked the following question: why be a Christian and a physicalist? Most Christian physicalists cite the recent successes of the neurosciences as their reason for being a Christian physicalist. Nancey Murphy, a Christian physicalist, writes the following: "[m]y argument in brief is this: all of the human capacities once attributed to the mind or soul are now being fruitfully studied as brain processes."[12] In short (and to put it in the vocabulary being employed in this chapter), Murphy and other Christian physicalists think that mental properties, once thought to be nonphysical properties of a nonphysical object (a soul), can now be studied as (metaphysically indistinct from) properties of the brain (or, perhaps, the human being: a wholly physical thing). While Murphy recognizes that this phenomenon doesn't prove[13] that these mental properties *are* (metaphysically indistinct from) properties of the brain or human person (physical properties), she thinks that the denial of this claim (or one very similar) is increasingly difficult to hold.[14] Where once we thought the mental properties of human beings were "nonphysical" we now have reason, so will say Murphy, to think this is not the case. Rather, we now have reason to think that they are physical properties, properties of brains, or (wholly physical) human persons.

It will be sufficient for our purposes to understand a "Christian" to be someone who would, for example, (C) affirm the Nicene Creed. Christian physicalists, for the purposes of this chapter, are those folk who affirm (C) and (LP).

SUBSTANCE DUALISM

Before I move on to discuss the medieval divines, I should also state what has undoubtedly been the dominant Christian view of the human person. I do this so that I can state where the medieval divines disagree with the dominant Christian view. The majority of Christians traditionally have been *substance dualists* with regards to human persons.

Substance dualists with regards to the human person typically believe the following:

> (SD) There are two kinds of substance (physical and nonphysical), and we, human persons, are either (i) nonphysical substances or (ii) composed, at least in part, by a nonphysical substance.[15]

Christian substance dualists are those who accept (SD) and (C). Typically, Christian substance dualists reject (LP). This is for a number of reasons. Primarily, however, this is because Christian substance dualists believe there are some properties that we instantiate that are nonphysical and that these properties are instantiated by a nonphysical substance. Take, for example, the property *what it is like to experience red.* According to the substance dualist, this property is a property that is instantiated by (at least part of) a human person, and this property is not a physical property. This property, so the substance dualist will argue, is instantiated by a nonphysical substance; namely, the soul.

It should be noted at this point that (LP) is inconsistent with another kind of dualism, namely, *property* dualism. Property dualists[16] think that we are physical objects but that we instantiate nonphysical properties. The reason why I rule out property dualism as a form of physicalism is because I understand most contemporary physicalists to reject property dualism,[17] Christian physicalists being among the contemporary physicalists.

Take, for example, the perhaps most liberal form of physicalism: nonreductive physicalism. Most nonreductive physicalists rule out property dualism as a form of physicalism. This is because they think that while the mental predicates of our sentences are essential for a full description of the world and that these predicates are not reducible to the predicates of physics, there are still no distinctly mental or nonphysical entities (such as nonphysical *properties*) that are required to make the sentences in which these mental predicates occur true. I know of no Christian physicalist who is also, at least explicitly, a property dualist. I know of Christian physicalists that reject property dualism. The Christian physicalist Peter van Inwagen explicitly rejects property dualism.[18] Furthermore, Murphy also makes statements that entail the falsity of property dualism. She writes, "statements about the physical nature of human beings made from the perspective of biology or neuroscience are about *exactly the same entity* as statements made about the spiritual nature of persons from the point of view of theology or religious traditions."[19] This claim is contrary to the claims of property dualists. Christian property dualists (if there are any) will likely hold that statements concerning the spiritual nature of human persons are about entities (namely, nonphysical properties) distinct from the entities that statements about the physical nature of human persons concern (namely, physical properties). Moreover, Murphy explicitly argues for "the acceptance of ontological reductionism."[20] Murphy's ontological reductionism is the view that "as one goes up the hierarchy of levels [e.g., from the level of physics to the level of the spiritual], no new metaphysical 'ingredients' need to be added."[21] This is precisely what property dualists reject. Property dualists hold that mental properties are entities that are not physical entities. If Murphy and other Christian physicalists are going to admit a form of dualism, it will be a "predicate dualism" as described above.[22]

Those physicalists who believe we are composed of substances will, I take it, believe we are composed of only physical substances: things like fermions, quarks, leptons, gauge bosons, and things that are wholly composed by fermions, quarks, leptons, and gauge bosons (if we human beings are, according to the physicalist, composed of anything we will be composed wholly from fermions, quarks, leptons, and gauge bosons).

I should note here that my primary concern, however, is not to try to show that the medieval divines (or at least some of them) were substance dualists (although some have found this view plausible[23]). My primary concern is to show that the medieval divines would not have endorsed (LP).[24]

THE MEDIEVAL DIVINES AND MEDIEVAL PERSONAL ONTOLOGY

The "medieval divines" consists of a wide variety of scholars, to name a few: Thomas Aquinas, Bonaventure, Duns Scotus, William of Ockham, and Peter John Olivi. Although the medieval divines, in broad, shared much the same view about the metaphysics of human persons, that view is shaded differently by each individual scholar. Here is what the medieval divines seemingly agreed upon. The medieval divines seemed to agree that we are *composite substances*. That is, we, human persons, are substances composed of matter and form. As Richard Cross notes, the medievals "were all convinced that body and soul are united in such a way as to form one (composite) substance."[25] This view is called hylemorphism with regard to the human person.[26] What is "matter" and what is a "form"? Very crudely, "matter" is the stuff from which a thing is made,[27] and "form" is the "dynamic configuration or organisation" that the matter takes.[28] In the case of a human person, the form of the human person is the substantial form: the human soul. The "body" of a human person is the prime matter that the human soul informs.

Given that, besides this point, medieval personal ontology was so nuanced, arguing convincingly that (LP) is inconsistent with *all* of the views around in the medieval period would require a book length treatment. In this chapter, I therefore restrict myself to demonstrating that (LP) is inconsistent with Thomas Aquinas's personal ontology. I take Aquinas as the starting point for two reasons. First, although medieval personal ontology is diverse, it seems so only insofar as it diverges from Aquinas's work. That is, most of the personal ontologies developed in medieval Europe consist of modifications to, or arguments against, aspects of Aquinas's personal ontology. Not only this but, second, out of all the medieval views, Aquinas's might be taken to be the view that is the closest to physicalism. The reason being that Aquinas thought the relationship between the body and the soul was tighter than some of his contemporaries thought it.[29] In consequence, if I can demonstrate that (LP) is inconsistent with Aquinas's personal ontology, it will likely be inconsistent with the thought of the other medieval divines.

Drawing a direct comparison between the personal ontology of the medieval divines (Aquinas in particular) and contemporary personal ontology is no mean feat.[30] As is true of most of the medieval divines, it is argued that their view about the nature of the human person does not easily fit into contemporary taxonomy. Richard Cross, for example, starts his chapter on the philosophy of mind in Duns Scotus's work by noting that "medieval views on the relation of mind and body occupy a strange territory somewhere between substance dualism, on the one hand, and some form of materialism, on the other."[31]

Likewise, Robert Pasnau and Christopher Shields claim in their chapter on Thomas Aquinas's philosophy of mind in particular that

> it is so far unclear whether Aquinas' hylomorphism is best regarded as a kind of materialism or a kind of dualism. On the one hand, in view of his orthodox Christianity, one might expect Aquinas to be attracted to some form of dualism, according to which the soul is separable from the body and capable of some postmortem existence. At the same time, it is not clear how dualism of any form could be reconciled with hylomorphism.[32]

The lack of clarity has meant that a number of contemporary commentators on Aquinas have him endorsing different contemporary accounts of the human person. J. P. Moreland and Scott Rae have Aquinas's view at least broadly consistent with substance dualism.[33] Brian Leftow puts Aquinas, in general, in the property dualism camp.[34] Robert Pasnau defends the view that Aquinas is neither a substance dualist nor a property dualist, but a kind of nonreductive materialist.[35] Eleonore Stump, thinks that Aquinas's position might best fit with a kind of "non-reductive materialism" or a "non-Cartesian substance dualism."[36] I hope to avoid the problems associated with positively formulating Aquinas's view in contemporary terms by merely arguing that Aquinas cannot be considered a physicalist as defined above.[37]

In the remainder of this chapter, I will outline the reasons why some contemporary thinkers take Aquinas's thought to be more in line with some sort of physicalism than substance dualism. I will also argue, however, that nevertheless Aquinas would have certainly rejected (LP).

THE CONSISTENCY BETWEEN CHRISTIAN PHYSICALISM AND THE MEDIEVAL DIVINES: WE ARE NOT SOULS

Put simply, the primary reason for the view that Christian physicalists and the medieval divines agree is that both the Christian physicalists and the medieval divines deny (i) of (SD); that is, they deny that we are immaterial substances.

Christian physicalists and the medieval divines deny (i), however, for different reasons. First, Christian physicalists, on the whole, deny (i) because they think that the findings of contemporary science render (i) implausible (see Murphy, mentioned earlier). Aquinas, of course, could not deny (i) for the same reason. Rather, Aquinas rejects (i) because he thinks that we are one substance (namely, we are instances of the substance "human being") and that this substance is composed of *physical and nonphysical* parts: matter and an immaterial substantial form. He writes, "body and soul are not two actually existing substances; instead one actually existing substance comes from the two."[38]

Aquinas does not simply assert this. He provides some arguments. In particular, Aquinas takes Plato's version of substance dualism (which posits (i)) as his target. According to Aquinas, Plato thought that "a human being is not something composed of soul and body; rather a human being is a soul using a body, so that the soul is understood to be in the body somewhat as a sailor is in a ship."[39] Aquinas gives a number of arguments in response to Plato. Here's one of them. Aquinas argues, in effect, that (i) is unsatisfactory in the light of our experience as embodied beings.[40] Put simply, Aquinas argues that because we experience ourselves as a thing that sees, touches, tastes, smells, and hears, and since these actions require a body (a material substance), we are not, therefore, an immaterial substance.[41] Whether or not one thinks this a good argument against (i) is beside the point. The point is that Aquinas thought one of the disjuncts of (SD) was false.

THE CONSISTENCY BETWEEN CHRISTIAN PHYSICALISM AND THE MEDIEVAL DIVINES: WE ARE NOT COMPOSITES OF TWO SUBSTANCES

As I have put it substance dualism is a disjunctive position. In this case, one can be a substance dualist and reject (i) so long as one accepts (ii); that is, we, human persons, are composed, at least in part, by a nonphysical substance. Aquinas, so it seems, would disagree with (ii) too, however. I say this because I think a certain set of Aquinas interpreters have accurately articulated his position as inconsistent with (ii). Even so, as will become apparent, Aquinas's view will not count as a form of physicalism. Here I will put forward an argument for the view that Aquinas would reject (ii).

Aquinas, as mentioned above, thought that we are composite objects and that we are composed of both body and soul. Importantly, however, the soul, according to Aquinas, is not a substance. In consequence, Aquinas is not a *substance* dualist with regards to the human person. Here, I recount a familiar argument for the view that Aquinas did not think, at least when he was thinking most carefully, that the soul was a substance.[42] Aquinas would have endorsed the following argument:

1. For any x, x is a substance only if x can (a) subsist per se and, is a (b) complete member of a particular species and genus.
2. Souls are able to (a) subsist per se, but are not (b) complete members of a particular species and genus.
3. Therefore,
4. Souls are not substances.

Aquinas believes premise 1. He writes, "[a]n individual in the genus of substance possesses not only per se subsistence, but is also something complete in a particular species and genus of substance."[43] It is also clear that Aquinas believes premise 2. Aquinas notes that while the rational soul can subsist when it is not informing matter, it is not complete in species per se.[44] Consider Amy, a human being. Amy's soul, while a subsisting thing, is not itself complete in species when separated from a body. This is because Amy's soul when separated from a body cannot carry out all of the functions necessary for its being completely human. It cannot, for example, consume, defecate, breathe, or exhale.

This argument renders Aquinas's view inconsistent with (SD). It is for these reasons that Aquinas's view may be considered closer to physicalism than any other view in the Church's history. That is, both Christian physicalists and Aquinas clearly reject (SD) where the dominant view throughout the Church's history has been (SD).

The question remains, however, how troubling should this be for our purposes? I do not think it's very troubling. Although Aquinas's views about personal ontology were not in keeping with the traditional substance dualism usually associated with Christianity, his view (a) is not prima facie in conflict with the Church's teaching with regards to the nature of human persons, and (b) his views are inconsistent with physicalism. I do not have time to defend (a); I will now turn to defending (b).

THE CONFLICTS BETWEEN CHRISTIAN PHYSICALISM AND THE MEDIEVAL DIVINES: WE INSTANTIATE MENTAL PROPERTIES THAT ARE METAPHYSICALLY DISTINCT FROM THE PHYSICAL PROPERTIES THAT WE INSTANTIATE

The first way by which Aquinas's views and the views of the Christian physicalists are inconsistent is that Aquinas (so I take it) thinks there are properties that are instantiated by human persons that are metaphysically distinct from the physical properties instantiated by those human persons. I will now attempt to argue for this.

In some passages, Aquinas recognizes there are some events that human persons engage in that, so it seems, involve the instantiation of both nonphysical and physical properties. Take, for example, the event of a human person's engaging in intellectual activity or "intellection" (that is, thinking about universals). Of all the acts of human cognition intellection is the one that we might think is the best candidate for an act that transcends the physical nature of human persons.[45] Having said this, two things should be noted.

First, it should be noted that according to Aquinas, even a human person's engaging in intellection during her terrestrial existence is, in part, a physical event which involves the instantiation of physical properties.[46] As Pasnau writes, according to Aquinas "even our intellect is unable to operate without the help of the body. All intellective cognition, [Aquinas] argues, requires the sensory images that he refers to as phantasms: 'It is impossible for our intellect, in its present state of life . . . , actually to cognize anything without turning towards phantasms' (84.7c . . .)."[47] This warrants the conclusion that the instantiation of physical properties by a human person is necessary for the instantiation of the mental properties (typical of intellection) of that human person (at least during that human person's terrestrial existence).

Second, this does not, however, warrant the conclusion that the mental properties of human persons instantiated upon intellection are metaphysically indistinct from the physical properties of those human persons. In fact, one may think that according to Aquinas the mental properties (typical of intellection) of human persons are metaphysically distinct from the physical properties of those human persons. As Pasnau notes, the "[i]ntellect is special because it transcends matter entirely; 'To the extent that it surpasses the existence of corporeal matter, being able to subsist and operate on its own.'"[48] If a human person instantiates mental properties (properties typical of intellection) and these properties surpass "the existence of corporeal matter," are these properties nonphysical? I think that the answer is "yes." Especially given that Aquinas admits that a thing is intelligent because it is immaterial.[49] In this context, in the very least "surpassing existence," I take it, means there exist properties instantiated upon some human person's engaging in the act of intellection that are not properties of corporeal matter. If properties that are not properties of corporeal matter are nonphysical, then there are properties typical of intellection that should be understood as metaphysically distinct from the physical properties of the human person that instantiates them.[50]

Pasnau, however, disagrees. Pasnau argues that Aquinas would not affirm either substance or property dualism. This is because Pasnau thinks that Aquinas has a

> deeper metaphysical account . . . according to which the only genuine reality in the world is actuality, and other things, even material things, are real only to the extent that they are actual. There is nonmaterial stuff in the mind—Aquinas calls it actuality—but he thinks that this actuality is spread throughout the created world. Some things, such as the senses, are higher on the scale of being because they have an operation that transcends the mere elements.[51]

While we may agree with Pasnau (for reasons that I have mentioned earlier) that Aquinas was not a substance dualist with regards to the human person, Pasnau still needs to give us a reason for believing that Aquinas should not be understood as a property dualist. While Aquinas may have thought that "the only genuine reality in the world is actuality," the question remains, why is it not sufficient for understanding Aquinas as a property dualist that Aquinas accepts that the properties that human persons instantiate when engaged in an intellective activity come in two different *kinds* of actuality: material and immaterial?[52] Here the material actuality involves the instantiation of physical properties and immaterial actuality involves the instantiation of nonphysical properties. I think this is sufficient for property dualism, and I can see no good reason as to why we should think that Aquinas would endorse this thesis.

THE CONFLICTS BETWEEN CHRISTIAN PHYSICALISM AND THE MEDIEVAL DIVINES: WE INSTANTIATE MENTAL PROPERTIES THAT ARE NOT NECESSITATED BY PHYSICAL PROPERTIES

Second, one can demonstrate that Aquinas would have disagreed with the claim that every property, *F*, instantiated by a human person is *necessitated* by some physical property, *G*, of that human person. Aquinas thinks there is at least one occasion when there are mental properties of human persons that are instantiated but there are no physical properties instantiated by that human person that necessitates them. [53] This occasion is during the intermediate state.

There is a lively and interesting debate between those who read Aquinas as a *corruptionist* and those who read Aquinas as a *survivalist*. Corruptionists read Aquinas as holding it to be the case that when a human person dies she ceases to exist even though her soul continues to exist. Survivalists read Aquinas as holding it to be the case that when a human person dies she continues to exist as her soul. Survivalists think that although the human person is never identical to her soul, she is composed of her soul during the interim state and so continues to exist. I, unfortunately, cannot here weigh in on this debate.

For the sake of the argument, however, in this chapter I assume that the survivalists have it right.[54] This point, however, should not matter too much. If one thinks that Aquinas was a corruptionist, the argument that I put forward can be adjusted slightly. One can argue that there are mental properties of souls (mere parts of human persons) that are instantiated during the intermediate state without there being any physical properties of a human person necessitating them. I take it that the Christian physicalist would disagree with this thesis too.[55]

Aquinas thinks that human persons (or, at least, the identity preserving part of human persons) exist immediately after the death of the body. As Christopher Brown notes, Aquinas thinks "the individual soul can preserve the being and identity of the human being whose soul it is. In other words, although the soul is not identical to the human being, a human being can be composed of his or her soul alone."[56] Brown summarizes Aquinas's view as follows:

> God creates the human soul such that it shares its existence with matter when a human being comes to exist (see, for example, SCG II, ch. 68, 3). Because the being of the human soul is numerically the same as that of the composite—again, the soul shares its being with the matter it configures whenever the soul configures matter—when the soul exists apart from matter between death and the general resurrection, the being of the composite is preserved insofar as the soul remains in existence (see, for example: SCG IV, ch. 81, 11; ST Ia. q. 76, a. 1, ad5; and ST Ia IIae. q. 4, a. 5, ad2).[57]

The souls of the deceased, however, do not remain in existence during the interim state devoid of any mental activity. As Aquinas notes, "when [the rational soul] is separated from the body, it has a mode of understanding, by turning to simply intelligible objects, as is proper to other separate substances."[58] If one grants that for a soul to have a "mode of understanding" it needs to instantiate mental properties, then there exist properties of a human person (namely, mental properties of a human person during the interim state) that are instantiated when there are no physical properties of that person instantiated.

In sum, at best, Christian physicalists and Thomas Aquinas share two beliefs to the exclusion of Christian substance dualists. That is, both Christian physicalists and Thomas Aquinas believe that human persons are not immaterial substances and that they are not a composite of two substances. These two beliefs, however, do not render Aquinas's view consistent with Christian physicalism. This is because the central claims of local physicalism (namely, the claims that we instantiate mental properties that are metaphysically distinct from the physical properties that we instantiate and the claim that we instantiate mental properties that are not necessitated by physical properties) are inconsistent with the view of Thomas Aquinas. In consequence, Christian physicalism marks a departure from, at least, the view of Thomas Aquinas and, no doubt, many of the medieval divines.[59]

NOTES

1. To refashion a claim made in Peter van Inwagen, "Dualism and Materialism," *Faith and Philosophy* 12, no. 4 (1995): 487.

2. I understand the *Catechism of the Catholic Church*, part 1, section 2, chapter 3, articles 11–12 (paragraphs 1005–1019) to be incompatible with physicalism. See also Oliver Crisp, *God Incarnate: Explorations in Christology* (New York: T & T Clark, 2009), 140–142. Crisp argues that the Council of Chalcedon throws up difficulties (to say the least) for the Christian physicalist. It also strikes me that it is very hard to make sense of the Westminster Confession of Faith, chapter 32, without rejecting physicalism (interestingly, it strikes me that this chapter is wholly consistent with Aquinas's hylemorphism).

3. The main points that are made in this chapter are not wholly original. They can be found elsewhere. I hope, however, that this chapter will (a) make the debates that concern Christian physicalism's departure from the thought of the medieval divines clear, and (b) provide a useful introduction for others wanting to explore the intricacies of the relationship between the thinking of the medieval divines and the Christian physicalists in more detail.

4. See Daniel Stoljar, *Physicalism* (New York: Routledge, 2010), 235. I use Stoljar's definition because I understand it to be the most up-to-date. I restrict Stoljar's

definition just to simple properties, however, to avoid some absurdities (thanks to Daniel Hill for alerting me to this point).

5. See Stoljar, *Physicalism*, 144.

6. I take it that a "nonphysical property" is any property that does not satisfy any of the necessary conditions for a property's being physical.

7. See, for example, Stoljar, *Physicalism*, 51–108.

8. See Stoljar, *Physicalism*, 57 and Barbara Montero, review of *Physicalism*, by Daniel Stoljar (June 2012) available at http://ndpr.nd.edu/news/physicalism/

9. Stoljar would likely count this property as a "nonphysical" property in virtue of the fact that it is a distinctive property of a wholly nonphysical being (d).

10. As mentioned, I restrict Stoljar's definition just to simple properties.

11. There may be other ways by which Christian physicalism conflicts with the thought of the medieval divines that is not to do with personal ontology. I restrict my focus here, however, to personal ontology.

12. Nancey Murphy, *Bodies and Souls, or Spirited Bodies?* (New York: Cambridge University Press, 2006), 56. Lynne Rudder Baker offers a slightly different but related reason for her accepting Christian physicalism. She writes, "[i]mmaterial souls just do not fit with what we know about the natural world. We human persons evolved by natural selection (even if God actualized this world on the basis of His foreknowledge of the outcome). Immaterial souls would simply stand out as surds in the natural world" (Lynne Rudder Baker, "Persons and the Metaphysics of Resurrection," *Religious Studies* 43, no. 3 [2007]: 341).

13. See Murphy, *Bodies and Souls, or Spirited Bodies?*, 112.

14. I recommend Moreland's "A Critique of and Alternative to Nancey Murphy's Christian Physicalism" (*European Journal for Philosophy of Religion* 8, no. 2 [2016]: 107–128) for a reply to Murphy.

15. Of course, one may be a substance dualist and think that human persons are wholly physical beings (Christian physicalists are substance dualists of this kind; they believe in two kinds of substance physical and nonphysical). This is not how the phrase is standardly used and I will not begin to use it this way.

16. *Local* property dualists to be precise.

17. Stoljar, *Physicalism*, 32, for example, seeks to rule out property dualism as a form of physicalism.

18. Peter van Inwagen, "A Materialist Ontology of the Human Person," in *Persons: Human and Divine*, eds. Peter van Inwagen and Dean W. Zimmerman (New York: Oxford University Press, 2007), 213–215.

19. Warren S. Brown, Nancey C. Murphy, and H. Newton Malony, "Preface," in *Whatever Happened to the Soul? Scientific and Theological Portraits of Human Nature*, eds. Warren S. Brown, Nancey C. Murphy, and H. Newton Malony (Minneapolis, MN: Fortress Press, 1998), xiii.

20. Nancey C. Murphy, "Nonreductive Physicalism: Philosophical Issues," in *Whatever Happened to the Soul? Scientific and Theological Portraits of Human Nature*, 130.

21. Murphy, "Nonreductive Physicalism," 129.

22. See also Howard Robinson, "Dualism," in *The Stanford Encyclopedia of Philosophy*, ed. Edward N. Zalta (Winter 2016), available at https://plato.stanford.edu/archives/win2016/entries/dualism/ (section 2.1).

23. It can be demonstrated that Aquinas himself, at times, explicitly identifies the rational soul as a substance. See *Summa Contra Gentiles; Book Two: Creation*, trans. James F. Anderson (Notre Dame, IN: University of Notre Dame Press, 1975), chapters 56 and 68, and the *proem* to *Summa Theologica*, Ia's "Treatise on Man."

24. I do this by showing that some of the (firmly held) claims of the medieval divines are inconsistent with the claims of the Christian physicalists. Of course, had the medieval divines been exposed to the arguments of the Christian physicalists they might have changed their minds. I assume in this chapter that they would not have changed their minds.

25. Richard Cross, "Philosophy of Mind," in *The Cambridge Companion to Duns Scotus*, ed. Thomas Williams (New York: Cambridge University Press, 2003), 263. An example of the different ways in which hylemorphism was shaded can be found by reading Olivi's work. Olivi rejects the Thomist claim that the *intellective part* of the soul is the form of the body. See Pasnau's, "Olivi on the Metaphysics of Soul," *Medieval Philosophy & Theology* 6, no. 2 (1997): 109–132.

26. More specifically, according to Aquinas we are human beings, human beings are rational animals and all human beings fall under the category of *person*. See *Summa Theologiae*, IIIa, q.16, a.12 *ad* 1. I will use the terms human being and human person interchangeably.

27. It should be stressed that matter in and of itself, according to Aquinas, is not a substance.

28. Eleonore Stump, *Aquinas* (London: Routledge, 2005), 36.

29. Take, for example, the disagreement between Aquinas and Olivi mentioned in footnote 19.

30. Although I do not think it impossible or naïve.

31. Cross, "Philosophy of Mind," 263.

32. Robert Pasnau and Christopher Shields, *The Philosophy of Aquinas* (Boulder, CO: Westview, 2004), 158.

33. James Porter Moreland and Scott B. Rae, *Body & Soul: Human Nature & the Crisis in Ethics* (Downers Grove, IL: InterVarsity Press, 2000).

34. Brian Leftow, "Soul, Mind and Brain," in *The Waning of Materialism*, ed. Robert C. Koons and George Bealer (New York: Oxford University Press, 2010), 411.

35. Robert Pasnau, *Thomas Aquinas on Human Nature: A Philosophical Study of Summa Theologiae 1a*, 75–89 (New York: Cambridge University Press, 2002).

36. Stump, at one point, considers Aquinas as a physicalist. She writes, "Patricia Churchland supposes it to be one of the main characteristics of physicalism to hold that 'mental states are implemented in neural stuff.' But if that characterization of physicalism is right, Aquinas should apparently be grouped with the physicalists" (Stump, *Aquinas*, 213). It should simply be pointed out that that characterization of physicalism is not right. That "mental states are implemented in neural stuff" is merely a necessary but not sufficient condition of physicalism. One can be a substance dualist and think that mental states are implemented in neural stuff, if one thinks that emergent substance dualism, for example, is true.

37. Although I should note that I find it plausible to think that Aquinas's view entails a kind of property dualism. This will become apparent. One may take my proposal here to be that Aquinas's view, in the very least, entails a position that

can broadly be called "property dualism" and that this view is inconsistent with physicalism.

38. *Summa Contra Gentiles*, II, 69.

39. *Quaestio disputata de spiritualibus creaturis, un.* 2.

40. Christopher Brown, "Thomas Aquinas," *Internet Encyclopedia of Philosophy*, available at http://www.iep.utm.edu/aquinas/ (section 7).

41. See *Summa Theolgiae*, I. Q. 76, a.1, *respondeo.*

42. Brian Leftow ("Soul, Mind and Brain,": 395–416), Christina van Dyke ("Not Properly a Person: The Rational Soul and 'Thomistic Substance Dualism,'" *Faith and Philosophy* 26, no. 2 (2009): 186–204), and Pasnau (*Thomas Aquinas on Human Nature*) all put forward arguments to the effect that Aquinas did not think that the soul was an immaterial substance. For a brief criticism of Pasnau's argument see Daniel Hill, "Thomas Aquinas on Human Nature," *British Journal for the History of Philosophy* 15, no. 2 (2007): 383–419. The argument put forward in this paper is drawn from van Dyke, "Not Properly a Person."

43. *Quaestiones Disputatae de Anima*, 1.co.

44. See *Summa Theologiae*, 75.2 ad 3 and 4.

45. See *Summa Theologiae*, Ia 75.2.

46. I think that Leftow's "Soul, Mind and Brain" is correct. Just because Aquinas thinks that the brain is not the "organ of thought" this does not sanction the conclusion that there is no organ of thought.

47. Pasnau, *Thomas Aquinas on Human Nature*, 68.

48. Pasnau, *Thomas Aquinas on Human Nature*, 71.

49. See *Summa Contra Gentiles*, I.44.

50. One might think that Stoljar would affirm the antecedent of this conditional. Properties instantiated upon intellection are not properties of an intuitively physical object or objects; namely, corporal matter.

51. Pasnau, *Thomas Aquinas on Human Nature*, 71.

52. The argument made here is a similar to an argument made in Hill's "Thomas Aquinas on Human Nature."

53. I understand that there is a lively and interesting debate between those who read Aquinas as a *corruptionist* and those who read Aquinas as a *survivalist*. Corruptionists read Aquinas as holding that when a human person dies it ceases to exist even though that human person's soul continues to exist. Survivalists think that although the human person is never identical to her soul, she is composed of her soul during the interim state and so continues to exist. I cannot weigh in on this argument here. For the sake of the argument, however, in this chapter I will assume that the survivalists have it right. This point, however, should not matter too much. If one thinks that Aquinas was a corruptionist the argument that I put forward can be adjusted slightly. One can argue that there are mental properties of souls (mere parts of human persons) that are instantiated during the intermediate state without there being any physical properties of a human person necessitating them. I take it that the Christian physicalist would disagree with this thesis too.

54. For a recent sophisticated "survivalist" view I recommend Jeffery E. Brower's *Aquinas's Ontology of the Material World: Change, Hylomorphism, and Material Objects* (New York: Oxford University Press, 2014).

55. I leave it to the reader to reformulate the arguments in this chapter with the assumption that corruptionists are right.

56. Brown, "Thomas Aquinas," section 7.

57. Brown, "Thomas Aquinas," section 7.

58. *Summa Theologia*, Ia. q. 89, a. 1, ad3.

59. My thanks go to Christopher Brown, Daniel Hill, and Greg Miller for their helpful comments on earlier drafts of this chapter.

SUGGESTED FURTHER READING

Aquinas, Thomas. *Summa Contra Gentiles.* Translated by James F Anderson. South Bend, IN: University of Notre Dame Press, 1976.

Brower, Jeffrey E. *Aquinas's Ontology of the Material World: Change, Hylomorphism, and Material Objects*. New York: Oxford University Press, 2014.

Cross, Richard. "Philosophy of Mind." In *The Cambridge Companion to Duns Scotus*, edited by Thomas Williams, 263–84. New York: Cambridge University Press, 2003.

Dyke, Christina van. "Not Properly a Person: The Rational Soul and 'Thomistic Substance Dualism.'" *Faith and Philosophy* 26, no. 2 (2009): 186–204.

Hill, Daniel. "Thomas Aquinas on Human Nature." *British Journal for the History of Philosophy* 15, no. 2 (2007): 383–419.

Leftow, Brian. "Soul, Mind and Brain." In *The Waning of Materialism*, edited by Robert C Koons and George Bealer, 395–416. New York: Oxford University Press, 2010.

Pasnau, Robert. "Olivi on the Metaphysics of Soul." *Medieval Philosophy & Theology* 6, no. 2 (1997): 109–32.

———. *Thomas Aquinas on Human Nature: A Philosophical Study of Summa Theologiae 1a*, 75–89. New York: Cambridge University Press, 2002.

Pasnau, Robert, and Christopher Shields. *The Philosophy of Aquinas*. Boulder, CO: Westview, 2004.

Sharpe, Kevin W. "Thomas Aquinas and Nonreductive Physicalism." *Proceedings of the American Catholic Philosophical Association* 79 (2005): 217–227.

Stoljar, Daniel. *Physicalism.* New York: Routledge, 2010.

Stump, Eleonore. *Aquinas*. London: Routledge, 2005.

van Inwagen, Peter. "Dualism and Materialism." *Faith and Philosophy* 12, no. 4 (1995): 475–488.

Chapter 3

Substance Dualism and the Diachronic/Synchronic Unity of Consciousness

J. P. Moreland

The great Presbyterian scholar, J. Gresham Machen, once observed, "I think we ought to hold not only that man has a soul, but that it is important that he should know that he has a soul."[1] From a Christian perspective, this is a trustworthy saying. Though not unique in this regard, Christianity is a dualist, interactionist religion in this sense: God, angels and demons, and the souls of men and beasts are immaterial substances that can causally interact with the world. Specifically, human persons are (or have) souls that are spiritual substances that ground personal identity in a disembodied intermediate state between death and final resurrection.[2] Clearly, this was the Pharisees' view in Intertestamental Judaism, and Jesus (Matt. 22:23–33; cf. Matt. 10:28) and Paul (Acts 23:6–10; cf. 2 Cor. 12:1–4) side with the Pharisees on this issue over against the Sadducees.[3]

Besides biblical teaching, property and substance dualism are the common-sense views held by the overwhelming number of humankind now and throughout history. As Charles Taliaferro points out, this is widely acknowledged by physicalists, including Michael Levin, Daniel Dennett, David Lewis, Thomas Nagel, J. J. C. Smart, Richard Rorty, Donald Davidson, and Colin McGinn.[4] Throughout history, most people have been substance and property dualists and most of them had no exposure to Greek thought. Thus, regarding the mind/body problem, Jaegwon Kim's concession seems right: "We commonly think that we, as persons, have a mental and bodily dimension. . . . Something like this dualism of personhood, I believe, is common lore shared across most cultures and religious traditions."[5] And regarding issues in personal identity, Frank Jackson acknowledges, "I take it that our folk conception of personal identity is Cartesian in character—in particular, we regard the question of whether I will be tortured tomorrow as separable from the question of whether someone with any amount of continuity—psychological, bodily, neurophysiological, and so on and so forth—with me today will be tortured."[6]

People do not have to be taught to be dualists like they must if they are to be physicalists. Indeed, little children are naturally dualists. Summing up the recent research in developmental psychology, Henry Wellman states that "young children are dualists: knowledgeable of mental states and entities as ontologically different from physical objects and real [non-imaginary] events."[7]

Given all this, one would think that everyone would be a property and substance dualist, or at least, every Christian would be one. But this is obviously not the case. So, in what follows, I want to show the inadequacy of physicalism regarding the bearer and unifier of consciousness, and the superiority of substance dualism in this regard. I will focus on three specific issues: (1) the irrelevance of neuroscience for grasping the nature and existence of consciousness and the soul; (2) the diachronic identity of the human person and the inadequacy of physicalism to account for it; (3) the nature and reality of the synchronic unity of consciousness and its ontological implications.

THE IRRELEVANCE OF NEUROSCIENCE FOR GRASPING THE NATURE AND EXISTENCE OF CONSCIOUSNESS AND THE SOUL

Since this volume addresses questions about the adequacy of Christian physicalism, it is obvious that some Christian thinkers have rejected the commonsense and historical Christian view (substance dualism of some form) in favor of a physicalist alternative. So far as I can tell, the main reason for the change of viewpoint is the idea that it is somehow required by advances in neuroscience. Thus, according to Nancey Murphy, "science has provided a massive amount of evidence suggesting that we need not postulate the existence of an entity such as a soul or mind in order to explain life and consciousness."[8] This evidence consists of the fact that "biology, neuroscience, and cognitive science have provided accounts of the dependence on physical processes of *specific* faculties once attributed to the soul."[9] Elsewhere, she claims: "My argument in brief is this: all of the human capacities once attributed to the mind or soul are now being fruitfully studied as brain processes—or, more accurately, I should say, processes involving the brain, the rest of the nervous system and other bodily systems, all interacting with the socio-cultural world."[10] Murphy acknowledges that "dualism cannot be *proven* false—a dualist can always appeal to correlations or functional relations between soul and brain/body—but advances in science make it a view with little justification."[11]

Addressing these claims is not the main purpose of this chapter. Nevertheless, I think something should be said in response, even if briefly, so the rest of the chapter can be taken more seriously. Murphy's appeal to neuroscience as justification for physicalism is constituted by two claims: (1) the evidence of neuroscience provides accounts of the dependence on physical processes of *specific* faculties once attributed to the soul. So, it is the specificity of these physical processes that makes it highly unlikely that the work is being done by mental faculties of the soul. (2) We no longer need to postulate a soul to explain life and consciousness because all the capacities once attributed to it are now being fruitfully studied as physical processes in the brain. So, physical brain processes have co-opted the role soulish capacities used to play.

I would like to address these in order. Regarding (1), a dualist can only scratch his/her head at this assertion. What, exactly, is supposed to be the problem here? It cannot be that we now know that the neurological correlations involve specific regions of the brain. As C. Stephen Evans notes regarding the findings of localization studies:

> What, exactly, is it about these findings that are supposed to create problems for dualism? [. . .] Is it a problem that the causal effects should be the product of specific regions of the brain? Why should the fact that the source of the effects is localized regions of the brain, rather than the brain as a whole, be a problem for the dualist? It is hard for me to see why dualism should be thought to entail that the causal dependence of the mind on the brain should only stem from holistic states of the brain rather than more localized happenings.[12]

In his 1886 lectures on the limitations of scientific materialism, John Tyndall claimed that "the chasm between the two classes of phenomena is of such a nature that we might establish empirical association between them, but it would still remain intellectually impassable. Let the consciousness of love, for example, be associated with a right-handed spiral motion of the molecules in the brain, and the consciousness of hate with a left-handed spiral motion. We should then know when we love that the motion is in one direction, and when we hate that the motion is in the other; but the 'WHY' would remain as unanswerable as before ."[13] Nothing substantial has changed since Tyndall made this remark. Specifically, no advance in knowledge of the specificity of detail regarding the correlations between mental and physical states provides any evidence against dualism. Aquinas knew that brain damage affected mental functioning. We have just added more details to what has always been known.

Regarding (2), many substance dualists do not believe in a substantial ego primarily because it is a theoretical postulate with superior explanatory power, certainly not superior *scientific* explanatory power. Rather, they take the ego to be something of which people are directly aware. Thus, for many dualists, belief in a substantial, simple soul is properly basic and grounded in self-awareness. Substance dualism is not a theory that could in principle be replaced by a better one.

Moreover, the neuroscientific empirical discovery of such correlations (causal or dependency relations both ways) provides no evidence whatsoever for physicalism. Indeed, the central issues regarding the mind—what is a thought, feeling, or belief; what is that to which my self is identical—are basically common-sense and philosophical issues for which scientific discoveries are largely irrelevant. Science is helpful in answering questions about what factors in the brain and body generally hinder or cause mental states to obtain, but science is largely silent about the nature of mental properties/states. Indeed, neuroscience must assume the common-sense nothing of the mental made aware to first-person introspection since neuroscientific correlations ultimately depend on first-person reports.

To see the basic irrelevance of empirical data, consider the following. We have discovered that if a certain type of neuron—*mirror neurons*—is damaged, then one cannot feel empathy for another. How are we to understand this? To answer this question, we need to get before our minds the notion of *empirically equivalent theories*. If two or more theories are empirically equivalent, then they are consistent with all and only the same set of empirical observations. Thus, an appeal to empirical data cannot be made in favor of one of such theories over the others.

Three empirically equivalent solutions to the discovery of the function of mirror neurons come to mind: (1) strict physicalism (a feeling of empathy is identical to something physical, for example, the firings of mirror neurons); (2) mere property dualism (a feeling of empathy is an irreducible state of consciousness *in the brain* whose obtaining depends on the firing of mirror neurons); (3) substance dualism (a feeling of empathy is an irreducible state of consciousness *in the soul* whose obtaining depends (while embodied) on the firing of mirror neurons).

Of these three, no empirical datum can pick which is correct, nor does an appeal to epistemic simplicity help. This is why three Nobel Prize winners working in neuroscience and related fields—John Eccles (substance dualist), Roger Sperry (mere property dualist), and Francis Crick (strict physicalist) could hold different ontologies regarding consciousness and the self even though they all knew the same neuroscientific data. Epistemic simplicity is a tie-breaker, and the substance dualist will insist that the arguments and evidence for substance dualism are better than those for the other two options mentioned earlier. Murphy implicitly acknowledges this when she admits that "dualism cannot be *proven* false—a dualist can always appeal to correlations or functional relations between soul and brain/body." But she apparently does not realize that this admission provides a defeater for her conclusion that "advances in science make it a view with little justification."

THE DIACHRONIC IDENTITY OF THE HUMAN PERSON AND THE INADEQUACY OF PHYSICALISM TO ACCOUNT FOR IT

So much for my view of the relevance of neuroscience regarding the crucial metaphysical topics before us. In this section, I turn to two issues that are supported by widespread, common-sense intuitions and by a classic, historical understanding of Christianity: diachronic identity and libertarian free will, and diachronic identity and personal responsibility.

Diachronic Identity, Physicalism, and Libertarian Free Will

Why are libertarian intuitions so pervasive that they clearly constitute the common-sense default position? Indeed, everyone is a libertarian unless their ideology requires them to embrace compatibilism. I think it is because through first-person introspection, people the world over are simply directly aware of themselves exercising active power in bringing about the effects they endeavor to achieve. When one sees a hammer driving a nail into lumber, they don't just see the hammer moving followed by the nail moving. They also perceive the hammer-moving-the-nail. Similarly, we are directly aware of our own endeavoring-to-raise-our-arm-in-order-to-vote. On the basis of such awarenesses, we form the properly basic belief that we exercise originative, free, active power for the sake of teleological goals.

Shortly, I shall unpack the formal elements of a libertarian view of free will that I take to be true and most obvious.[14] Most philosophers are agreed that libertarian freedom and a theory of agency it entails are incompatible with the generally accepted depiction of physicalism. Thus, Roderick Chisholm claimed that "in one very strict sense of the terms, there can be no science of man."[15] Along similar lines, John Searle says that "our conception of physical reality simply does not allow for radical [libertarian] freedom."[16] And if moral (and intellectual) responsibility has such freedom as a necessary condition, then reconciling the natural and ethical perspectives is impossible. In what may be the best naturalist attempt to accomplish such a reconciliation, John Bishop frankly admits that "the idea of a responsible agent, with the 'originative' ability to initiate events in the natural world, does not sit easily with the idea of [an agent as] a natural organism. . . . Our scientific understanding of human behavior seems to be in tension with a presupposition of the ethical stance we adopt toward it."[17]

Stated formally, a person, P, exercises libertarian agency and freely performs some intentional act, e, just in case 1) P is a substance that has the active power to bring about e; 2) P exerted his/her active power as a first, unmoved mover (an "originator") to bring about e; 3) P had the categorical ability to refrain from exerting his/her power to bring about e; 4) P acted for the sake of reasons which serve as the final cause or teleological goal for which P acted. Taken alone, 1–3 state necessary and sufficient conditions for a pure voluntary act, for example, freely directing my eyes toward a specific desk upon entering a room. Propositions 1–4 state necessary and sufficient conditions for an intentional act, that is, a voluntary act done for a reason (for example, raising my hand to vote).

There are five features of a free act that makes it difficult and, indeed, virtually impossible to reconcile with a strictly physicalist standpoint. First, the free agent is a substance and not an event, a bundle of events or an ordered aggregate of separable parts. But according to physicalists, all causes and effects are events. The laws of nature govern causal processes in which a temporal state (an event) of an object (an electron, water molecule, storm cloud, location or temperature of the earth) brings about a different temporal state (a subsequent event) according to a natural law.

This is called "event-event causation" and it is the only sort of efficient cause recognized in physics, chemistry, geology, biology, neuroscience or other physicalistically certified hard sciences. Event-event causation governs changes of state in or among objects—nothing more, nothing less. Substances as substances—essentially characterized particulars, substantial things—do not cause things. Strictly speaking, it is not the first billiard ball that moves the second. It is the moving-of-the-first-ball (the causal event) that causes the moving-of-the-second-ball (the effect event).

By contrast, it is the agent as a substantial self or I, not some state in the agent, that brings about a free act. By "substance" I mean a member of a natural kind, an essentially characterized particular that sustains absolute sameness through (accidental) change and that possesses a primitive unity of inseparable parts (a.k.a., modes), properties, and capacities/powers at a time. This strong view of substance is required for libertarian agency for at least three reasons: (1) libertarian agency is possible only if there is a distinction between the capacity to act or refrain from acting and the agent that possesses those capacities. (2) The type of unity present among the various capacities possessed by an agent is the type of unity (that is, a diversity of capacities within an ontologically prior whole) that is entailed by the classic Aristotelian notion of substance. (3) Ordinary free acts take time and include subacts as parts, and an enduring agent is what gives unity to such acts by being the same self who is present at the beginning of the action as intentional agent, during the act as teleological guider of means to ends, and at the end as responsible actor.

But this is not countenanced by physicalism. Thus, naturalist John Bishop frankly admits, "the problem of natural agency is an ontological problem—a problem about whether the existence of actions can be admitted within a natural [i.e., physicalist for Bishop] scientific perspective. . . . [A]gent causal-relations do not belong to the ontology of the natural perspective. Naturalism does not essentially employ the concept of a causal relation whose first member is in the category of person or agent (or even, for that matter, in the broader category of continuant or 'substance'). All natural causal relations have first members in the category of event or state of affairs."[18]

Second, the ontology of physicalism knows nothing of active powers. The particulars that populate that ontology are, one and all, exhaustively characterized by passive liabilities with regard to their causal powers. A passive liability is such that, given the proper efficient cause, it is and, indeed, must be actualized. As such, the actualization of a passive liability is a passive happening, not an *action*. This fact about passive liabilities is what makes them, along with their causes, fitting entities for subsumption under law. And it is precisely as passive and so subsumable that makes their owners bereft of the sort of first-moving, active spontaneity that is a necessary condition for the exercise of free will.

All physical objects with causal powers possess them as passive liabilities. Again, these liabilities are triggered or actualized if something happens to the object and, once triggered, they can produce an effect. For example, dynamite has the power (passive liability) to explode if something is first done to it. And so on for all causes. They are, one and all, passive potentialities. Their actualizations are mere happenings to the relevant object.

But active power is different. In virtue of possessing active power, an agent may *act*, initiate change or motion, perform something, bring about an effect with nothing causing it to do so. Active power is not something admitted in the ontology of the hard sciences, period.

Third, a "first mover" is a substance which has active power. As such, it is the absolute originator of its actions. It is not just another caused cause, just one more event (or bundle of events) in a chain of events in which earlier causes bring about later effects which, in turn, bring about later effects to form one big series of passive happenings governed by natural law. No, a first mover is not subject to laws in its initiation of action. Since such an initiation is a first, spontaneous, action not caused by a prior event, it amounts to the absolute origination of initiatory movement. Such an origination comes into being instantaneously and spontaneously, and while the effect it produces (for example, the earliest stages in the raising of one's arm) may well be subject to natural laws, the initiating event is not since there is nothing prior to its coming-to-be on which a law may operate. Moreover, such a first mover is an unmoved mover, that is, it has the power to bring about an action without having to change first before it can so act.

By contrast, since all events in a physicalist ontology are passive happenings, they all are examples of moved movers, that is, something has to happen to an object first, namely an event that triggers and actualizes its causal powers, before it can cause something else to happen. In this sense, all strictly physicalist causation involves changed changers. But a first mover can produce change without having to change first to do so. It should be obvious why such an agent is not an object that can be located in a physicalist ontology. Unmoved movers are quintessentially nonphysical!

Fourth, the notion of "categorical ability" in (3) has two important aspects to it. First, it expresses the type of ability possessed by a first mover that can exercise active power and, as such, it contrasts with the conditional ability employed by compatibilists. Second, categorical ability is a dual ability: if one has the ability to exert his power to do (or will to do) A, then one also has the ability to refrain from exerting his power to do (or to will to do) A. This means that the circumstances within (for example, motives, desires, reasons) and outside (environmental conditions) the agent at the time of action are not sufficient to determine that or fix the chances of the action taking place. Given those circumstances, the agent can either exercise or refrain from exercising his/her active power, and this ability is the essential, causal factor for what follows. Among other things, this implies that libertarian acts cannot be subsumed under natural laws, whether construed as deterministic or probabilistic.

But all the particulars in the physicalist ontology are so subsumable. In fact, all of them are subject to diachronic and synchronic determinism in the following sense: regarding diachronic determinism, at some time, t, the physical conditions are sufficient to determine or fix the chances of the next event involving the object and its environment. Regarding synchronic determinism, at any time, t, the object's states and movements are determined or have their chances fixed by the microphysical states of the object and its environment. This latter determination is bottom-up.

Fifth, (4) expresses a view of reasons as irreducible, teleological goals for the sake of which a person acts. In general, we may characterize this by saying that person S F'd (for example, went to the kitchen) in order to Y (for example, get coffee or satisfy S's desire for coffee). This characterization of action, according to (4), cannot be reduced to a causal theory of action that utilizes belief/desire event causation such that reasons amount to efficient causes (or causal conditions) for action. To see this, consider these two sentences:

1. The water boiled because it was heated.
2. Smith went to the kitchen because he wanted to get coffee.

(1) is a straightforwardly (efficient) causal assertion. The event cited after "because" (the water's being heated) is the efficient cause for the water's boiling). But while grammatically similar, (2) doesn't employ reasons as causes. This can be seen by paraphrasing (2) as follows:

(2) Smith went to the kitchen in order to get coffee.

Here, getting coffee is the goal, purpose, end of Smith's free action. Every step he takes (getting out of his chair, walking toward the kitchen, opening the pantry) are means to this end.

If there is anything that physicalists agree upon, it is that there is no such thing as teleology. Matter is mechanistic, not in the sense that it only engages in action by contact and is bereft of forces, but in that it only behaves according to chains of efficient causes. As philosophers Joshua Hoffman and Gary Rosenkrantz note, attempts to slap teleology onto a naturalist framework really amount to abandonments of naturalism:

> Aristotle's account [of natural function and teleology] does *not* provide a naturalistic reduction of natural function in terms of efficient causation. Nor do characterizations of natural function in terms of an irreducibly emergent purposive principle, or an unanalyzable emergent property associated with the biological phenomenon of life, provide such a reduction. Theistic and vitalistic approaches that try to explicate natural function in terms of the intentions of an intelligent purposive agent or principle are also nonnaturalistic. Another form of nonnaturalism attempts to explicate natural function in terms of nonnatural evaluative attributes such as intrinsic goodness. . . . We do not accept the anti-reductionist and anti-naturalistic theories about natural function listed above. Without entering into a detailed critique of these ideas, one can see that they either posit immaterial entities whose existence is in doubt, or make it utterly mysterious how it can be true that a part of an organic living thing manifests a natural function. . . . [T]he theoretical unity of biology would be better served if the natural functions of the parts of organic life-forms could be given a reductive account completely in terms of nonpurposive or nonfunctional naturalistic processes or conditions.[19]

For these reasons, the ontology of libertarian agency and the diachronic unity/identity it entails are inconsistent with a common-sense and biblically respectable view of free agency. Of course, if one is a compatibilist, this section may have no relevance to that person. But it seems to me that more and more Christian philosophers and philosophically informed theologians are embracing libertarianism. If so, they need to think carefully about the ontology in which such a move is intelligible.

Diachronic Personal Identity and the Inadequacy of Physicalism to Account for It

If anything about Christianity is self-evident it would be the idea that a number of its entailments—especially those about punishment/rewards for past actions and punishment/rewards in the future judgment—require human persons to be literal continuants who sustain absolute identity through their lives. Their diachronic "identity" is not partial, arbitrary or conventional. It cannot be reduced to mere sortal dependent persistence conditions. The Christian view of punishment and rewards makes sense only if it really is I myself who did such and such an act in the past, who now is contemplating my past and future, and who will be punished or rewarded in the future for that (and other) past act(s). As Geoffrey Madell noted long ago, if absolute diachronic personal identity is set aside, "then the notion of responsibility for past wrong loses its foundation. . . . [O]ur present notion of responsibility wold be destroyed."[20]

Unfortunately, I do not believe there is a physicalist view of the human person (or animal) that can avoid rejecting absolute diachronic identity and putting in its place some weaker relation constituted by persistence conditions, even sortal dependent ones, that fails to ground and comport with the metaphysical entailments of the Christian view.

To see why strict identity is not an option for the extant versions of physicalism, consider the following argument:

1. If something is a physical object composed of parts, it does not survive over time as the same object if it comes to have different parts.
2. My body and brain (or subregion of the brain) are physical objects composed of parts.
3. Therefore, my body and brain do not survive over time as the same object if they come to have different parts.
4. My body and brain are constantly coming to have different parts.
5. Therefore, my body and brain do not survive over time as the same object.
6. I do survive over time as the same object.
7. Therefore, I am not my body or my brain.
8. I am either a soul or a body or a brain.
9. Therefore, I am a soul.

Premise (2) is commonsensically true, given the scientific image. Premise (4) is obviously true as well. Our bodies and brains are constantly gaining new cells and losing old ones, or at least, gaining new atoms and molecules and losing old ones. So understood, bodies and brains are in constant flux. I will assume that (8) represents the only live options for most ordinary people. This leaves premises (1) and (6). I am granting (6) as the common-sense and Christian requirement for the intelligibility of responsibility for past acts and future rewards or punishments.

That leaves us with (1). Why should we believe that ordinary material objects composed of parts do not remain the same through part replacement?[21]

To see why this makes sense, let's start with some simple illustrations: Consider five scattered boards, a–e, each located in a different person's back yard. Commonsensically, it doesn't seem like the boards form an object. They are just isolated boards. Now, suppose we collected those boards and put them in a pile with the boards touching each other. We would now have, let us suppose, an object called a pile or heap of boards. The heap is a weak object, indeed, and the only thing unifying it would be the spatial relationships between and among a–e. They are in close proximity and are touching each other. Now, suppose we took board b away and replaced it with a new board f to form a new heap consisting of a, c–f. Would our new heap be the same as the original heap? Clearly not, because the heap just is the boards and their relationships to each other, and we have new boards and a new set of relationships. What if we increased the number of boards in the heap to 1000? If we now took one board away and replaced it with a new board, we would still get a new heap. The number of boards does not matter.

Now imagine that we nailed our original boards a–e together into a makeshift raft. In this situation, the boards are rigidly connected such that they do not move relative to each other; instead, they all move together if we pick up our raft. If we now took board b away and replaced it with board f, we would still get a new object. It may seem odd, but if we took board b away and later put it back, we would still have a new raft because the raft is a collection of parts and bonding relationships to each other. Thus, even though the new raft would still have the same parts (a–e), there would be new bonding relationships between b and the board or boards to which it is attached.

We cannot go beyond these simple illustrations and put the argument in a more precise, technical way. Premise (1) expresses a commitment to mereological essentialism for physical objects composed of separable parts. Why is mereological essentialism a problem for virtually all versions of physicalism besides those who identify us with an atomic simple? Because, at the end of the day, these versions of physicalism identify us as mereological aggregates, and mereological essentialism cannot be avoided for such wholes (if there are such wholes.)

Here is a definition of a mereological aggregate: it is a particular whole that is constituted by (at least) separable parts and external relation-instances between and among those separable parts (there is a debate as to whether or not one should add an additional constituent, such as a surface or boundary to the analysis). Mereological aggregates are very different to genuine Aristotelian substances. Jonathan Schaffer characterizes the difference in terms of grounding (ontological dependency or priority):

> The notion of grounding may be put to further use to capture a crucial mereological distinction (missing from classical mereology) between an integrated whole

which exhibits a genuine unity, and a mere aggregate which is a random assemblage of parts. Thus, Aristotle speaks of "that which is compounded out of something so that the whole is one—not like a heap, however, but like a syllable." (1984: 1644; Meta.1041b11–2). This intuitive distinction may be defined via:

Integrated whole: x is an integrated whole =df x grounds each of its proper parts.
Mere aggregate: x is a mere aggregate =df each of x's proper parts ground x.[22]

Shaffer's integrated whole and mere aggregate are the same as my substance and mereological or ordered aggregate, respectively. Later, we shall see that the "proper parts" of a mere aggregate are separable parts and those of an integrated whole are inseparable parts/modes.

Why think that mereological essentialism characterizes mereological aggregates? Because a proper metaphysical analysis of such wholes does not provide an entity adequate to ground their literal identity through part alteration. To see this, suppose we have some mereological aggregate W, say a car, in the actual world, w, at some time, t, and let "the ps" refer distributively to all and only the atomic simples (assuming such) that make up W. Now, given that the ps just are a specific list of simples taken distributively without regard to structure, it would seem obvious that if we have a different list of simples, the qs, it is not identical to the ps even if the two lists share all but one part in common. This same insight would be true if we took "the ps" and "the qs" collectively as referring to some sort of mereological sum. In either case, there is no entity "over and above" the parts or sum members that could serve as a ground of sameness through part alteration.

Now, W has different diachronic identity conditions than, and, thus, is not identical to the ps. W could be destroyed and the ps (taken in either sense) could exist. Let S stand for all and only the various relations that stand between and among the ps. S is W's structure. Is W identical to S and the ps? I don't think so. W has its own structure, say in comparison to some other whole W* that is exactly similar in structure to W. W and W* have their own structures. Given that S is a universal, it is not sufficient for individuating W's specific structure. For that we need SI, W's structure-instance, W's token of S and SI will consist of all and only the specific relation-instances that are instantiated between and among the ps.

Let "the rs" stand for all and only the relevant relation-instances that compose SI. I think it is now obvious that SI is a mereological aggregate composed of the rs. If the rs undergo a change of relation-instances, it is no longer the same list of relation-instances. Given that SI just is a mereological aggregate or, perhaps, a specific ordering of the rs, if the rs undergo a change of relation-instances, SI will cease to exist and a different structure (perhaps exactly similar to SI) will obtain since there is no entity to serve as a ground for SI's sameness through part replacement. If W is the ps plus SI, it seems to follow that W is subject to mereological-essentialist constraints. Adding a surface/boundary to W won't help avoid these constraints.

Now, consider our bodies and brains and assume they are mere physical objects composed of billions of parts. From our daily vantage point—the manifest image—they appear to be solid, continuous objects. But if we could shrink down to the level of an atom, we would see that, in reality, they are like a cloud—gappy, largely containing empty space and composed of billions of atoms (molecules, cells) that stand in various external (bonding) relation-instances between and among those parts. If we were to take a part away and replace it, we would have a new object, assuming there is an object there in the first place. The body and brain are mereological aggregates. Besides the parts and the relation-instances between and among them, there is nothing in the body or brain to ground its ability to remain the same through part alteration.

This is the fundamental insight behind the claim that the body and brain cannot remain the same if there is part alteration.[23] Since the body and brain are constantly changing parts and relationships, they are not the same from one moment to the next in a strict philosophical sense (though, for practical day-to-day purposes, we regard them as the same in a loose, popular sense.) Thus, physicalist views that identify us with our bodies or brains (and adding lives or psychological criteria won't help because these are constantly in flux just as much as bodies and brains) fail to ground the strict diachronic identity of human persons that is the ground of deeply ingrained intuitions about rewards and punishment and of the Christian view of such.

THE NATURE AND REALITY OF THE SYNCHRONIC UNITY OF CONSCIOUSNESS[24]

Perhaps more than anyone else in the literature, Tim Bayne and David Chalmers have thought carefully about the synchronic unity of consciousness (hereafter, simply the unity of consciousness), although in my view, Bayne is the chief current thinker on these matters. They distinguish several different types of unified consciousness, but only three of them are relevant to my present concerns.[25] To understand these, I should point out that a phenomenal conscious state, ϕ, is one such that there is a what-it-is-like to be in ϕ.

The Types of Consciousness's Unity

First, there is *objectual phenomenal unity*. Two or more states are so unified if they are experienced as being of the same object. For example, the state of feeling a desk and the state of seeing the desk's color, or the state of seeing the desk's shape and seeing its color are objectually phenomenally unified just in case they are experienced as being of the same object, the desk. This type of unity generates the binding problem. How is it that we experience, say perceptually, separate pieces of information as bound together in pertaining to the same object?

Second, there is *subject phenomenal unity*. This occurs when all of one's phenomenal states are had by the same subject. According to Bayne and Chalmers, this sort of unity is irrelevant for investigating the nature of consciousness's unity because it is trivially true by definition and tells us nothing about consciousness. Unfortunately, this claim is all too convenient for Bayne because without it his own solution, as we shall see below, is not intellectually motivated. Moreover, it is just plain false. Many thinkers, including Bayne himself, believe there is no self or subject that unifies consciousness. And some thinkers like me think that a simple self or soul is the best explanation for the unity of consciousness, especially when a careful metaphysical account is given for how conscious states are in the self in the first place. More on this later.

Third, there is *subsumptive phenomenal unity*. Two (or more) states are subsumptively phenomenally unified just in case there is something it is like to be in both states simultaneously and conjointly. All of one's phenomenal states are subsumed within a single (totalizing) phenomenal state. Bayne and Chalmers insist, rightly in my view, that one's total phenomenal field is not built up atomistically as a complex conjunction of individual phenomenal states. Rather, one's totalizing state is a whole, and the various individual phenomenal states are aspects of that ontologically prior whole.

For example, suppose you are having two different phenomenal experiences—an awareness of ϕ and an awareness of ψ. Then there will be a totalizing state, T, in its own right with its own what-it-is-like to be in T. T is "over and above" and not merely an atomistic conjunction of the two awarenesses of φ and ψ, taken individually. According to Bayne and Chalmers, any subject of experience will have a single, total phenomenal state at a point in time.[26] Coming from a different direction, Cleeremans makes roughly the same point this way: "[There is] the intuitive idea that consciousness requires unity of consciousness, that is, that there is no sense in which one could simultaneously have separate conscious experiences that failed to present themselves as integrated in a single phenomenal field."[27]

Bayne and Chalmers go on to formulate what they call the Total Phenomenal Unity Thesis: necessarily, the set of all phenomenal states of a subject at a time is phenomenally unified.[28] According to them, this thesis captures the central insight behind unity of consciousness: there is always a single phenomenal state that subsumes all of one's phenomenal states at a time, and this state is one's totalizing phenomenal state.

It is worth briefly making two observations. First, the necessity of the Total Phenomenal Unity Thesis should be explained (if possible) and not taken as a primitive. Second, the notion of subsumption wherein one state subsumes another is unclear to say the least. Bayne and Chalmers claim that it is "something of an intuitive primitive."[29] In my view, this is an unfortunate situation, and if we can ground the totalizing unity of consciousness in a clearer, metaphysically available entity, then we should seek to do that.

In sum, these three sorts of unity are all relevant to what follows. Indeed, I will try to show that proper analysis of one of these theses provides the solution to the other two. In any case, when speaking of the unity of consciousness, I will use the specific thesis that I have in mind.

The Unity of Consciousness as a Defeater for Physicalism about the "Self"

As we shall see, the unity of consciousness presents a significant difficulty for physicalism regarding the possessor of consciousness: given such physicalism, the various entities to which a human person is identical—the organism, the brain, an object constituted by an organism, a 4-dimensional physical object synchronically composed of its physical parts and diachronically composed of its stages—do not have the sort of unity needed to account for or ground the unity of consciousness because they are all mereological aggregates. Rather, the best explanation/ground for the unity of consciousness is a simple (not composed of separable parts) soul.

More than anyone else, William Hasker has championed this argument for substance dualism.[30] To remind us of what we noted earlier, by the unity of consciousness, say, of one's visual field, I mean (at least) two things. First, there is what Bayne and Chalmers call subsumptive phenomenal unity: all of one's experiences are subsumed within a single, totalizing state of consciousness. This totalizing state is a conscious state in its own right, and there is a what-it-is-like to be in that state.[31] The Total Phenomenal Unity Thesis says that, necessarily, there is always a single phenomenal state that subsumes all of one's other phenomenal states at a time.

The notion of "subsuming" is a bit unclear, so let me state the second thing I mean by the unity of consciousness, also from Bayne, and add my own metaphysical clarification to it. According to Bayne, an atomistic theory of consciousness states that the phenomenal field is composed of "atoms of consciousness"— independent conscious states.[32] Among other things, this would mean that one's field of consciousness is like a mereological aggregate—it contains and is built up by separable parts ("atoms of consciousness")— placed into various external relation-instances which constituted the field's structure—a mereological aggregate of relation-instances.[33]

By contrast, Bayne—and I—accept holism: the components of the phenomenal field are conscious only as components of that field (it is interesting to note that diachronically, consciousness changes as a continuous flow, but the brain changes states in a discrete, atomistic way). I add the qualification that the phenomenal field is a whole in which subsumptive components are modes or inseparable parts of something—either the whole field, or much more likely, of the grounding entity for subsumptive unity.

Now consider the following principle:

> (F) For any complex object (one with a plurality of separable parts) O, if O performs function F, then O's performing function F consists in parts p_1–p_n and subfunctions/activities f_1–f_n, such that p_1 performs f_1 . . . p_n performs f_n.

For example, a computer performing function F just is a certain set of its parts performing their own subfunctions. Principle F can also be stated in terms of properties such that an object, O, having some property, P, consists in each part having some property or other. This is clearly the case with additive properties, for example, mass. It does not, however, rule out emergent properties. Given the reasonable assumption that supervenience for simple, emergent properties is local (the supervenient simple property obtains and is dependent on what is going on right there at the subvenient base), the principle disallows emergent properties exemplified by complex objects like O taken as an irreducible whole. But it does not disallow each of the relevant parts of O to have an emergent property as long as these parts are mereological simples.

The following argument, then, is an attempt to show that the unity of consciousness cannot be explained if one is a brain (or any of the other naturalist candidates mentioned earlier), because a brain is just an aggregate of different physical (separable) parts. It is only if the self is a single, simple subject that we have an adequate account for the unity of consciousness.

To illustrate, consider one's awareness of a complex fact, say one's own visual field consisting of awareness of several objects at once, including a number of different surface areas of each object. One's entire visual field contains several different experiences, for example, being aware of a desk toward one's left side and being aware of a podium in the center of one's visual experience of an entire classroom. Corresponding to such an experience, numerous different light waves bounce off different objects (and off different locations on the surface of the same object, say different areas of the desk's top side), they all interact with the subject's retinas, and they all spark signals that terminate in a myriad of locations in the brain, breaking objects down into constituents.[34] If we add local emergence, then we could hold that each relevant part of the brain instantiates an atomistic sensory experience.

Accordingly, a physicalist may claim that such a unified awareness of the entire room by means of one's visual field consists in a number of different physical parts of the brain each terminating a different wavelength, each of which is aware only of part—not the whole—of the complex fact (the entire room). But this cannot account for the single, unitary awareness of the entire visual field. There is a what-it-is-like to have the whole visual field. If we terminate our search for an explanation for this with a holistic phenomenal field, then two problems arise. First, it is hard to see how a myriad of atomistic parts could give rise to a single, nonatomistic, holistic field; we are owed an account of this within the constraints of subject physicalism.

Second, a basic datum of our experience is not simply this or that item of awareness in the room, but that I *have* and *am not identical to* the totalizing state. In the history of philosophy, classic substances have served to unify things in this way, and Hasker and LaRock believe this ontology provides the best answer for how we could have a totalizing, unified field of consciousness. The very same substantial soul is aware of the desk to the left, the podium at the center, and, indeed, each and every distinguishable aspect of the room. But no single part of the brain is correspondingly activated as a terminus for the entire visual fields. Only a single, uncomposed mental substance can adequately account for the unity of one's visual field or, indeed, the unity of consciousness in general.

The most prominent physicalist rejoinder attempts to explain objectual phenomenal unity in terms of synchronicity: all the different locations of the brain processing electrical signals associated with different aspects of the object of perception (for example, color, size, shape, etc.) fire together at the very same time, and this explains objectual unity. Unfortunately, a growing amount of empirical evidence refutes this thesis.[35] And, philosophically, the connection between synchronicity and objectual unity is unclear. Consider LaRock's analogy: "If five chefs are located in separate kitchens and each chef is consciously aware of only part of the same recipe, it does not follow that any one chef is consciously aware of the recipe as a whole—*even if all of the chefs are consciously aware of their respective recipe parts at the same time*."[36] The synchronicity solution, then, fails to be adequate.

What, If Anything, Grounds the Three Types of Unity, Especially Subsumptive Phenomenal Unity?

As I mentioned earlier, the leading thinker on these matters is Tim Bayne. Because of this, and due to space limitations, I will focus my attention on Bayne's approach,[37] explain why I think it is inadequate, and provide what I take to be a better solution.[38]

As Bayne sees it, there are three roles the unifier of consciousness (which he calls the self or subject of experience) must play: (1) Ownership—that which has conscious experiences. (2) Referential—objects of I-thoughts involved in first-person reflection. (3) Perspectival—"selves" have a perspective, a first-person point of view. In my view, there are other key roles for the self (for example, being the agent of libertarian acts, being that which is [metaphysically] possibly disembodied), but since, with qualifications, I accept Bayne's three, I shall not quibble with his list. However, I do want to make a few brief comments about it. Regarding (1), there should be a clear, plausible metaphysical analysis of the sort of "having" it involves. Regarding (2), it should be expanded to say that it is not only the object of the self-referring use of "I" but also that which employs "I" to self-refer. I also think it is question-begging at this early stage of analysis to use "I-thoughts" to characterize the nature of the role of reference the unifier must satisfy. Regarding (3), it is unclear what metaphysical notion of "having" is being used or what it is that does the having. More on this later.

Bayne begins by defending the phenomenal unity thesis on empirical grounds by beginning with the self as a biological organism, looking at specific cases of animalism and psychological views of unity to see if phenomenal unity prevails. He finds that it does. But then he shifts to a stronger a priori conceptual claim: necessarily, x is a self iff x has phenomenal unity. According to Bayne, what we need is a phenomenalist (not a functionalist) conception of the self that allows us to construct selves out of streams of consciousness and affirm, as a matter of conceptual necessity, that no self can possess simultaneously two phenomenally unified streams of consciousness. Thus, if we entertain a thought experiment in which there are two functionally interactive, isomorphic streams of consciousness (that is, the two streams exhibit functional unity) that, nevertheless, are not phenomenally unified, then we have two minds, not one.

This seems right to me. David Barnett offers a thought experiment that undergirds this intuition.[39] Consider two people, Fred and Ted, who have trained for years such that Fred and Ted can completely imitate the functional activities of the left and right hemisphere, respectively. Now suppose we take a third person, Joe, remove his brain, shrink Fred and Ted down to hemisphere size, and put Fred in Joe's left hemisphere and Ted in his right. After the operation, from a third person perspective, there is only one person and one stream of consciousness present since Fred and Ted are completely functionally unified. But, says Barnett, given two streams of integrated phenomenal consciousness, there are two persons present, not one.

So, Bayne's project becomes one of finding a view that allows us to construct a self out of a phenomenally unified stream of consciousness. He considers and rejects two views—naïve phenomenalism and substrate phenomenalism—and concludes by proffering his own view—virtual phenomenalism. Before we look at virtual phenomenalism, it will be instructive to see the way Bayne handles the two positions he rejects.

Naïve Phenomenalism. According to naïve phenomenalism, it is no great mystery as to why there is a 1:1 ratio between selves and streams of conscious, *because selves just are streams of consciousness.* While Bayne ultimately rejects this view, he provides defeaters for a number of objections raised against it. Why would Bayne take the time to do this? In my view, it is because the metaphysics of his overall position entails that a person is synchronically identical to "his" totalizing phenomenal field and diachronically identical to (or, perhaps, the "same" as) "his" stream of consciousness. As we will see shortly, his own view (virtual phenomenalism) simply adds a Kantian-like twist to the metaphysics. In any case, in what follows I shall state the objection to naïve phenomenalism, present the response I imagine Bayne would offer, and follow that with my own reply to Bayne.

Argument 1: Selves can't *be* streams of consciousness—selves are things in their own right and streams are modifications of selves.
Bayne's Reply: In some sense, streams *are* things in their own right, for example, they have their own principle of unity. The forces that knit together the components of a stream of consciousness are no less robust than those that knit together a single mind or animal.
My Reply to Bayne: Nowhere has Bayne demonstrated that a stream of consciousness has its own principle of unity, and the thin metaphysical framework within which he works—the totalizing phenomenal state "subsumes" its substates—employs an obscure metaphysical notion that provides no insight whatsoever as to how this is supposed to take place. Moreover, his analogy with "forces that knit together a single . . . animal" is quite revealing. When forces "knit" together independently existing entities upon which those forces work, the result is an atomistic building up of an ordered aggregate—the parts "knit" together are separable parts and the forces that bring and hold them together are external relations. You simply don't get holism out of this; the analogy is atomistic, a view Bayne rejects.

Argument 2: Naïve phenomenalism does not do justice to the sense in which streams are owned/had by selves.
Bayne's Reply: This will be cashed out mereologically. I suppose by this Bayne means that, not unlike bundle theories of substance, we can give a reductive analysis of sentences like "(1): I am exemplifying the property of being-an-appearing-of-red to form a mode of me—being-appeared-to-redly." To "(2): A particular phenomenal state—being-appeared-to-redly—is a part of 'my' totalizing phenomenal state at that time." Thus, ownership is a part/whole relation.

My Reply to Bayne: I will provide my alternative view of ownership later, but for now, I offer three brief replies. First, it seems difficult to read this view in any other way than atomistically: there is a separable part/whole relationship going on and the whole is simply a group of independent phenomenal parts standing in various external relations to each other. The problem here is that while Bayne merely and correctly points to the holistic nature of consciousness's unity, he does not give a supporting metaphysical analysis of how this could be. Absent such an analysis, it is hard to avoid bringing to bear fairly standard metaphysical notions when it comes to evaluating his position, even when those notions entail propositions Bayne explicitly rejects. Second, a notorious difficulty for bundle or mereological theories of substance is that they seem to lack the ontological resources to ground absolute identity through standard changes. For many, this will be a problem. Finally, Bayne's theory leaves opaque why most of us do not think our mental states are parts of us; rather, we take ourselves to be wholes that are not composed of our mental states. As we will see later, we think of ourselves as simple substances that "have" mental properties in that we exemplify them.

Argument 3: Naïve phenomenalism can't make sense of locutions such as "I weigh 185 pounds" or "I will die."
Bayne's Reply: Fairly obvious paraphrases are available.
My Reply to Bayne: I basically agree, except for one thing. I do not believe naïve phenomenalism has paraphrases available that properly handle the indexical "I."[40] For example, in handling objection 2, Bayne has to appeal to a part/whole relation between an independent phenomenal state and a totalizing state. But which totalizing state? I don't think this can be answered without saying "mine," or something indexically equivalent. But this is most naturally interpreted as claiming that I am one thing, and my totalizing phenomenal state is another, namely, a complex property I exemplify to form a mode of me.

Argument 4: Naïve phenomenalism cannot account for modal properties of the self, for example, I could have had different experiences, but this would not be possible if I just am a specific stream of consciousness. Indeed, one can conceive of a case in which there are no experiences in common between the actual and a merely possible self.
Bayne's Reply: Streams of events do not have their parts essentially. For example, World War II could have started as it did, but could have taken a different direction. Moreover, *any* conception of the self will have to deny at least some modal intuitions. The psychological view must reject the intuition that I could have had massive brain damage as a child. Animalism must reject the intuition that I could have been a different animal or disembodied.
My Reply to Bayne: It is hard to see how a stream of events held together by external relations could, in fact, have had different events. In the loose and popular sense, World War II could, indeed, have gone off in a different direction. But in the

strict metaphysical sense, this war (call it World War II*) would not be identical to World War II, but it could be treated as such for various purposes. And Bayne's claim that any view of the self must deny some intuitions is false. He only considers the psychological and animalist views. But a version of substance dualism that includes a mereologically simple, spiritual substance does not require abandoning basic intuitions in this area, including the three Bayne mentions. In fact, it seems pretty obvious that a substance-dualist conception of the person is actually the source of modal intuitions regarding the self, and advocates of alternative positions must tweak their views so as to be as close intuitively to substance dualism as possible without collapsing into substance dualism.

Argument 5: If your stream of consciousness fissions into two with psychological continuity, naïve phenomenalism cannot tell us which stream I am.
Bayne's Reply: Maybe consciousness is so deeply unified that fission is impossible. Maybe the physical basis of consciousness is such that continuity is broken and neither is the original self. In any case, animalism and the psychological view have troubles here too.
My Reply to Bayne: Our intuitions that fission is possible are much stronger and better justified than Bayne's claim that "maybe" fission is impossible. But more importantly, Bayne admits that naïve phenomenalism, animalism, and the psychological view have problems here. Elsewhere, Richard Swinburne has identified why this is the case: substance dualism resolves this issue and physicalist views do not.[41]

Bayne concludes that naïve phenomenalism is too naïve, and we must look elsewhere for a more adequate model. Bayne very briefly considers and rejects a second view—substrate phenomenalism—the view that the self is identical to the material substrate that underlies and generates consciousness. But this won't work, says Bayne, because there is no a priori guarantee that a single generative, underlying mechanism will produce only one stream of consciousness, and he is looking for a view that makes the unity of consciousness and its 1:1 relationship with a self a matter of a priori necessity. We now turn to a presentation of Bayne's own view.

Virtual Phenomenalism. For Bayne, the "self" is merely an intentional entity, one whose identity is determined by the cognitive architecture underlying the stream of consciousness; a sort of brain architecture that generates a fictitious entity like a character in a novel. So, the "self" is a virtual center of phenomenal gravity. In *de se* reference, the "subject" represents itself as itself; conscious states are automatically *de se*. Streams of consciousness are constructed around a single intentional object like a narrative is unified around the novel's main character. So, the cognitive architecture underlying your stream of consciousness represents that stream as if it were had by a single self—the virtual object that is "brought into being" by *de se* representation.

So, the cognitive architecture underlying consciousness creates a unifying single subject/center of consciousness as a projected, virtual reality due to the *de se* nature of the constructed conscious states. A unified field projects one and only one virtual self. Other approaches go wrong in thinking that there must be a real entity that plays the role of the self, but the self is a mere intentional object (like Zeus?). The self isn't really real, but self-talk is still "legitimate" in the way we talk of a character in a novel.

What should we say about Bayne's virtual phenomenalism? To begin with, I must confess that when I first read it, I thought that Bayne must be kidding. In my view, his position amounts to a bunch of mere assertions that allow him to avoid an obvious solution: substance dualism. How could we ever tell whether virtual phenomenalism is true? What possible evidence could be marshaled for it? Ontologically speaking, I think his view is really just a version of naïve phenomenalism (thus, Bayne's defense of the view) with a Kantian-style just so "as if" story added to it.

Second, we don't start with a role that needs to be filled—if we did, we would have no idea what roles to choose—and posit a self to fill it. We begin, rather, with knowledge by acquaintance of our own simple self, and then upon reflection, we see that it plays various roles, for example, the unifier of consciousness.[42] Moreover, why do people throughout human history and all over the world take themselves to be indivisible, disembodiable souls? I think the answer is that people simply are able to be aware of themselves.

Third, if the phenomenal field just is unified in and of itself, what would be the need for the brain to project a (virtual) self? Why would conscious states automatically and of necessity be *de se*? I think Bayne just posits these as brute facts—but surely, they are so odd, given his view, that we would be better off trying to find a different position that makes more sense of these and related issues.

Fourth, on Bayne's view, *de se* reference is systematically false. I-thoughts have no real, veridical intentional object. But surely this is far too extreme and skeptical. Part of what allows Bayne to get away with his view of *de se* reference is his inadequate characterization of the second role for the self, namely, to be an object of reference for I-thoughts. A more adequate characterization is this: the self is the object of the self-referring use of "I" but also that which employs "I" to self-refer. Once we see this, it becomes clear that there are no irreducible I-thoughts. Rather, there are substantial souls, selves, Is, with the power of self-awareness and self-reflection that can be expressed indexically.

Finally, I believe Bayne's view gives inadequate analyses for the other two roles he claims the self must play: ownership and having a perspectival point of view. Regarding ownership, there is no real self that owns anything. There is simply the totalizing, holistic phenomenal field that has individual states as parts. Given his rejection of atomism, it is hard to see what kind of parts these are. At the very least, this aspect of his view is in need of considerable clarification. Regarding the next role, what exactly is it that has this perspectival point of view? Given the arguments by Hasker and LaRock considered earlier, it can't be the brain because it is not a simple. Nor can it be the totalizing phenomenal field because for Bayne, that field just is the perspectival point of view in and of itself. In my view, for two reasons there is no such thing as the property of being a (first-person) perspectival point of view that something exemplifies.

For one thing, if there is such a property, it is an impure one. An impure property, for example, being identical to Socrates or to the left of a desk, requires reference to a particular to be described. Such a property cannot constitute such a referent without being circular—the property presupposes and, therefore, cannot constitute the particular—Socrates, the desk, or an individual person—to which reference is made. Similarly, being a first-person perspectival point of view presupposes the I.

For another thing, there most likely is no such property. In general, one may give a reductive analysis of the first-person perspective as follows: S has the property of being a first-person perspectival point of view iff S is a personal, viewing kind of point, that is, S is a kind of substance (point), a sentient (viewing) substance, with the properties (including ultimate potentialities) characteristic of persons (e.g., self-awareness and so on). The first-person perspective is not a property persons have, it is the thing persons are: centers of a personal kind of consciousness. Persons qua substantial, unified centers, exemplify ordinary mental properties—being-a-thought-that-P, being-a-sensation-of-red, being painful. But they do not have in addition to these the property of being a first-person perspectival point of view. A substantial personal ego's exemplifying an ordinary mental property is ipso facto a first-person perspectival point of view. There is no additional fact that needs grounding in a superfluous property—being a first-person perspective. The "first-person perspective" is just a way of describing/referring to an ontologically prior substantial, sentient person with ordinary mental properties to which that perspective can be reduced.

At the end of the day, Bayne has provided no clear, explanatory model of what it would be for consciousness to be unified or how this can be. Talk about subsumption, forces that knit together the components of a stream of consciousness (while avoiding atomism!) and Kantian-like fictitious selves, will hardly do. For Bayne, the unity of consciousness turns out to be a brute fact. But I think that unity can be further analyzed by appropriating ontological categories that are clear and that have long been with us.

Before I briefly present my alternative ontology of the unity of consciousness, it would be helpful to spell out some ontological notions I shall use to cash out my position.

Properties. A property is a universal, that is, something that can be nonspatially in, exemplified, or possessed by many things at the same time. As I have defended elsewhere, constituent realism is the best view of how properties relate to the ordinary particulars that "have" them.[43] According to constituent realism, properties are universals that, when exemplified, become constituents of the ordinary particulars that have them. Thus, if the mind exemplifies a mental property, say, the property of being-a-thought-of-London, then that property enters into the very being of the mind as a metaphysical constituent.[44]

> *Parts*. There are two kinds of parts relevant to our discussion: separable and inseparable. p is a *separable part* of some whole W $=_{\text{def.}}$ p is a particular, p is a part of W and p can exist if it is not a part of W.
>
> p is an *inseparable part* of some whole W $=_{\text{def.}}$ p is a particular, p is a part of W and p cannot exist if it is not a part of W. Inseparable parts get their existence and identity from the whole of which they are parts.

The paradigm case of an inseparable part in this tradition is a (monadic) property-instance or relation-instance. Thus, if substance s has property P, the-having-of-P-by-s is (1) a property-instance of P; (2) an inseparable part of s which we may also call a mode of s. For example, let s be a chunk of clay, P be the property of being round, and the-having-of-P-by-s be the-clay's-being-round. The clay could exist without being round, and the property of being round could exist without there being clay (e.g., a baseball could have that property), but the-clay's-being-round could not exist without the clay. The-clay's-being-round is a mode or inseparable part of the clay.

A *substance* $=_{\text{def.}}$ an essentially characterized particular that (1) has (and is the principle of unity for its) properties but is not had by or predicable of something more basic than it; (2) is an enduring continuant; (3) has inseparable parts but is not composed of separable parts; (4) is complete in species.[45]

A *spiritual substance (self or soul)* $=_{\text{def.}}$ (1) a substance; (2) metaphysically indivisible in being (though it may be fractured in functioning); (3) not spatially extended (though some characterizations hold that it may be spatially located); (4) essentially characterized by the actual and potential properties of consciousness.

Internal Relations. If something, A (say the color yellow), stands in an internal relation (brighter than) to B (say the color purple), then anything that did not stand in that relation to B could not be A. So, if any color was not brighter than purple, it could not be the color yellow. If a thing, X, stands in an internal relation to another thing, Y, then part of what makes X the very thing it is, is that it stands in that relation to Y.

Given this framework, it is fairly straightforward to spell out the nature of the unity of phenomenal consciousness and the nature of its ground. The substantial self (soul, I) is spatially unextended and not composed of separable parts. The self's having a mental state, say, an awareness of a table, occurs when the self exemplifies the property being-an-awareness-of-the-table—call this property P, and this forms a mode of the self, call it the-having-of-P-by-the-self. This mode may be described as the self's being-appeared-to-tablely. It is an inseparable part of the self and it stands in an internal relation to the self. This is what a particular phenomenal state is. Synchronically, the various phenomenally conscious modes of the self are unified into one totalizing phenomenal mode (state) by being modes of the same simple self and by being internally related to that self. Finally, the self is a unique kind of substance in that it has the power of self-awareness and self-reference.

It is not my purpose here to argue directly for substance dualism. I have done that elsewhere.[46] Rather I want to show how that view addresses and solves the problems that I have claimed Bayne's view does not.

1. It solves the binding problem and Hasker's argument considered earlier by employing a simple, substantial self.
2. It employs clear, standard ontological notions that have constituted the heart of ontology for a long time.
3. It provides a solid analysis of Bayne's three roles for a self: *Ownership*: that which has conscious states is the simple self, and the "having" amounts to the self's exemplifying various properties of consciousness to form modes of the self. *Referential*: substantial, simple selves simply have the power of self-awareness and self-reference. Thus, the I both employs "I" in linguistic acts of self-reference and is the object of those acts. *Perspectival*: the substantial, simple self exemplifies and unifies various properties of consciousness that, in turn, have the property of intentionality. In this way, the self/I has an irreducible, unified totalized conscious state/mode that is about intentional objects. Because modes are inseparable parts internally related to the self, this totalizing state is unique to one self and cannot be shared.
4. It provides a way of relating the three types of unity: (i) *Objectual phenomenal unity*: two or more states are so unified if they are experiences as being of the same object. (ii) *Subject phenomenal unity*: this occurs when all of one's phenomenal states are had by the same subject. (iii) *Subsumptive phenomenal unity*: two (or more) states are subsumptively phenomenally unified just in case there is something it is like to be in both states simultaneously and conjointly. Subject phenomenal unity occurs when all of one's phenomenal states are modes of the same, simple I. This view is neither irrelevant nor true by definition. Rather, it is a substantive (!) thesis with a developed ontology. It is because of subject phenomenal unity that objectual phenomenal unity obtains: two or more states are experienced

as being of the same object (e.g., the color and shape of a table) because they belong to the same object and the substantial I is simply aware of the table as a whole, including its various aspects. Finally, the reason subsumptive phenomenal unity obtains is because it is the same self that exemplifies the conscious property constituting each phenomenal mode/state and the self's unification of these into a totalizing mode is due to the self's simplicity.

5. My model provides a simple explanation for why Frank Jackson's observation is correct: "I take it that our folk conception of personal identity is Cartesian in character—in particular, we regard the question of whether I will be tortured tomorrow as separable from the question of whether someone with *any* amount of continuity—psychological, bodily, neurophysiological, and so on and so forth—with me today will be tortured."[47] As we saw earlier, people don't have to be taught to be dualists like they must if they are to be physicalists. Indeed, little children are naturally dualists. Summing up research in developmental psychology, Henry Wellman states that "young children are dualists: knowledgeable of mental states and entities as ontologically different from physical objects and real [non-imaginary] events."[48] The reason human persons all over the world and throughout history have overwhelmingly believed in a substantial self is because they *are* substantial selves and they have the ability to be aware of themselves.

FINAL PLEA TO CHRISTIAN PHYSICALISTS

In light of what I have argued so far, I urge my Christian-physicalist brothers and sisters please to reconsider their views. Let's be honest. The findings of neuroscience have virtually nothing to do with the nature of consciousness or the self. Those findings provide insight about correlations, causal relations, or dependency relations between mental states and physical states and vice versa. Given this, I believe my solution to the unity-of-consciousness problem provides a number of advantages over physicalism. I offer this list for reflection.

1. The best analysis of the nature of the unity of consciousness requires a form of holism that is not consistent with a Darwinian naturalist account of how life came about. This strengthens the design argument for God's existence as many naturalists acknowledge. Thus, they attempt to reduce, eliminate, or appeal to magic (emergence) to avoid providing evidence for the design argument. Christians should be excited about this rather than being resistant.

2. Given the nature of consciousness's diachronic and synchronic unity, the best explanation for it is the existence of a spiritual substance or soul. Despite contemporary Christian physicalism, this sort of dualism is clearly what the Bible teaches and what everyone in the history of the church—whether exposed to the Greeks or not—have held. It also comports with what 99 percent of the human race worldwide have known by direct introspection since Neanderthals.
3. The reality of the soul provides the clearest and best ground for absolute diachronic and synchronic personal identity in this life and into the next.
4. The reality of the soul provides the clearest and best ground for libertarian freedom, whether the soul is taken to be an emergent individual or not. This, in turn, provides a model of the human person that sheds light on how there could be a synchronic and diachronic unity to responsible moral actions such that the enduring subject is an appropriate object of praise and blame.
5. The reality of the soul comports well with the massive and ever-growing evidence that near death experiences are real and involve the soul leaving the body.
6. The nature of the unity of consciousness implies that there is only one stream of consciousness and one self per body, setting aside the demonic. If the brain is the bearer of consciousness, it is hard to explain how (i) a mereological aggregate could give rise to one, unified, holistic stream of consciousness; and (ii) one can avoid several selves in the brain/body, for example, a visual self, an auditory self, and so forth.

NOTES

1. J. Gresham Machen, *The Christian View of Man* (New York: Macmillan, 1937), 159.

2. See John Cooper, *Body, Soul & Life Everlasting*, rev. ed. (Grand Rapids, MI: Eerdmans, 2000).

3. N. T. Wright, *The Resurrection of the Son of God* (Minneapolis, MN: Fortress, 2003), 131–34, 190–206, 366–7, 424–26.

4. See Charles Taliaferro, "Emergentism and Consciousness," in Kevin Corcoran (ed), *Soul, Body and Survival* (Ithaca, NY: Cornell University Press, 2001), 60.

5. Jaegwon Kim, "Lonely Souls: Causality and Substance Dualism," in *Soul, Body and Survival*, 30.

6. Frank Jackson, *From Metaphysics to Ethics* (Oxford: Clarendon, 1998), 45.

7. Henry Wellman, *The Child's Theory of Mind* (Cambridge, MA: MIT Press, 1990), 50. I owe this reference to Stewart Goetz and Mark Baker.

8. Nancey Murphy, "Human Nature: Historical, Scientific and Religious Issues," in Warren S. Brown, Nancey Murphy, and H. Newton Malony (eds), *Whatever Happened to the Soul?* (Minneapolis, MN: Fortress Press, 1998), 18.

9. Murphy, "Human Nature," 17; cf. 13, 27.

10. Nancey Murphy, *Bodies and Souls, or Spirited Bodies?* (Cambridge, UK: Cambridge University Press, 2006), 56.

11. Murphy, *Bodies and Souls*, 112.

12. C. Stephen Evans, "Separable Souls: Dualism, Selfhood, and the Possibility of Life After Death," *Christian Scholar's Review*, 34 (2005): 333–34.

13. John Tyndall, "Scientific Materialism," in his *Fragments of Science: A Series of Detached Essays, Addresses and Reviews*, vol. 2 (New York: D. Appleton and Co., 1896).

14. See J. P. Moreland and Scott Rae, *Body and Soul* (Downers Grove, IL: Inter-Varsity Press, 2000), chapter 4.

15. Roderick Chisholm, "Human Freedom and the Self," reprinted in *On Metaphysics* (Minneapolis: University of Minnesota Press, 1989), 14.

16. John Searle, *Minds, Brains, and Science* (Cambridge, MA: 1984), 98.

17. John Bishop, *Natural Agency* (Cambridge, UK: Cambridge University Press, 1989), 1. Bishop's own solution eschews libertarian agency in favor of a version of compatibilism.

18. John Bishop, *Natural Agency*, 40. An interesting implication of Bishop's view is that naturalism cannot allow for there to be a first event in the absolute sense of not being preceded by other events because all events are caused by prior events or else they are simply uncaused. In the latter case, the coming to be of the event cannot be "natural" since it is just a brute fact. In the former case, this means that if the kalam cosmological argument is correct and there was a beginning to the universe, then the beginning itself was not a natural event nor was its cause if it had one. For more on this, see William Lane Craig and Quentin Smith, *Theism, Atheism, and Big Bang Cosmology* (Oxford: Clarendon Press, 1993).

19. Joshua Hoffman and Gary S. Rosenkrantz, *Substance: Its Nature and Existence* (London: Routledge, 1997), 98–99.

20. Geoffrey Madell, *The Identity of the Self* (Edinburgh: Edinburgh University Press, 1981), 16. Cf. Georg Gasser, Matthias Stefan, eds., *Personal Identity: Complex or Simple* (Cambridge, UK: Cambridge University Press, 2012).

21. For more on problems of material composition, see Michael Rea, ed., *Material Constitution: A Reader* (Lanham, MD: Rowman & Littlefield, 1996); Christopher M. Brown, *Thomas Aquinas and the Ship of Theseus* (London: Continuum, 2005).

22. Jonathan Schaffer, "On What Grounds What," in David Manley, David J. Chalmer, and Ryan Wasserman (eds), *Metametaphysics: New Essays on the Foundations of Ontology* (Oxford: Oxford University Press, 2009), 28.

23. The view I am advancing is called mereological essentialism (from the Greek word "meros" which means "part." Mereological essentialism is the idea that an object's parts are essential to its identity such that it could not sustain its identity to itself if it had alternative parts. Animalists and constitutionalists deny mereological essentialism. For a brief exposition of these views, see Eric Olson, *What are We? A Study in Personal Ontology* (Oxford: Oxford University Press, 2007), chapters 2 and 3. In different ways, each view claims that, under certain circumstances, when parts come together to form a whole, as a primitive fact, the whole itself just is the sort of

thing that can survive part alteration. In my view, this is just an assertion. The whole just is parts and various relations, and neither the parts nor the relations can sustain identity if alternatives are present. The whole is not a basic object—it is identical to its parts and relations.

24. A somewhat fuller presentation of the material in this section may be found in my "Substance Dualism and the Unity of Consciousness," in *The Blackwell Companion to Substance Dualism* edited by Jonathan Loose, Angus Menuge, and J. P. Moreland (Oxford: Blackwell, forthcoming, 2018).

25. Tim Bayne and David J. Chalmers, "What is the Unity of Consciousness?" in Axel Cleeremans (ed) *The Unity of Consciousness: Binding, Integration and Dissociation* (Oxford: Oxford University Press, 2003), 23–58.

26. Bayne and Chalmers, "What is the Unity of Consciousness?" 32–33.

27. Axel Cleeremans, "Introduction," in *The Unity of Consciousness*, 2.

28. Bayne and Chalmers, 33.

29. Bayne and Chalmers, 40. They go on to show that, under certain conditions, the Subsumptive Unity Thesis (for any set of phenomenal states of a subject at a time, the subject has a phenomenal state that subsumes each of the states in that set) is materially equivalent with the Logical Unity Thesis (for any set of phenomenal states of a subject at a time, the subject has a phenomenal state that entails each of the states in the set). But material equivalence is not identity, and subsumption is still left as an alleged intuitive primitive.

30. William Hasker, *The Emergent Self* (Ithaca, NY: Cornell University Press, 1999), 122–44. See also William Hasker, "On Behalf of Emergent Dualism," in *In Search of the Soul*, ed. Joel B. Green and Stuart L. Palmer (Downers Grove, IL: InterVarsity Press, 2005), 75–100.

31. Bayne and Chalmers, 26–27.

32. Bayne, *The Unity of Consciousness*, 225–29.

33. According to constituent realism, when an object exemplifies a property or a collection of objects exemplify a relation, the property or relation (universals) constitutes an immanent essence of the individuating property or relation instances, which in turn, particularize the individual structure in view, keeping it from being an abstract universal. I have defended this view elsewhere. See my "Exemplification and Constituent Realism: A Clarification and Modest Defense," *Axiomathes* 23 (2013): 247–59.

34. Eric LaRock, "From Biological Naturalism to Emergent Substance Dualism," *Philosophia Christi* 15 (2013): 97–118.

35. Eric LaRock, "Emergent Dualism is Theoretically Preferable to Reductive Functionalism," unpublished manuscript, 31 March 2015.

36. LaRock, 15.

37. The best exposition of Bayne's views is found in his *The Unity of Consciousness*, chapter 12.

38. The other significant attempt to solve the unity of consciousness problem is offered by Lynne Rudder Baker. See her *Persons and Bodies: A Constitution View* (Cambridge, UK: Cambridge University Press, 2000), especially Part I. I have criticized her view elsewhere: J. P. Moreland, *The Recalcitrant Imago Dei* (London: SCM

Press, 2009), 131–37. It may be worth mentioning that split-brain issues have been raised as a significant defeater for advocates of the sort of unity of consciousness like Bayne and me. In my opinion, Bayne has provided an adequate response to this problem. See Tim Bayne, "The Unity of Consciousness and the Split-Brain Syndrome," *The Journal of Philosophy* 105, no. 6 (2008): 277–300.

39. David Barnett, "You Are Simple," in Robert Koons and George Bealer (eds), *The Waning of Materialism* (New York: Oxford University Press, 2010), 161–74.

40. For more on this, see Geoffrey Madell, *The Essence of the Self* (New York: Routledge, 2015).

41. See Richard Swinburne, *The Evolution of the Soul* (Oxford: Clarendon Press, 1986), 147–51; *Mind, Brain & Free Will* (Oxford: Oxford University Press, 2013), 152–58.

42. See J. P. Moreland, "Substance Dualism and the Argument from Self-Awareness," *Philosophia Christi* 13, no. 1 (Summer 2011): 21–34.

43. See J. P. Moreland, "Exemplification and Constituent Realism: A Clarification and Modest Defense."

44. See Dallas Willard, "How Concepts Relate the Mind to its Objects: The 'God's Eye View,'" *Philosophia Christi* 1, no. 2 (1999): 5–20.

45. Two helpful treatments of substances and related entities are Joshua Hoffman and Gary S. Rosenkrantz, *Substance: Its Nature and Existence* (London: Routledge, 1997); Christopher M. Brown, *Aquinas and the Ship of Theseus* (London: Continuum, 2005).

46. Besides the source cited in note 20 above, see my *Recalcitrant Imago Dei*, chapter 5; "A Conceptualist Argument for Substance Dualism," *Religious Studies* 49 (March 2013): 35–43.

47. Frank Jackson, *From Metaphysics to Ethics* (Oxford: Clarendon: 1998), 45.

48. Henry Wellman, *The Child's Theory of Mind* (Cambridge, MA: MIT press: 1990), 50. I owe this reference to Stewart Goetz and Mark Baker.

SUGGESTED FURTHER READING

Baker, Lynne Rudder. *Persons and Bodies: A Constitution View*. Cambridge, UK: Cambridge University Press, 2000.

Barnett, David. "You Are Simple." In *The Waning of Materialism*, edited by Robert Koons and George Bealer, 161–74. New York: Oxford University Press, 2010.

Bayne, Tim. *The Unity of Consciousness*. Oxford, UK: Oxford University Press, 2010.

———. "The Unity of Consciousness and the Split-Brain Syndrome." *The Journal of Philosophy* 105, no. 6 (2008): 277–300.

Bayne, Tim, and David J. Chalmers. "What is the Unity of Consciousness?" In *The Unity of Consciousness: Binding, Integration and Dissociation*, edited by Axel Cleeremans, 23–58. Oxford, UK: Oxford University Press, 2003.

Gasser, Georg, and Matthias Stefan, eds. *Personal Identity: Complex or Simple?* Cambridge, UK: Cambridge University Press, 2012.

LaRock, Eric. "From Biological Naturalism to Emergent Substance Dualism." *Philosophia Christi* 15 (2013): 97–118.

Moreland, J. P. *Consciousness and the Existence of God*. New York: Routledge, 2008.

———. "Exemplification and Constituent Realism: A Clarification and Modest Defense." *Axiomathes* 23 (2013): 247–59.

———. *The Recalcitrant Imago Dei*. London: SCM Press, 2009.

Willard, Dallas. "How Concepts Relate the Mind to its Objects: The 'God's Eye View.'" *Philosophia Christi* 1, no. 2 (1999): 5–20.

Chapter 4

Christian Physicalism and Our Knowledge of God

Angus Menuge

Christian physicalism (henceforth, CP) claims that human persons do not possess immaterial souls, but are either strictly identical to,[1] emergent from,[2] or constituted by,[3] material objects—namely, living human beings. Many of the theological criticisms of CP question its ability to account for basic Christian teachings like the Incarnation[4] and the Resurrection.[5] In this chapter, I focus on the epistemological adequacy of CP. Several philosophers have argued that standard (i.e., non-Christian) physicalism faces a variety of general problems in accounting for knowledge.[6] Here I argue that, although it recognizes God's existence, CP inherits similar difficulties and, in particular, seems incompatible with biblical teachings about our ability to know God through general and special revelation.[7]

I begin by considering a general requirement for our having knowledge of God—that we can form appropriate, valid concepts of God (henceforth, *divine concepts*)—and note that there seem to be just three main ways that this requirement might be fulfilled. Divine concepts are either (1) acquired, (2) innate, or else (3) constantly emanated by God. I then argue that this presents a trilemma for CP. Regardless of which of these three options is selected, CP does not give a credible account of our possession of divine concepts, and therefore cannot explain our knowledge of God. Along the way, I consider several natural responses that a proponent of CP might make, and show that they are inadequate

But is the dualist—who affirms that the human person is an integrated union of a material body and an immaterial soul—any better off? At the end of the chapter, I outline some reasons to think that immaterial souls are more apt to possess the divine concepts required to know God.

KNOWLEDGE OF GOD

Historic, orthodox Christianity clearly teaches that humans can have knowledge of God. They can come to know *about* God—that He exists and has certain transcendent attributes—through an understanding of *nature*: the general revelation noted in such passages as Psalm 19:1–2 and Romans 1:19–20.[8] And they can come to know God personally through Jesus Christ as revealed in the *scriptures* (special revelation). But such knowledge assumes that humans have adequate concepts of God and His nature, so that we can think that God exists, that He is infinite and perfect, and that He is incarnate in the person of the God-man, Jesus Christ.

This reflects a general requirement for knowledge: we cannot know what we cannot think. Consider a mundane example. To know that there is a bird in the shrub requires one to have concepts of birds and shrubs. If one does not have these concepts one cannot even *think* that there is a bird in the shrub. But if one cannot think that something is the case, one cannot have a true belief about it, and so there is no question of knowing it. But even if one does have the concepts, this is not enough to make knowledge possible. The concepts must also be *valid*, that is, they must be appropriately grounded in the way reality is. If I have somehow acquired a fictitious concept, such as the concept of a leprechaun, I can certainly believe that there is a leprechaun in the shrub, but I cannot know it, because this concept is not appropriately grounded in the nature of reality: there are no leprechauns. Valid concepts must derive from a reliable source, objects that actually exemplify the properties a concept presents to the mind. Suppose that we lived in a world void of birds and shrubs (as our actual world is void of leprechauns), and that an evil genius had implanted the concepts of birds and shrubs in someone's mind: that person could think, but could not know, that a bird is in the shrub, because the way those concepts present the world as being does not derive from the way the world is. A valid concept must derive its presentational features (how it represents the world as being) from properties the world actually exemplifies (how the world is).

It is this basic requirement for knowledge—that one cannot know unless one has relevant, valid concepts—which makes it difficult to see how CP can explain knowledge of God. If we can know God, then we must have valid concepts of God (*divine concepts*). It seems that there are only three main ways we can acquire basic concepts (and our other concepts derive from complex combinations of these). Empiricists like Thomas Aquinas and John Locke argue that our basic concepts are *acquired* (for example, by abstraction from experience), and they offer a posteriori arguments for God based on divine concepts we have derived from experience. By contrast, rationalists like Anselm and René Descartes emphasize our possession of innate concepts, and they offer a priori arguments for God's existence based on divine concepts engrafted into our being. A third possibility is that divine concepts are neither acquired nor innate, but constantly emanated by God as He calls us to know Him.[9]

But this, I argue, presents a trilemma for CP, since a physicalist anthropology makes all three of these options problematic. I will begin by sketching the trilemma, then develop each of its horns in more detail in subsequent sections.

A TRILEMMA FOR CHRISTIAN PHYSICALISM

Consider the first horn: divine concepts are *acquired*. On a physicalist anthropology, it is not plausible that we can acquire divine concepts due to limitations imposed by the brain. To qualify as physicalist, CP must assert that our psychological capacities (in particular, our ability to form concepts) are dependent on the physical state of the brain, but there are serious limitations on the kind of information that the brain can transmit to our thoughts. On a physicalist anthropology, I argue that the brain is an informational "bottleneck," that will not permit us to form the transcendent concepts of God—as an infinite, perfect, eternal being—that we need to know Him.

Now consider the second horn: divine concepts are *innate*. This may seem compatible with CP, since there is no obvious contradiction in the idea of God engrafting divine concepts in a physical being.[10] But knowledge of God requires not merely the possession of divine concepts, but also a subject that can access those concepts. This, I argue, is a problem for CP because it fails adequately to account for the existence of a subject at and over time. So innate divine concepts of God would be like those library books students don't read—material for potential knowledge, but not actual knowledge, because there is no knowing subject.

Finally, consider the third horn: God tirelessly refreshes our brains by constantly emanating divine concepts. This is not a good fit with CP because it is not really a physicalist proposal: the intervention of God, an immaterial being, is routinely required to account for our psychological capacities, and so these capacities are not exclusively determined by our physical states. In fact, on this view, God becomes a very busy surrogate for the soul, personally doing all of the work that dualists attribute to creaturely human souls.

We will now take a closer look at each of the three horns of the trilemma in turn, and also consider likely rejoinders from proponents of CP.

Acquired Divine Concepts

Proponents of CP may affirm or deny that mental properties reduce to properties of the brain. If they affirm it, call this Reductive CP. If they deny it, call this Nonreductive CP. We will consider each of these alternatives in turn.

Reductive CP

Reductive CP holds that psychological capacities really are just physical capacities of the brain. In that case, our ability to think must be limited by the causal powers of the brain: for, on the plausible principle that we can individuate properties in terms of their distinctive causal powers, if our psychological capacities exceed the causal powers of the brain, there must be mental properties that do not reduce to properties of the brain.[11] However, it does not seem possible that the brain's operations can account for important divine concepts, for example, the concepts of an *infinite*, *perfect*, and *eternal* being.

Consider infinity. Consistently with his materialism, Thomas Hobbes argues that the scope of our thought is limited to what we can imagine and that our imagination is determined solely by physical interactions between our brain, our senses, and the environment. But we only ever experience finite objects and qualities, and both the senses and our brain are capable of only a finite number of operations on that sensory input. Finite operations (e.g., of cutting and pasting, and various forms of combination) on finite input can only produce a finite output, so representations (ideas, mental images, and models, etc.) are themselves obviously finite. And, Hobbes argues, finite operations on sensory input that represents finite objects and qualities cannot yield a representation that *presents* an infinite object or quality: "Whatever we imagine is *finite*" and hence that "there is no idea or conception of anything we can call *infinite*."[12]

This second claim is less obvious, however. One may suppose that even though our representations (ideas) are, as objects or states, finite, still they might have infinite *representation content*, and there might be a finite route to their acquiring it (to their being able to represent something as being infinite). In particular, even if our concepts derive from finite sensory input, surely there might be a *via negativa* (way of denial) that takes us to a presentation of the infinite by the finite operation of denial that some quality is present to a finite, limited degree.[13] While this proposal has some initial plausibility, further reflection shows that the concept of the infinite cannot be constructed in this way. If we look closely at what is involved in a denial of finite limits, we will see that this denial presupposes that we already have a concept of the infinite, and therefore cannot explain the origin of that concept.

A proponent of CP who thinks that the concept of infinity is acquired via operations on sensory input might claim that all we need is to abstract the concepts of finite limits and negation. The concept of the infinite is then formed simply by a finite application of the concept of denial to the concept of limits: to say that something is infinite is simply to say that there is no finite limit to it. However, this argument collapses once we examine its meaning in terms of first order quantified logic. For whether a denial of finite limits entails infinity depends on the domain over which we are quantifying. If that domain is finite—if it contains a finite number of limits—saying that something *x* has no finite limit in that domain does not imply that *x* is infinite, for it is possible that *x* is simply finite but larger (it has a greater limit) than any member of the domain. Therefore, in order for a denial of finite limits to entail that something is infinite, it must be that the domain contains an *infinite* set of finite limits: when we say that an entity is not bounded by any of these limits, that object must be infinite. But that alerts us to the fact that in the sense required, "*x* has no finite limit" means "for any limit L in an *infinite* set of limits, *x* is not limited by L." And therefore, in order for me to understand what "*x* has no limit" means *in the appropriate sense*, I must already have the concept of infinity. I have to grasp the concept of an infinite set of limits before the denial of limits can yield a valid concept of the infinite. But if so, this denial cannot be the means by which I originally acquired the concept of infinity.

Further, if our concept of limits is acquired from our sensory experiences, the fact is that we have only ever experienced finite entities, and so any denial that an entity is limited by what we have experienced would only entail that it is something larger, but finite. We clearly cannot get the concept of the infinite just from the idea of something beyond the limits we have actually experienced. It is therefore only by smuggling in the idea that we have a grasp of *limits in general*—of all possible limits in an infinite domain—that the *via negativa* proposal gains any plausibility.

Interestingly, though not a materialist, Descartes agrees with Hobbes that we cannot derive the concept of infinity from experience. For Descartes, infinity and other divine concepts are innate, and he counters the *via negativa* by challenging its assumption that our concept of the finite is unproblematic:

> And I must not think that I perceive the infinite not through a true idea, but rather only through the negation of the finite, just as I perceive rest and shadows through the negation of movement and of light. For—on the contrary—I manifestly understand that there is more reality in an infinite substance than there is in a finite one, and therefore that the perception of the infinite is in me in some mode prior to the perception of the finite, that is, that the perception of God is in me in some mode prior to the perception of me myself.[14]

Descartes argues that it is only through a prior grasp of God's infinity that he is—by way of negation—able to grasp his own finitude. The *via negativa* from our concept of the finite to the concept of the infinite has it exactly backward: we have a concept of the finite only by way of denying our prior concept of the infinite. Our tendency to prioritize the finite over the infinite may only reflect a linguistic illusion, since the word "infinite" is constructed by adding the negation "in" to "finite." Ontologically, though, being itself is not limited, so my primary concept of being does not imply any limitation. However, I discover my own existence by distinguishing it from being in general, by seeing that I have boundaries and deficiencies that being in general does not possess. If Descartes is right, and I can know anything finite only by a prior grasp of God as an infinite being, there is no question of my being able to construct a concept of God's infinity from any finite source.

Similar arguments would appear to show that, on Reductive CP, we can have no concept of perfection or eternity either. The operations of imperfect brains on the imperfect objects and qualities presented to the senses do not explain our ability to imagine (or subsequently conceive of) a perfect being. Indeed, Descartes argues that here again, the *via negativa* runs in just the opposite direction:

> For how would I understand that I doubt and that I desire, that is that something is lacking in me and that I am not completely perfect, if there were no idea of a more perfect being in me from whose comparison I might recognize my defects?[15]

Likewise, the inputs and operations of brains are exclusively temporal, so they do not seem able to account for our ability to think of an eternal being. A remarkable feature of the human mind is that it possesses the concept of the *cosmos*—of the totality of space and time—and this is what makes it possible to develop the cosmological argument, which infers some being beyond the cosmos. But how can we think about *all* of time unless we have some implicit counter-perspective, some grasp of what is *not* conditioned by time? Just as our ability to think of limits in general presupposes a concept of the infinite, it seems that our ability to think of time in general presupposes a concept of the eternal.

Reductive CP fails to give a plausible account of our acquisition of divine concepts such as infinity, perfection, and eternity, and therefore this popular combination (a reductive materialist anthropology and an empiricist view of concept acquisition) fails to explain how knowledge of God is possible.[16]

Nonreductive CP

Reductive materialism has taken a beating in recent times, and it is increasingly fashionable for proponents of CP to take the nonreductionist option, for example, by claiming that mental properties emerge from the brain and bring new causal powers into the world, maybe even the power to act back on the brain and body (so-called "downward mental causation").[17] Despite many subtle differences in their positions, Lynne Baker, Kevin Corcoran, and Nancey Murphy all affirm that persons have mental properties that do not reduce to the physical states from which they emerge. If the powers of the mind transcend those of the brain, perhaps the mind can overcome the limitations of the brain so that it can acquire divine concepts.

However, if CP is to qualify as a physicalist theory, it must limit the extent to which mental properties transcend physical properties. By definition, physicalism is committed to the view that emergent mental states and properties are exclusively determined by physical base states and properties. Thus psychophysical *supervenience*, the thesis that there can be no mental difference without a physical difference, is typically cited as the lowest common denominator commitment of any variety of physicalism. A physicalist may affirm *property dualism*, claiming that mental properties do not *ontologically* reduce to physical properties.[18] But it is arguably not compatible with physicalism to assert that these mental properties can operate in complete *causal independence* of their physical bases. The assertion of such a strong form of mental independence sounds like the dualist claim that (whether the mind is a strict substance or not)[19] the mental is a realm with its own set of powers, powers not exclusively determined by physical objects and properties. If so, it risks trivializing physicalism to affirm such mental independence.[20] But this requirement of the dependence of the mental on the physical makes it hard to see how Nonreductive CP fares better than Reductive CP as an explanation of our acquisition of divine concepts.

If our physical sensory input always presents finite, imperfect, and temporal objects and qualities, and our brain is capable only of finite, imperfect, and temporal operations, any physical base state it yields as output does not represent or indicate an infinite, perfect, and eternal being. But if our thoughts are determined by these physical base states, this surely has implications for the information these thoughts can contain. If our thoughts are limited by their informational input from the brain to being thoughts of the finite, imperfect, and temporal, and if, as we argued, there is no *via negativa* that uses finite operations of negation to convert these into thoughts of an infinite, perfect, and eternal being, Nonreductive CP does not explain our acquisition of divine concepts.

But perhaps our thoughts have a power to represent that brains lack: thus, proponents of Nonreductive CP emphasize that thoughts but not brains have intentionality[21] and that persons but not brains are capable of a first-person perspective.[22] However, this is highly contentious, since it is arguable that both of these claims are really dualist ones: they appear to affirm the independence of the mental. Nonreductive physicalists (Christian or not) may retort that one can be a property dualist without being a substance dualist, understanding intentionality and a first-person perspective as emergent mental properties, while denying that these properties belong to an immaterial substance. But, on inspection, this kind of "mere property dualism," which affirms independent mental properties, but not independent mental subjects, is unstable, and arguably untenable.

As J. P. Moreland has pointed out, if we understand having a first-person perspective as a mental property, it is clearly an *impure* property. An impure property is one that presupposes the existence of a particular. Thus a tree's property of *being rooted* is impure because it presupposes the existence of particular roots: a tree without roots cannot be rooted. Likewise, the property of *having a first-person perspective* (or subjectivity) presupposes the existence of a mental subject that has that perspective.[23] It is therefore not coherent to substitute the property of having a first-person perspective for the existence of a mental subject. First-person perspectives are ontologically dependent on mental subjects, as the roundness of a ball (that property instance or mode) is dependent on the ball.

Less obviously, if we understand the intentionality of thoughts as merely a property, it is arguably also an impure one, one that presupposes the existence of a mental subject. Our evidence for this is that we confidently and literally attribute nonderived intentionality only to mental subjects. While books and computers may contain information, most philosophers agree that inscriptions in books and states of computers are not intrinsically about anything: they can be said to exhibit intentionality only in the derived sense that persons whose mental states have nonderived intentionality can *take* inscriptions or computer states to be about other things. And, as the brain is physically described by neuroscience, it does not exhibit intentionality either. This is because any power the brain has to represent the world must be limited to what arises from its physical interactions with the world, but intentionality clearly transcends physical causation. We can think of future events (a vacation), unrealized possibilities (a vacation that is cancelled), actually nonexistent objects (leprechauns), and necessarily nonexistent objects (round squares). It is safe to say that future events, unrealized possibilities, and nonexistent objects have no physical causal power to make us think of them. And so intentionality does not appear to result from any causal power of the brain, but is rather a power of mental subjects.

So, if first-person perspectives and intentionality are ontologically dependent on mental subjects, they presuppose that mental subjects really exist and so the existence of mental subjects cannot be explained away simply by viewing first-person perspectives and intentionality as emergent mental properties. If this is right (and it will take another chapter to defend it in detail), the mere property dualism espoused by Nonreductive CP is an unavailable position.

Setting those objections to the side, we may also notice that in order for Nonreductive CP to remain physicalist, it must claim that the informational content of our thoughts derives from the physical base states of the brain. If it does not, it is false that our thoughts are exclusively determined by physical properties of the brain, and if nonphysical determinants of our thoughts are allowed, we do not have a physicalist theory of mind. On physicalism, our ability to think that an object has some property *F* must be determined by the fact that our brain has some property *G*, and so our power to represent objects and qualities must be limited by the resources in the brain. As we have seen, these resources only indicate finite, imperfect, and temporal objects and qualities; they do not contain any information about what is infinite, perfect, and eternal that could be used to form divine concepts. What I am suggesting here is a principle of *representational inheritance* (RI):

> (RI) For any thought T, if T's representing an object O1 as *F* derives from an object O2's being *G* (where O2 may be an extra-mental object or an intermediary state of the senses or the brain), then the information required for T to represent O1 as *F* must be inherited from O2's being *G*.

Notice it is not required by RI that O1 actually is *F*, since we are capable of misrepresentation (e.g., when we think that our mother's patience is infinite, or that some Danish bakery was perfect, etc.). But our ability to think (truly or falsely) that O1 is *F* does require that we can present something as being *F*, and the informational content of the thought, "that's an *F*" must have a source. Now, if Nonreductive CP is true, then the *proximal* cause of the thought that O1 is *F* is some brain state (O2's being *G*). But as we have seen, no physical state of the brain indicates the existence of an infinite, perfect, or eternal being, so in this case, there is no information present in O2's being *G* that could account for the ability to present O1 as *F* (infinite, perfect, eternal).

This, I believe, reflects the fact that our representational capacities are limited by general principles governing the transmission of information. As Fred Dretske argues, the flow of information obeys a basic conservation law. He calls it the *Xerox principle* because it is illustrated by the process of photocopying: a copy can preserve all of the information in the original(s) or it may lose some (a bad copy, low toner, etc.); but a copy can never contain more information than is found in the original(s) from which it is derived.[24] The Xerox principle expresses the simple idea that the transmission of information is conservative: the informational signal that a system receives can contain no more information that the sum of its informational sources. Applied to nonreductive physicalism, if thoughts are entirely determined by physical base states of the brain, then a thought cannot contain any information about *F*-ness (so that it can represent O1 as *F*) that does not derive from the properties of its physical base state (O2's being *G*). If this is correct, then Nonreductive CP fares no better than Reductive CP in accounting for our possession of divine concepts, since information about infinite, perfect, and eternal beings cannot be inherited by our thoughts from their physical base states in the brain.

Proponents of both Reductive and Nonreductive CP may protest that my argument is unfair. After all, CP is not standard physicalism, which assumes the materialist thesis that there are *no* immaterial substances. CP affirms that God (an immaterial substance) exists and causally interacts with the world, so even if persons are identical with (or emerge from, or are constituted by) living physical organisms, still it seems that God could bring it about that physical states of the brain are (or generate) representations of divine qualities. And obviously, if God's divine qualities are causally responsible for our having representations of them, our divine concepts are valid (rooted in objective reality).

The Bottleneck Argument

My response is what I call the *bottleneck argument*. If our divine concepts are acquired, then physicalism requires that God works through physical means to generate those concepts. For otherwise, God, an immaterial being, is required as a *direct* explanation of some of our psychological capacities and so the physicalist thesis that our psychological capacities are exclusively determined by physical properties of the brain is false. However, although God Himself is infinite, perfect, and eternal, all of the physical means through which, on physicalism, He must work, are finite, imperfect, and temporal. So these means do not appear able to bear the information required to form divine concepts. To use an analogy with modern digital communication, the physical links between God and the brain do not have the "bandwidth" to transmit information about a divine being. If so, there is still an informational "bottleneck" between God and our thoughts: brain states cannot contain the information necessary to be (or generate) divine concepts. As an analogy, suppose someone sends you a jpeg file of a Rembrandt painting. The painting contains information in continuous form (smooth areas of paint), but the jpeg can only capture a digital approximation using discrete color points (pixels). Even if the jpeg does derive from the original artwork, its limited resolution means it cannot fully capture the original, and on close inspection, the jpeg misrepresents the original as being made up of pixels. Likewise, whatever states and operations God may cause in the brain, they cannot adequately represent God as infinite, perfect, or eternal, and they would in fact misrepresent God as finite, imperfect, and temporal. (Why we should think finite, imperfect, temporal souls are any better off is a question to which we will return.)

Now one might balk at this, and argue that God's omnipotence means that He can simply make it the case that we have divine concepts. But there are problems with this response. If the bottleneck argument is correct about the limitations of physical states to bear information, then if God does make it the case that we have divine concepts, the information that these concepts bear is *not* derived from the physical base states of the brain (their limitations have simply been overridden). But if that is the case, physicalism is false: it is not true that all aspects of our thoughts depend on our physical states. If God simply makes it the case that whenever we are in brain state B, our thoughts involve divine concepts, what makes it possible for us to have those concepts derives directly from God and not from B. So there are aspects of our thought that can be explained only by the direct action of an immaterial being, God, which is not a physicalist view.

Now a theist might claim that causation is simply a reliable, regular connection decreed by God, and that if God decrees that certain brain states are followed by thoughts involving divine concepts, then the former cause the latter, even if the information in the concepts does not derive from the brain states. One problem with this response is that it appears to endorse *occasionalism*: being in brain state B is not really sufficient to produce a divine concept; rather, on the occasion of being in brain state B, God produces a divine concept. This, again, is hardly a physicalist view, since the content of our thoughts depends on the direct, special action of an immaterial being. And it is questionable that B really is a cause. B might be a cause of something mental, but regarding our divine mental concepts in particular, it seems rather to be a noncausal condition of divine causation, and a "cause" only in the vulgar sense that humans conditioned by Humean association will continue to call it one.

But even if this response did account for divine concepts, it would not explain *knowledge* of the divine. If the physical "causes" of the divine concepts do not contain any information about a divine being's attributes, those "causes" would be just as they are even if no such attributes existed. So, it is only a lucky coincidence that our divine concepts happen to be valid, because *in fact*, we have those concepts only because there is a being with divine attributes who produces divine concepts on the occasion that the physical "causes" obtain. In terms of their intrinsic information-bearing capacity, there is no reason that the physical base states that "generate" our divine concepts could not exist even though the base states were not caused by a being with those attributes. So if one wants to maintain an attenuated sense in which the brain states "cause" divine concepts, still they do not provide knowledge of God, since it is only a coincidence that any beliefs about God employing those concepts are true. (One cannot know something if one is only right by coincidence in believing it.)[25]

True, one can plausibly argue that our possession of divine concepts independently justifies belief in God since the informational content of these *concepts* is derivable only from God Himself. But since that informational content is not present in the physical base states of our brain but derives directly from God, an immaterial being, this is not a *physicalist* account of the mind: it again implies that our thoughts are not exclusively directly determined by physical causes.[26]

I conclude that physicalist anthropology does not adequately explain how we can *acquire* knowledge of God. Either the divine concepts cannot be constructed, because of the bottleneck of the brain, or if they are, it is only by routine appeal to the intervention of an immaterial being, God, and this is not a physicalist theory of mind.

Innate Divine Concepts

But perhaps we do not need to acquire knowledge of God. Perhaps God engrafts divine concepts into our being, so that we have an innate capacity to come to know Him under the right conditions (e.g., through contemplating nature or scripture). I have no problem with this proposal per se: in fact, I have defended it.[27] But it is very difficult to reconcile it with physicalism.

For one thing, if God engrafts divine concepts into us by bypassing the bottleneck of the brain, then our psychological capacities do not exclusively depend on physical base states of the brain. In fact, God becomes a sort of surrogate soul for each person doing work that dualists attribute to those persons' creaturely souls. And there are several other serious problems.

If God engrafts divine concepts into our being, then we must be the sort of unified, persistent entities capable of having these (and other) concepts *at* and *over* time (i.e., persons). Encoding information about God in a book or on a computer would do no good because both the book and the computer are impersonal aggregates of parts: they lack substantive unity at and over time and cannot be the subject of any concepts about God. But the problem for physicalism is that the human brain is also an impersonal aggregate of parts, constantly in flux, and this does not explain the emergence of a single, persistent person capable of understanding concepts at and over time.

This problem is more acute when we look at the brain in more detail, and consider its capacity to generate our personal mental life. As we saw, proponents of CP may agree with dualists that persons require a "first-person perspective," a subject that has and understands concepts at and over time. On this view, the "I" is a nonsubstantial mental subject, perhaps the property of having a stream of consciousness, or the potential for one, one that reduces to, or emerges from, the brain (or particular states of the brain). But, setting aside the worry that having a first-person perspective is an impure property,

Dean Zimmerman raises a further difficulty. If we look in the brain for plausible physical candidates for the purpose of being, or generating, the "I," they all "appear surprisingly like clouds on close inspection: it is not clear where they begin and end, in space or time. Many particles are in the process of being assimilated or cast off; they are neither clearly 'in,' nor clearly 'out.'"[28] This creates a problem of vagueness about just what matter constitutes a brain at and over time. Just as there are many sets of particles with an equal claim to be a particular cloud, there are many sets of particles with an equal claim to be a particular brain. As Igor Gasparov has argued, this vagueness about what constitutes a brain is inherited by any mental life that emerges, with the result that we have no reason to say that there is a single person either at or over time.[29] And as Joshua Farris has pointed out in response to Lynne Rudder Baker, this matters, because "The first-person perspective requires a metaphysical grounding, a 'what' and a thing that persists in and through time to account for the unity of consciousness,"[30] but the physical resources of the brain do not seem adequate to generate it.

Gasparov points out that *at* a time, there is a set of many clouds of particles in the brain, C_1, C_2, . . . C_n, such that each of the C_i ($1 \leq i \geq n$) has an equal claim to be the brain. This creates a *synchronic* problem of identity: if at time *t*, any one of the C_i generates a personal subject[31] capable of understanding concepts, then all of the C_i should do so. But it would be an implausible example of massive, systematic overdetermination if all the C_i generated the *same* subject.[32] So it is most probable that the C_i would generate multiple psychological subjects, S_1, S_2, . . . S_n. But then we do not have a plausible account of how just one subject per human being comes to understand divine concepts.

A similar problem surfaces when we consider the brain *over* time. The brain is a complex, dynamic system in a state of constant flux, with particles constantly being added or lost. So, over time, any particular brain exhibits vagueness: even if we could specify one and only one cloud C* that is the brain at each time, over time there would still be a sequence of different clouds of particles, C^*_1, C^*_2, . . . C^*_n and each of the C^*_i ($1 \leq i \geq n$) would have an equal claim to be the brain. This is a *diachronic* problem of identity: it would be an implausible case of massive, systematic overdetermination if each of the C^*_i generated the same subject. Instead, it is more probable that there would be a sequence of nonidentical, instantaneous self-stages, S^*_1, S^*_2, . . . S^*_n. But if that is so, even if divine concepts are engrafted innately, it seems that the psychological subject that initially owned those concepts would not persist over time and so those concepts could not be recollected, or if they were, still would not count as a subject's recollecting its *own* concepts.

Thus, it is not clear why an adult self-stage S_k would be able to access divine concepts engrafted into an earlier, nonidentical self-stage S_j that no longer exists. But even if this is possible, because divine concepts are somehow transmitted from one self-stage to the next, this would not be an account of a single self *recollecting* its own innate concepts because S_k is not identical to S_j (S_k was "born" more recently than S_j), so, bizarrely, S_k would be recollecting someone else's (S_j's) innate concepts. To overcome this, God would apparently need to engraft divine concepts separately in each self-stage. This is essentially the "constant emanation" theory of divine concepts that we will consider in the next section.

To his credit, Kevin Corcoran is aware that physicalism cannot plausibly ground personal identity in the mere physical aggregate of parts in an organism. Corcoran agrees with Zimmerman and Gasparov that, viewed as physical aggregates, human bodies are indeed like clouds (or storms): "Human bodies . . . are storms of atoms moving through space and time. They take on new stuff . . . and throw off old stuff as they go."[33] But he denies that human bodies find their identity merely in the atoms themselves. Rather, following John Locke, Corcoran maintains that the identity of a living organism is not determined by a particular aggregate of parts, but only requires that some aggregate or other is united by a common life. He defines this life as a "biological event . . . that is remarkably stable, well individuated, self-directing, self-maintaining and homeodynamic."[34] What maintains the identity of the living human body over time is immanent causation, in which "a state x of thing A brings about a consequent state y in A itself."[35] Clearly, if organisms derive their identity from such a common life which unites various clouds of particles at a time and connects them over time via immanent causation, concerns about the vagueness of the human body (including human brains) as physical aggregates are irrelevant. In that case, it may be that the identity of a human person is grounded in the identity of the living human organism that constitutes it. And then there is no problem with a person recollecting its own divine concepts.

However, a major problem with Corcoran's solution to the problem of personal identity is that it does not seem to be a physicalist one. Corcoran's description of life as "self-directing" and "self-maintaining" is surely teleological: these do not describe simple physical states of a system, but goals that it has (living systems are goal-directed). Now either this teleology is a fundamental part of the physical world or it emerges. If the former, this hardly sounds like physicalism: physicalism generally holds that at the base (nonemergent) level—the world described by physics—there is only undirected efficient causation. If the latter, and teleology emerges from the nonteleological, this is scarcely less puzzling than the claim that conscious persons emerge from an unconscious, impersonal world. And if teleology is emergent, then the arguments of Zimmerman and Gasparov can be adapted to show that it is unlikely that just *one* teleological system (one life) emerges from so many different "storms of atoms" at and over time, each storm having an equal claim to belong to a human body and an equal capacity to generate that body's life.

A second problem is that Corcoran's description of a human life seems question-begging. To talk of a human life as "self-directing" and "self-maintaining" assumes that there is a single unified, persistent human life, and how such a thing emerges is what needs to be explained. Reapplying Zimmerman and Gasparov's argument shows that this is not trivial and cannot be taken for granted. The same problem besets Corcoran's appeal to immanent causation, since it simply assumes that there is some well-defined, persistent entity A, such that one of A's states, x, produces a further state y in A. Let A be a particular living human organism. What is it about the underlying clouds of particles that make it the case that there is just *one* persistent A?

If the answer is that a life is radically emergent or sui generis, in the sense that it has a kind of unity and persistence not predicted by the underlying physical constituents of the body, then the account appears to be vitalist, since it relies on a fundamental difference of kind between living and nonliving systems. But if so, Corcoran's attempt to ground the identity of human persons in the living human bodies that constitute them avoids psychological (substance) dualism only by embracing *biological* dualism. This no longer sounds like physicalism. Even supposing Corcoran's account successfully grounds the identity of single, persistent persons (one per body, at and over time) capable of having and recollecting divine concepts, it appears to do so by abandoning core physicalist doctrines.

I conclude that if divine concepts are innate, CP fails to provide a plausible, physicalist account of the personal identity required for single persistent persons to possess and recollect those concepts.

Constant Emanation of Divine Concepts

Perhaps divine concepts are neither acquired nor innate. Instead, perhaps God refreshes our brains (and hence minds) with these concepts by a constant emanation of them. Then it does not matter if our brains are incapable of acquiring divine concepts through physical means (such as our brains interacting with nature and physical copies of the Bible), and perhaps it does not matter if we cannot recollect innate divine concepts, because God directly and continually forms in us the concepts necessary to know Him.

To be sure, God could do it this way. But the emanation theory is not a good fit for physicalism because it makes at least some of our psychological capacities depend on the continual direct intervention of an immaterial being. As an explanation, emanation also seems ad hoc and to violate Occam's razor, as it is far more complex than necessary. Surely an omnipotent God who wishes to achieve a regular result (all human beings can access divine concepts) can so make us that this is a necessary, lawful consequence of our natures. Dualists will argue that God can achieve this result by creating us as embodied souls, because souls, as unified persistent immaterial subjects, do not have the same limitations as material objects like the brain. If these souls are immaterial substances created with innate divine concepts as modes, then the same soul can access these concepts at and over time. This is because souls are substantively simple and persistent: their mental states are inseparable parts (modes) so that one and the same soul can access many divine concepts at the same time, and they remain the same substance over time, despite changes in the particular content of their thoughts and in the physical states of their embodiment.

To the extent that the dualist can flesh out such an account, the emanation theory can justly be criticized for committing a real "God of the gaps" fallacy: it multiplies beyond necessity the number of gaps in nature which God must fill.[36] And it calls into question whether God really made us in His image, as creatures designed to know Him, so that we could carry out our primary vocation to be stewards of the rest of creation. In matters of divine knowledge, it seems we are all afflicted by Alzheimer's disease and must each day learn what a previous self-stage knew about God. Before accepting this conclusion, one would need a very strong argument to show that no other alternative that is more elegant and in keeping with our status as image bearers was open to God, and it seems unlikely such an argument is forthcoming.

Finally, there are also biblical concerns about the emanation theory. Both the Old and New Testament describe knowledge of God as an achievement, something we can gain and retain, and something which can increase. This implies that the same person can go from not knowing to knowing God and from knowing something about God to knowing Him better. For example, Proverbs 2:4–5 says: "if you seek it like silver and search for it as for hidden treasures, then you will understand the fear of the Lord and find the knowledge of God." Again, God tells us He wants us to acquire this knowledge: "For I desire steadfast love and not sacrifice, the knowledge of God rather than burnt offerings" (Hosea 6:6). And Paul exhorts us to "walk in a manner worthy of the Lord . . . bearing fruit in every good work and increasing in the knowledge of God" (Colossians 1:10). This knowledge is important to God's plan of salvation for mankind, for He "desires all people to be saved and to come to the knowledge of the truth" (1 Timothy 2:4). But on the emanation theory, it is difficult to understand why knowledge of God is an achievement, and why there is apparently such wide variation in people's knowledge of God.

If divine concepts are acquired, it is easy to see that some would more fully master these concepts than others and how greater diligence in thinking about

God and searching for greater understanding of Him could lead to greater knowledge. Likewise, if divine concepts are innate, they might either be neglected and lay idle or be exercised frequently, leading to lesser or greater knowledge. But on the emanation theory, God's continually rewriting our brains with divine concepts would serve as a constant reminder, making it difficult to see how anyone could avoid having significant knowledge of God. Despite the natural man's willful suppression of the divine, on the emanation theory, one could not easily avoid thinking about the transcendent attributes of God, and so it seems likely that resistance to knowing God would, sooner or later, be overcome.

And one might reasonably predict that all people without brain deficits would at some time or other eventually acquire about the same knowledge of God, making it difficult to understand why some appear to know God much better than others ever will and how there can be people (e.g., Thomas à Kempis, Martin Luther, John Bunyan, Dietrich Bonhoeffer, C. S. Lewis, Richard Swinburne) whose knowledge of God appears to have grown enormously over time.

More generally, the emanation theory suggests that God literally brainwashes us with knowledge of Him, making it hard to see why there are any atheists. At the very least, a defender of the emanation theory would seem to need an elaborate theory that explains the wide distribution of beliefs about God that we actually find. It is true that scripture speaks of our willful suppression of the knowledge of God (Rom. 1:18 and Rom. 3:23), but this is a general problem for all mankind and so does not explain why the resistance to knowing God is greater in some people than others. A defender of the emanation theory would therefore need to appeal to factors other than possession of divine concepts that vary sufficiently among people to account for the diversity of belief about God.

DUALISM AND DIVINE CONCEPTS

Some may respond to my argument with a *tu quoque*: CP may not give a plausible account of our possession of divine concepts, but why suppose that substance dualism fares any better? This challenge must be taken seriously, and it deserves another whole chapter. Here, I can only outline my response.

First, I will make a concession. On my view, appeal to creaturely souls does not make it any easier to see how we could *acquire* divine concepts. The finite, imperfect, temporal interactions of finite, imperfect, temporal souls do not account for the formation of the concepts of infinity, perfection, and eternity. Thus, my sympathy is with Descartes and Leibniz: divine concepts are innate, engrafted into our souls by God Himself. Notice, that because the dualist is committed to finite immaterial beings (souls) doing causal work, it is not an embarrassment if we sometimes must directly appeal to an infinite immaterial being to causally explain some aspect of our psychological capacities.

Still, a critic may push back: why think that any concepts God engrafts into a soul are capable of presenting God as infinite, perfect, and eternal? Won't the limitations of the soul (a finite, imperfect, and temporal being) present the same difficulties as the limitations of the physical brain? Here, my response appeals to what we know about intentionality. First, we know that intentionality is a *self-transcendent property* of persons: our ability to think of things transcends our own limitations. Thus, someone who is five feet tall has no difficulty in thinking of something much taller, like Mount Everest and a fool has no difficulty in thinking of someone wiser than himself. More impressively, as Pascal noted, while physically I am like a speck in a vast cosmos, "through thought I grasp it."[37] But if we can grasp the totality of space-time (something presupposed by the naturalistic statement that space-time is all there is), this requires our understanding to transcend the temporal. Although we are time-bound creatures, our thought contacts the eternal. Likewise, whether or not infinite sets really exist in a Platonic sense (a vexed question in the philosophy of mathematics), certainly finite mathematicians can think of them and prove theorems about them. And Descartes seems right that in order for us to know that we are imperfect, we must have a prior concept of perfection. So if it is a fact that intentionality is self-transcendent property of persons, the real question is whether dualism provides a better explanation of that fact than does physicalism. I submit that it does.

For second, as we saw earlier, intentionality cannot be explained by the physical causal powers of organisms, and since the only noncontroversial cases of nonderived intentionality are mental subjects, it is reasonable to think that intentionality is a power that mental subjects have qua mental beings. It is therefore more plausible that souls, rather than brains, have the self-transcending power of intentionality, and with that power, finite, imperfect, and temporal souls could think of an infinite, perfect, and eternal being *if* they were equipped with the right concepts. But if God wishes all people to come to know Him, it is plausible that He could and would provide us with the particular self-transcendent concepts required to do this. This is one way in which we clearly reflect the image of God: we alone among creatures are made so that we can come to know God.

Third, once creatures are granted the gift of souls with nonderived intentionality and the innate concepts required to think of God, a further advantage of souls is that, being simple mental substances, they can maintain a mental subject's identity at and over time, despite the constant flux of the physical organism. As a result, a mental subject stocked at birth[38] with divine concepts can later gain knowledge of God. Thus, given innate divine concepts, one and the same mental subject might one day employ those concepts to conclude that nature is the handiwork of God or that the Jesus of the Scriptures is God in human flesh.

CONCLUSION

There seem to be three main ways of explaining our possession of divine concepts: they are acquired, they are innate, or they are emanations of God. But each alternative is difficult to square with CP. It does not seem that transcendent concepts can be acquired by any interaction of our brains with the physical environment; if they are innate, that hardly helps unless there at mental subjects which can access those concepts at and over time; and if they are constant emanations of God, CP hardly counts as a physicalist theory of human beings, since our psychological capacities depend on interventions by an immaterial being.

Moreover, to the extent that any of these theories rely on the actions of an immaterial God, it makes it hard to see why the idea of an immaterial soul is found so problematic by proponents of CP. If immaterial agency is objectionable in general, then this conflicts with the *Christian* part of CP, since Christianity is a religion of divine providence and miracles. But if it is not, there seems no compelling reason to say that souls are any more troubling than God, which undercuts the *physicalist* part of CP. And if CP makes it difficult to account for the knowledge of God that Christianity assumes is possible, but the capacity for such knowledge is at least plausible if we have souls, we have good reason to reaffirm the traditional view that human persons are embodied souls.

NOTES

1. The identity view is defended by Peter van Inwagen and Trenton Merricks. See van Inwagen, "A Materialist Ontology of the Human Person," in *Persons: Human and Divine*, eds. Peter van Inwagen and Dean Zimmerman (New York: Oxford University Press, 2007), 199–215, and Trenton Merricks, "How to Live Forever without Saving Your Soul: Physicalism and Immortality," in *Soul, Body and Survival: Essays on the Metaphysics of Human Persons*, ed. Kevin Corcoran (Ithaca, NY: Cornell University Press, 2001), 183–200.

2. See Nancey Murphy, *Bodies and Souls, or Spirited Bodies?* (New York: Cambridge University Press, 2006).

3. The constitution view is defended by Kevin Corcoran and Lynne Rudder Baker. See Corcoran, "Physical Persons and Postmortem Survival without Temporal Gaps,"

in *Soul, Body and Survival: Essays on the Metaphysics of Human Persons*, ed. Kevin Corcoran, 201–217, and his *Rethinking Human Nature: A Christian Materialist Alternative to the Soul* (Grand Rapids, MI: Baker Academic, 2006); also see Baker, *Persons and Bodies: A Constitution View* (Cambridge, UK: Cambridge University Press, 2000), her "Materialism with a Human Face," in *Soul, Body and Survival: Essays on the Metaphysics of Human Persons*, ed. Kevin Corcoran, 159–180, and her *Naturalism and the First-Person Perspective* (New York: Oxford University Press, 2013).

4. See Luke Van Horn, "Merricks's Soulless Savior," *Faith and Philosophy* 27/3 (July 2010): 330–341.

5. See Jonathan Loose, "Constitution and the Falling Elevator," *Philosophia Christi* 14, no. 2 (Winter 2012): 439–449.

6. Michael Rea argues that, on physicalist (naturalist) assumptions, there are no well-defined, persistent objects of knowledge; see Rea's "Naturalism and material objects," in *Naturalism: A Critical Analysis*, eds. William Lane Craig and J. P. Moreland (New York: Routledge, 2000), 110–132, and his *World without Design: The Ontological Consequences of Naturalism* (New York: Oxford University Press, 2002). Robert Koons further argues that, on physicalism, even if such objects of knowledge existed, their being persistent objects is not a causal power they have, since a physical aggregate at a time has just the same causal powers whether or not it composes a persistent physical object, and so we still could not know them; see Koons's "Epistemic Objections to Materialism," in *The Waning of Materialism*, eds. Robert C. Koons and George Bealer (New York: Oxford University Press, 2010), 291. R. Scott Smith argues that physicalism fails to explain our acquisition of concepts that correspond to extra-mental reality. See Smith's *Naturalism and our Knowledge of Reality: Testing Religious Truth Claims* (Farnham, UK: Routledge, 2016).

7. Obviously, I cannot consider each epistemological theory in detail, but my argument does show that CP faces difficulties if it relies on any of the following: (1) foundationalist theories of knowledge that depend on empirically acquired or innate concepts; (2) standard causal theories of knowledge, for example, the reliability theory; and (3) reformed epistemology. These theories all rely on some acquired or innate concepts to make knowledge possible. It is conceivable that a proponent of CP will propose some alternative epistemology which is immune to my objections.

8. While the apophatic way emphasizes our knowing God indirectly, by what He is not, texts like Romans 1:19–20 appear to support our having *positive* knowledge about God: "For what can be known about God is plain to them, because God has shown it to them. For his invisible attributes, namely, his eternal power and divine nature, have been clearly perceived, ever since the creation of the world in the things that have been made."

9. It is of course possible that there is a fourth alternative I have not thought of. But for this to be a serious objection, one would have to flesh out this alternative, make it independently plausible as a path to knowing God, and show that, on this account, CP could avoid the epistemic difficulties facing the other three alternatives.

10. Actually, I think there are serious problems here. It is not at all obvious that a physical object can have a concept. Notice that we are generally willing to allow

that books and computers contain information, but not that they contain concepts or thoughts. As J. P. Moreland has argued, concepts (and thoughts) appear to be inseparable parts of simple, mental subjects, and not the sort of thing that can be located in an aggregate of separable parts like a book, computer, or brain. See J. P. Moreland, *The Recalcitrant Imago Dei* (London: SCM Press, 2009), chapter 5.

11. In general, property *F* is not reducible to property *G* if *F* has causal powers lacked by *G*. So a mental property is not physically reducible if it has causal powers not possessed by any properties of the brain.

12. Thomas Hobbes, "Of the Consequences or Train of Imaginations," *Leviathan*, ed. Marshall Missner (New York: Pearson Longman, 2008), 13. For further development of this argument, see my "Knowledge of Abstracta: A Challenge to Materialism," *Philosophia Christi* 18, no. 1 (Summer 2016): 7–27.

13. Thanks to Charles Taliaferro for pressing this point in his comments on an earlier draft of this chapter.

14. René Descartes, "Third Meditation," in *Meditationes de prima philosophia / Meditations on First Philosophy*, ed. and trans. by George Heffernan (Notre Dame, IN: University of Notre Dame Press, 1990), 137, 139.

16. Descartes, "Third Meditation," 139.

17. It is conceivable that a reductive materialist would not opt for an empiricist epistemology. But both historically and today, most reductive materialists have been empiricists, as they believe materialism finds its justification in empirical science, which they see as the most reliable source of knowledge.

18. For example, see Robert C. Koons and George Bealer's *The Waning of Materialism* (New York: Oxford University Press, 2010) and Thomas Nagel's *Mind and Cosmos: Why the Materialist Neo-Darwinian Conception of Nature is Almost Certainly False* (New York: Oxford University Press, 2012).

19. A nonreductive physicalist can also deny property dualism. Searle's position is that mental properties are higher order *physical* properties of the brain. They are causally, but not ontologically, reducible to the lower level physical properties of the brain, but causal reduction is enough to make the higher level properties physical.

20. Some philosophers claim to be emergent *subject* dualists without affirming that they are substance dualists. It is fair to say, though, that the mental "subject" does most of the same work as a mental substance.

21. In practice, this is easily done by the process of "minding up" the brain, attributing distinctively mental properties and powers to the brain without showing how they derive from the brain's physical capacities.

22. See, for example, Kevin Corcoran, *Rethinking Human Nature*, 67

23. See, for example, Kevin Corcoran, *Rethinking Human Nature*, 68 and Lynne Baker's *Naturalism and the First-Person Perspective*.

24. For discussion, see J. P. Moreland, *The Recalcitrant Imago Dei*, 132–133.

25. Fred Dretske, *Knowledge and the Flow of Information* (Cambridge, MA: MIT Press, 1981), 57.

26. Thus, to use the classic example, if one looks at a broken clock that stopped at 2 o'clock, concludes that it is 2 o'clock, and, by coincidence, it is 2 o'clock, one has

a true belief but not knowledge, because there is no reliable connection between the time and the representation of the time by the clock.

27. My view is that to qualify as physicalist, CP must assert that the *direct* causal determinants of our psychological states are physical properties of the brain. Thus, on physicalism, an immaterial being like God can only influence psychological states *indirectly*, by influencing the physical properties of the brain.

28. Angus Menuge, "Knowledge of Abstracta: A Challenge to Materialism."

29. Dean Zimmerman, "From Experience to Experiencer," in *The Soul Hypothesis*, ed. Mark C. Baker and Stewart Goetz, 168–196 (New York: Continuum, 2011), 187.

30. Igor Gasparov, "Emergent Dualism and the Challenge of Vagueness," *Faith and Philosophy* 32, no. 4 (October 2015): 432–438.

31. Joshua Farris, "Bodily-Constituted Persons, Soulish Persons, and the *Imago Dei*: The Problem from a Definite I," *Philosophy and Theology* 28, no. 2 (2016): 455–468, 463.

32. Note that this subject is not being construed in dualistic terms as an immaterial substance: for nonreductive materialists like Baker, Corcoran and Murphy, the "subject" is just a unified cluster of emergent psychological properties capable of a first-person perspective, etc.

33. Overdetermination can be plausible in particular cases. If 100 soldiers in a firing squad fire on a prisoner simultaneously, then the prisoner's death is overdetermined because it has many individually sufficient causes. But overdetermination is implausible, and a violation of Occam's razor, when it is systematic or ubiquitous. Thus, if it is claimed that *both* a mental act of volition and a simultaneous state of the brain are individually sufficient for *every* intentional action we perform, we feel that one or other of the causes must be redundant. Indeed reductive physicalists like Jaegwon Kim uses such an argument to show that irreducible mental causes are excluded from independent causal power by their subvenient physical bases (the "exclusion argument"). See Jaegwon Kim, *Philosophy of Mind*, 3d ed. (Boulder, CO: Westview Press, 2011), chapter 7.

34. Corcoran, *Rethinking Human Nature*, 72.

35. Corcoran, "Physical Persons and Postmortem Survival without Temporal Gaps," 206.

36. Corcoran, *Rethinking Human Nature*, 72.

37. See, for example: William Hasker, *The Emergent Self* (Ithaca, NY: Cornell University Press, 1999); E. J. Lowe, *Personal Agency: The Metaphysics of Mind and Action* (Oxford: Oxford University Press, 2008); J. P. Moreland, *The Recalcitrant Imago Dei*; Angus Menuge, "The Ontological Argument From Reason: Why Compatibilist Accounts of Reasoning Fail," *Philosophia Christi* 13, no. 1 (Summer 2011): 59–74, and "Neuroscience, Rationality and Free Will: A Critique of John Searle's Libertarian Naturalism," *Philosophia Christi* 15, no. 1 (Summer 2013): 81–96; and Richard Swinburne, *Mind Brain and Free Will* (Oxford: Oxford University Press, 2013).

38. Blaise Pascal, *Pensées*, trans. A. J. Krailsheimer (New York: Penguin Books, 1966), #113, 59.

39. Or whenever it is that God does this: the matter has been disputed.

SUGGESTED FURTHER READING

Baker, Lynne Rudder. *Naturalism and the First-Person Perspective.* New York: Oxford University Press, 2013.

Corcoran, Kevin. *Rethinking Human Nature: A Christian Materialist Alternative to the Soul.* Grand Rapids, MI: Baker Academic, 2006.

Gasparov, Igor. "Emergent Dualism and the Challenge of Vagueness." *Faith and Philosophy* 32, no. 4 (October 2015): 432–438.

Koons, Robert C. "Epistemic Objections to Materialism." In *The Waning of Materialism*, edited by Robert C. Koons and George Bealer, 281–306. New York: Oxford University Press, 2010.

Menuge, Angus. "Neuroscience, Rationality and Free Will: A Critique of John Searle's Libertarian Naturalism." *Philosophia Christi* 15, no. 1 (Summer 2013): 81–96.

———. "Knowledge of Abstracta: A Challenge to Materialism." *Philosophia Christi* 18, no. 1 (Summer 2016): 7–27.

Moreland, James Porter. *The Recalcitrant Imago Dei.* London: SCM Press, 2009.

Murphy, Nancey. *Bodies and Souls, or Spirited Bodies?* New York: Cambridge University Press, 2006.

Rea, Michael. "Naturalism and material objects." In *Naturalism: A Critical Analysis*, edited by William Lane Craig and J. P. Moreland, 110–132. New York: Routledge, 2000.

Smith, R. Scott. *Naturalism and our Knowledge of Reality: Testing Religious Truth Claims*. Farnham, UK: Ashgate, 2012.

Swinburne, Richard. *Mind Brain and Free Will.* Oxford, UK: Oxford University Press, 2013.

Zimmerman, Dean. "From Experience to Experiencer." In *The Soul Hypothesis*, edited by Mark C. Baker and Stewart Goetz, 168–196. New York: Continuum, 2011.

Chapter 5

Physicalism, Divine Eternality, and Life Everlasting

R. Keith Loftin and R. T. Mullins

Contemporary Christian reflection on anthropology can sometimes be a funny thing. At times, we cannot help but suspect that there is a disconnect between the doctrine of God and certain claims about human nature. For instance, consider the often quoted, but rarely fleshed out, interaction problem raised against substance dualism: how can an immaterial substance interact with a physical substance? If this is a problem—and we doubt that it is—it should be a problem for the doctrine of God, as well.[1] The Christian God is a personal (indeed tripersonal) immaterial substance that causally interacts with the physical universe.[2] If the Christian theologian wishes to affirm that an immaterial God can interact with the physical universe, she cannot deny the possibility that an immaterial soul could interact with a body. In other words, a Christian who is a physicalist ought not to raise the interaction problem for substance dualism without recognizing the problem this causes for her doctrine of God.

We point out this example as one, among many, possible disconnects we see between the doctrine of God and theological-philosophical anthropology. In this chapter, we wish to focus on another possible area of disconnect between the doctrine of divine eternality and life everlasting. We maintain that the eternal God wishes to grant humanity life everlasting. We will argue that physicalism is not compatible with the Christian claim that the eternal God is going to grant redeemed humanity life everlasting.

In the first section, we shall outline the key theological assumptions that are guiding our thoughts. In the second section, we shall articulate the doctrine of divine eternality, drawing out its implications for the philosophy of time, personal identity through time, theological anthropology, and the doctrine of life everlasting. In the third section, we shall argue that endurantism is needed for God to grant us eternal life. In the fourth section, we shall argue that physicalism is not compatible with endurantism.

THEOLOGICAL ASSUMPTIONS

There are four theological assumptions that we wish to identify before moving forward. For the sake of transparency, we think it is helpful for readers to have an understanding of our starting points as it should help readers better grasp some of the arguments and moves that we make in later sections. Those four assumptions are as follows: 1) the centrality of immortality to Christian hope, 2) discontinuity and continuity between this life and the life to come, 3) the temporality of creation, and 4) constraints on theological adequacy.

Immortal Life

The promise of life everlasting is central to the message of Christian hope. Whereas the way of the unrighteous leads to separation from the creator, according to Christian teaching, the redeemed in Christ are granted life everlasting (Jn. 3:14–16, 11:25–26; Matt. 25:46). What does this mean? Jesus states: "Now this is eternal life—that they know you, the only true God, and Jesus Christ, whom you sent" (Jn. 17:3). In virtue of knowing God through Christ, the redeemed are united into communion with God (Jn. 17:22–23), that they may "remain in the Son and in the Father" (1 Jn. 2:24–25). As Irenaeus explains, "the Lord thus has redeemed us . . . and has also poured out the Spirit of the Father for the union and communion of God and man, imparting indeed God to men by means of the Spirit, and, on the other hand, attaching man to God by His own incarnation, and bestowing upon us at His coming immortality durably and truly, by means of communion with God."[3] Life everlasting, then, involves being granted a share in God's own life. Echoing Irenaeus, in his sermon on 1 John 5:20, John Wesley describes life everlasting as "the happy and holy communion which the faithful have with God the Father, Son and Holy Ghost."[4] Integral to the Christian hope of life everlasting is the promise of immortality (2 Tim. 1:10; 1 Cor. 15:53–54). God cannot give humanity a gift that He does not have, but thankfully we worship an immortal God. We believe that this gift of eternal life is grounded in the immortal and eternal nature of God (1 Tim. 1:17, 6:14–16). We shall have more to say on divine eternality below.

Discontinuity and Continuity

There is a soteriological theme in the New Testament that scholars often refer to as the "now" and the "not yet." Like many other theologians, we affirm that the facets of this New Testament theme are important to any nuanced conception of life everlasting.[5] An interesting question that arises from this New Testament theme is to what extent life everlasting is a present versus an as yet future reality. To be sure, the story of Christian salvation is of a hope that includes a future resurrection of the dead where God will give us glorified bodies that are made to last forever in a redeemed creation (1 Cor. 15:12–49). However, the New Testament witness also speaks of experiencing this everlasting life in the here and now. Paul writes, "if anyone is in Christ, he is a new creation; what is old has passed away—look, what is new has come!" (2 Cor. 5:17). The fourth Gospel tells us "the one who believes in the Son has eternal life" (3:36, 6:46), with John's first epistle similarly stressing believers' present possession of life everlasting (1 Jn. 5:11–13). In short, scripture gives us reason to think that our experience of life everlasting is indeed a present reality that is had in our pre-resurrection life. This present reality is granted upon new birth in Christ. But we further believe that the redeemed look forward to a post-resurrection experience of life everlasting that encompasses but also exceeds our present experience.

It is beyond our purpose fully to describe the sense(s) in which our pre-resurrection experience of life everlasting is different from post-resurrection experience of the same, yet we believe the scripture affirms that there is an important continuity between the two. The typical claim from biblical scholars and theologians is that the life to come will be characterized by the presence of God's Holy Spirit and an absence of sin, sickness, and death.[6] So there is a discontinuity between our present state and the life to come since our present lives include sin, sickness, and death. However, there is important continuity because the believer currently experiences the presence of the Holy Spirit. Further, there is continuity between the two in that the one who currently believes, the one who currently experiences the Holy Spirit, will be the one who will receive immortality and enjoy life everlasting without sin, sickness, and death. As Jesus makes clear, blessed are those who currently seek righteousness, for they will inherit the kingdom of heaven (Matt. 5:2–12).

To conceive life everlasting as something radically disconnected from life "here and now" would be to fail to appreciate the biblical emphasis on continuity in God's program—a program of *redeeming* creation, rather than annihilating and replacing it.[7] This of course includes the Christian experience of life everlasting. God is not interested in replacing those who presently mourn. Instead, God is interested in comforting those who presently mourn. God is not interested in replacing the meek. He is interested in giving the meek a redeemed earth (Matt. 5:2–12). Again: while our pre- versus post-resurrection experiences of life everlasting are not qualitatively identical, there is an essential continuity between the two. As Hans Weber notes, "if there was no continuity whatsoever between both lives, the New Testament idea of an 'eternal life' would be proven wrong, since it is not confined to the absolute future but rather reaches from there to the present life before death."[8]

The Temporality of Creation

This might seem like an odd theological affirmation since it is quite obvious that the created order is temporal. We are making a stronger claim merely than that God merely made a temporal universe. What we are affirming is that the temporality of creation is a good thing, and that God is going to redeem the temporality of creation. God did not make temporal life only to scrap it in the eschaton. God made temporal life because it has value and makes several of His purposes for creation possible. This is in stark contrast to claims that God is going to bring about a timeless eschatological state of affairs.[9] We think that such a thing is not only impossible, but that it is unbiblical.[10] Further, we believe that a timeless life after death offers a far too radical kind of discontinuity between the present life that God seeks to redeem and the life to come.

If the life everlasting of Christian hope is conceived as timeless, then one's understanding of God's program of redeeming creation shall demand a radical discontinuity with our temporal experience here and now. Against this, Miroslav Volf argues, "*ultimate fulfillment is not only compatible with temporality but also unthinkable without it*, partly because any intelligible notions of both reconciliation and contentment in fact *presuppose* change . . . with the erasure of temporality in the 'life' of the world to come, it takes away the possibility of communal peace and personal joy," and this seems to us correct.[11]

Theological Adequacy

We readily acknowledge that there are different ways to articulate Christian doctrines, but not all articulations are created equal. Our focus in this chapter is the connection between the doctrines of God, theological anthropology, and life everlasting. So there will be various nuances and systematic connections at play in any given articulation of these doctrines. We believe that some accounts of anthropology are theologically inadequate. For the purposes of this chapter, we affirm a simple rule of thumb for judging if a particular account of Christian doctrine is theologically adequate. It goes as follows: if a particular rendering of the doctrine of anthropology cannot account for the hope of life everlasting, something is wrong with that rendering. Perhaps it is false, or perhaps it merely needs to be revised a bit. Given space constraints, we shall argue that two popular versions of physicalist understandings of personal identity over time cannot adequately account for the hope of life everlasting.

DIVINE ETERNALITY

We need to be clear about a peculiar stance that we take on divine eternality. We reject the traditional claim that God is timeless. We modify the doctrine of divine eternality in a way that we believe better reflects the basic teachings of scripture and Christian doctrine. In order to understand our modification, one must first understand the traditional view.

There are at least three questions about the philosophy of time that must be answered in order properly to understand the traditional doctrine of divine eternality, as well as our modification. These three questions are as follows. 1) The metaphysical question: what is time? 2) The ontological question: what moments of time exist? 3) The persistence question: how does a person persist through time?

Metaphysical Question

Traditionally, Christian theologians overwhelmingly have held to what is called a relational theory of time.[12] This view says that time exists if and only if change occurs. This is because change creates a before and after. Christians also overwhelming have held that one key characteristic of a temporal object is that it undergoes change. Further, a key characteristic of an atemporal object is that it does not undergo any kind of change.[13]

Near the end of the scholastic era and moving into the scientific revolution, various thinkers in the West began to reject the relational theory of time, affirming instead a view called the absolute theory of time. On the absolute theory, time is thought to exist without change (one might think that time is dependent upon the existence of a substance that possibly can undergo change, with this being all that is needed for time to exist).[14] During this period a fair number of thinkers, such as Samuel Clarke, believed God to be temporal—or at least the substratum of time[15]—and capable of undergoing certain kinds of changes. In their view, the eternal existence of God guaranteed the eternal existence of time, since God is a being that can possibly undergo change.[16] (More on divine temporality later.)

Ontological Question

Traditionally, Christians have held to a view called presentism.[17] This view says that only the present moment of time exists. The past no longer exists, and the future does not yet exist. On presentism, the present moment of time exhausts reality.[18] There are no nonpresent objects in existence. This can be contrasted with the view called eternalism according to which the past, present, and future equally exist.[19] By way of example, on eternalism, even though we are not currently located in the year 1763, the year 1763 and its occupants exist. On presentism, the year 1763 no longer exists since the present moment of time exhausts reality.

Most contemporary proponents of eternalism grant that presentism is the default view because of its intuitive nature. The intuitive pull of presentism can be witnessed throughout the history of Christian thought.[20] Augustine, for example, takes presentism to be obviously true. Augustine's most famous, though not most important, treatment of time is found in his *Confessions*. Throughout his discussion, he examines a series of puzzles that arise on presentism related to the measurement of time and how prophets might be able to know the future. Near the end of his discussion he says, "It is now, however, perfectly clear that neither the future nor the past are in existence."[21] Even though various philosophical puzzles arise from presentism, Augustine sees no reason to give it up.

Despite the intuitive pull of presentism, eternalism has grown in popularity since the advent of Albert Einstein's special theory of relativity. Proponents of eternalism often claim that the special theory of relativity entails eternalism. However, this is far from obvious since a fair number of philosophers and scientists see no reason to give up presentism on the basis of a theory that fails to capture the fundamental laws of nature. Unfortunately, we do not have space to go into those issues here as it will take us off topic.[22]

Persistence Question

Traditionally, Christians have held to a view called endurantism, or three-dimensionalism. A person endures through time by living her life as a whole, all at once, in the present moment. There is numerically one thing (the person) that persists as a whole from one moment to the next. She does not have parts at different times. She exists wholly in the present since, given presentism, there are no nonpresent objects in existence.[23]

Endurantism is contrasted with a view called four-dimensionalism.[24] Four-dimensionalism is a doctrine about temporal parts, and is typically held in conjunction with an eternalist ontology. On eternalism, all moments of time exist. On four-dimensionalism, the entire space-time world can be cut up into temporal parts—numerically distinct objects that exist at each instant of time. A temporal part does not change its temporal location because its location is eternally fixed in the space-time world. Unlike endurantism, objects do not persist as a whole from one moment to the next. There is no numerical identity across time on four-dimensionalism. Instead, objects persist by having temporal counterparts at later times.[25]

Divine Eternality

With these three questions laid out, it is worth recalling how classical Christians have answered them. Classical Christian theism has overwhelmingly affirmed a relational theory of time, a presentist ontology of time, and an endurantist theory of persistence through time. With these answers, we can offer an insight into the traditional view of divine eternality. The traditional claim is that God is timeless, which means that God exists without beginning, without end, and without succession.[26] Yet this claim needs to be unpacked a bit.

Recall that, on the relational theory of time, time exists if and only if change exists. Classical theists hold that God is strongly immutable in that God cannot undergo any changes whatsoever, be they intrinsic or extrinsic.[27] Alongside their commitment to the relational theory of time, classical theists maintain that immutability entails divine timelessness.

Given their commitment to presentism and endurantism, classical theists maintain that God lives as a whole, or all at once, in a timeless present that lacks a before and after.[28] Creatures endure in an ever fleeting present, with moments of their lives fading away into the nonexistent past. Classical theism maintains that God is not like that. God's life never fades away into the past, nor does God experience new moments in His life. Since God cannot change, God exists in a present that lacks a before and after.

At this point, one might wonder how classical Christians came to believe that God is timeless, especially since the Bible contains no hint of timeless existence.[29] Classical theism derives its understanding of God from the definition of God as a perfect being. What must be understood is that all human persons have varying intuitions about what makes a being perfect. These intuitions have a significant impact on how one formulates one's understanding of God, and how one interprets scripture. Due to space limitations, we will focus on one example of this from Anselm.

For Anselm, the present is the only moment of time that exists.[30] Further, creatures exist as a whole, or all at once, at each moment of their existence. A human person exists as a whole through individual times.[31] In light of these assumptions, Anselm considers whether it is better for God to endure through all times as a whole (i.e., divine temporality), or to exist as a whole in a timeless present that lacks a before and after (i.e., divine timelessness). If God endures through all times, God's life can be conceptually divided up into parts.[32] What this means is that part of God's life will be over and done with as it fades away into the nonexistent past. Anselm regards such a claim as inappropriate for the eternal God.[33] Why?

Anselm, like all classical theists, is committed to the doctrine of divine simplicity. The assumption here is that a perfect being must be an absolutely simple being. We do not have the space to delve into all of the complex nuances of this doctrine, so we must stick to a quick definition. Peter Lombard offers the following definition of divine simplicity: "The same substance alone is properly and truly simple in which there is no diversity or change or multiplicity of parts, or accidents, or of any other forms."[34] This doctrine is actually much stronger than most contemporary Christian philosophers and theologians realize. It entails that God does not have any kind of metaphysical complexity of any sort. A simple God does not have any intrinsic or extrinsic properties because a simple God does not have any properties at all. What is relevant for Anselm is the extreme extent to which God lacks any kind of complexity or parts. Anselm explicitly denies that God can be divided into either actual parts or conceptual parts.[35] If God endured through time, His life would be divided into conceptual parts such as before and after. That would violate divine simplicity. So Anselm maintains that we should affirm divine timelessness.

We deny divine simplicity, so we see no reason to follow Anselm to his conclusion that God is timeless. Unlike the classical tradition, we do not see divine simplicity as a possible perfection. We do not think that a perfect being can be absolutely simple because we think that it is metaphysically impossible for a simple being to have free will.[36] We wish to affirm that God freely creates and sustains the universe, and that God the Son freely became incarnate for our salvation. So we are unable to affirm that God is simple, and we thus see no motivation to follow Anselm to the conclusion that God is timeless.

On our view, God is not timeless. God is eternal in that God exists without beginning and without end, but God is not timeless because God experiences succession in the divine life. In order to understand this, we must make it clear how we answer the three questions about time. First, we differ over how to answer the metaphysical question. However, we agree that if there is a change, there is clearly time because there will be a before and after. Second, we agree with the tradition that presentism is true. Third, we agree with the tradition that endurantism is true.

So, what do we say about God's eternality? We say that God exists as a whole, or all at once, in the present moment of time. It is the same present as ours. It would be better, though, to say that we exist in God's present, since God is actively sustaining us in existence from moment to moment. In other words, we believe that God is an endurant being who persists through time.[37] We disagree with the traditional way of making this divine duration timeless because we cannot understand how such a thing could possibly be true.[38] The God of the Bible is immutable in certain respects, but mutable in others. God is immutable in that He is a necessarily existent, omnipotent, omniscient, perfectly good, tripersonal, and perfectly free being. However, this God changes through the exercise of His free will, and thus undergoes a succession of moments. God was not always creating a universe, but freely brought a universe into existence. This marked a new moment in the life of God. Having brought the universe into existence, God freely continues to maintain and interact with it in rather astonishing ways. He was not always in a covenantal relationship with Abraham because Abraham did not always exist. Yet at one point in time, God freely chose to enter a covenant with

Abraham, and He promised to bless the entire world through Abraham's offspring. Again, this marked a new moment in the life of God. One of the most dramatic moments in history is the incarnation. The incarnation of God the Son is a free and gracious act of the triune God. The universe had never seen anything quite like it before. The incarnation marked a new moment in the life of God as He sought to establish an everlasting relationship with His creatures. This is a God who endures through time, and who seeks to ensure that His redeemed creatures endure with him forever.

PERSONAL IDENTITY OVER TIME AND LIFE EVERLASTING

We believe that God and creatures are both endurant beings. Further, we believe that God is going to grant creatures a life without end. Since God is an endurant being who will never cease to exist, He can give us a life that will never cease.

In order to get clear on the nature of personal identity over time as it relates to life everlasting, we need to distinguish two related but distinct questions.

1. What is a human person?
2. What does personal identity consist in?

The first question is about the metaphysical make up of human persons. To understand this question, consider your friend Sally. Sally is a human person, but you want to know exactly what kind of thing Sally is. A physicalist is going to say that a human person like Sally is a purely physical being. A substance dualist will say that a person like Sally is an immaterial substance, or a mind. The dualist will go on to say that what makes Sally a *human* person is that she is an immaterial mind that is appropriately related to a human body. This appropriate relationship is often called *embodiment.* (For more on embodiment, see the chapter, "Physicalism and the Incarnation" in this volume.)

The second question is about the necessary and sufficient conditions for personal identity over time. Consider Sally again. Say that Sally is 30 years old, and that you have known her since she was 5 years old. What makes the Sally at age 5 the same person as the Sally at age 30? To be clear, this is not an epistemic question. We are not asking, "How do I know that this is the same Sally?" We are asking a metaphysical question. We want an account that explains how this is the same Sally. To answer this question, we need to distinguish several different accounts of personal identity over time. There are two kinds of accounts that one can offer: complex and simple.[39] On the complex view, there are substantive conditions for personal identity over time. The complex view says that personal identity can be explained in nonpersonal or subpersonal terms. What the complex theorist says is "that a person persists over time is nothing more than some other facts which are generally spelled out in either biological or psychological terms, or both."[40] A simple theorist will deny that these complex conditions capture personal identity. On the simple view, there are no nontrivial or noncircular conditions for personal identity over time. This is because personal identity is a primitive notion that is not subject to a deeper analysis.

Consider again Sally. What makes the Sally at age 5 the same person as the Sally at age 30? The simple theory says that personal identity is an ontologically primitive notion. So nothing makes the Sally at age 5 the same person as the Sally at age 30. It simply is the same Sally. That is just what personal *identity* is. To be clear, the simple view affirms a strict numerical identity. This is the sort of numerical identity that endurantism wishes to affirm. The complex view will disagree that personal identity is an ontologically primitive notion. The complex view will say that personal identity is reducible to some deeper biological or psychological relationships between objects or mental states.

The proponent of the complex view has several options available to her to unpack this. Due to space limitations, we shall restrict ourselves to discussing two popular accounts that physicalists about human persons often adopt. The first account we shall examine is the combination of physicalism, four-dimensionalism, and the psychological continuity account of personal identity. The second view we shall examine is a combination of physicalism, endurantism, and the biological account of personal identity.

Physicalism, Four-Dimensionalism, Psychological Continuity, and Life Everlasting

Some complex theorists affirm both physicalism about human persons and four-dimensionalism about personal identity over time. With regards to personal identity over time, a four-dimensionalist will say that certain temporal parts are united, or fused together, in interesting ways to form a space-time worm. On one version of four-dimensionalism, called perdurantism, a human person is a space-time worm. What unites the different temporal parts together such that they constitute a particular space-time worm? This is where the complex account of personal identity can come into play. On one version of the complex theory, the temporal parts are psychologically continuous with one another.

On this option for the complex view, one will say that what makes the Sally-at-age-5 the same person as the Sally-at-age-30 is the fact that the later Sally stands in some sort of psychological continuity with the earlier Sally.

There are many numerically distinct temporal parts that can be referred to as Sally (one for each instant in fact!). The Sally-at-age-5 has later temporal counterparts, and she is connected to these later temporal counterparts by immanently causing them to have certain psychological states.[41] What this means is that the Sally-at-age-5 passes her psychological states on to the next temporal counterpart, who in turn passes on her psychological states to the next temporal counterpart, who in turn . . . and so on till we reach the temporal part at age 30. Each of these temporal parts, or person stages, is a numerically distinct Sally. Yet they are unified through this interesting psychological relationship and immanent causation. According to the psychological view, this interesting relationship is what personal identity over time consists in.

We do not like this physicalist view. Why? On four-dimensionalism, the object that is fused together through psychological continuity forms a space-time worm, but this space-time worm does not enjoy numerical identity. The only objects that enjoy numerical identity are the temporal parts themselves. Each temporal part is identical to itself. What the space-time worm enjoys is a continuity relation cut in terms of psychological continuity.[42] This continuity relation is explicitly not numerical identity, and is sometimes referred to as the gen-identity relation in order to get that fact across.[43] As proponents of four-dimensionalism often say, personal identity is not what matters. What really matters is that we persist by having later temporal counterparts.[44]

We disagree with this four-dimensionalist stance because numerical personal identity matters quite dearly to us. We want to be the numerically same people who enjoy life everlasting with God and all of redeemed humanity. We believe that this four-dimensionalist account causes problems for our hope in life everlasting.

Here is where problems arise for life everlasting. Each temporal part of a human person space-time worm is called a *person stage*. Each person stage is a thinking thing with free will, and each person stage of a space-time worm is psychologically continuous with particular person stages that have come temporally before and that come temporally after the person stage via immanent causal relations. Consider the person stage of the apostle Peter that exists at some time t_x which is temporally prior to Christ's return. This person stage is thinking, "I sure look forward to Christ's return." Things get rather unfortunate at this point for this particular person stage. The complex theorist will say that this person stage is psychologically continuous with later person stages that exist at Christ's return. So, the Peter stage that exists at t_x has later temporal counterparts that are able to say, "Wow, I really am enjoying the return of Christ." However, that poor person stage back at time t_x is saying, "I sure look forward to Christ's return." This person stage never gets to enjoy the return of Christ. We are of the opinion that this is not the sort of life everlasting in which the apostle Peter has placed his hope. In fact, it seems to us that this Peter stage has nothing to hope for since this Peter stage is eternally located at time t_x, and so can never experience the glorious return of Christ.

The four-dimensionalist might reply that this isn't so bad. The person stage at t_x is eternally located at that time. The Peter that exists at that time never ceases to exist given an eternalist ontology of time. Surely that is a kind of everlasting life.

We concede that this is a kind of eternal life. On eternalism, there is no state of affairs where God exists without the eternal space-time world.[45] All moments of time simply do exist, and are sustained by, God eternally. However, we maintain that this is not the sort of life everlasting that the Bible speaks of. In Jesus's sermon on the mount, He makes certain promises to those who seek healing, forgiveness, mercy, righteousness, and peace. Those who seek such blessings eventually will come to enjoy those very things at the eschaton (Matt. 5:1–12). In the situation that the four-dimensionalist is describing, none of this pans out. The Peter stage that exists at time t_x is seeking the kingdom of heaven, but it is a numerically distinct Peter stage who enjoys the kingdom of heaven. That is not what Jesus has promised. Jesus did not promise, "Blessed are those who seek righteousness, for your later temporal counterparts who are numerically distinct from you shall find it." To be sure, the author of Matthew does not state Jesus's underlying metaphysical assumptions about personal identity over time. However, we maintain that four-dimensionalism is a far cry from a natural reading of the text. Further, four-dimensionalism seems to make Jesus's promises on the sermon on the mount unintelligible.

Much more could be said here, but we shall rest our case against the combination of physicalism, four-dimensionalism, and the complex view. We wish to turn our attention to the combination of physicalism and endurantism.

PHYSICALISM, ENDURANTISM, AND LIFE EVERLASTING

As unrepentant substance dualists, we fail to see the compatibility of physicalism with Christian doctrine. So far, we have argued that a physicalist cannot account for life everlasting if she adopts four-dimensionalism and some sort of psychological criterion of personal persistence over time. It seems to us that endurantism is essential to the biblical notion of life everlasting. It is, after all, the *redeemed* who receive life everlasting—but that notion requires the numerical identity of the one who first walked in unrighteousness but later is united into communion with God. In this section, we shall argue that physicalism does not sit well with endurantism.

Someone who affirms endurantism will be able to hold either a simple or a complex view. Most, but not all, physicalists affirm a complex view.[46] Most, but not all, dualists affirm the simple view.[47] We are dualists who affirm the simple view. We say that the persistence conditions of a human person are different from the persistence conditions of her body. A person, the immaterial mind, just does persist with numerical identity over time. Her body, however, persists in a complex way, and we do not think that this complex way allows for strict numerical identity. It is to this problem that we now turn.

We focus our attention on physicalists who are endurantists and who also affirm a complex view of personal persistence over time. This is often referred to as the biological approach.[48] This version of the complex view says that what makes Sally at age 5 the same person as the Sally at age 30 is that there is some sort of biological continuity between each Sally. A popular version of this is called animalism. On animalism, a person is numerically identical to a human organism.[49]

Animalism so defined still needs to answer some important questions in order to shed light on this biological continuity. For instance, what makes one human organism the same organism at later times? It cannot be (on pain of raising the specter of mereological essentialism)[50] the physical stuff that makes up the organism since a human organism is constantly losing and gaining parts over time through mitosis and other such events. The animalist readily acknowledges this and looks elsewhere for an answer. The typical answer is that there is a persistence of some sort of underlying biological process called Life. Life is the "self-organizing biological event that maintains the organism's complex internal structure" amidst the perpetual need to "take in new particles, reconfigure and assimilate them into its living fabric, and expel those that are no longer useful to it."[51] On this view, what makes one organism identical to an organism at a later time is the fact that each organism is caught up in the same Life: so long as the same biological event of Life continues, Sally the organism persists.[52]

If Sally at age 5 is caught up in a Life distinct from the Life of Sally at age 30, then clearly Sally's endurance through time cannot be grounded in a Life for there would in fact be numerically distinct Lives. Can animalism account for Sally's endurance through time? That, it seems, will depend upon whether Life can endure, and that will depend upon the endurance of the relevant biological activities. However, as Brandon Rickabaugh has pointed out,

> A life is an event composed of a collection of separate relation instances and atomic parts, and as parts are replaced, so are the relation instances. Because of this inherent process, a life at *t*1 is not numerically identical to a life at *t*2. Although the relation types and part types may remain, the specific relation tokens and part tokens are expelled and replaced. That is, the life at *t*2 might have the same type of structure and same type of parts as the life at *t*1, although the life at *t*2 does not have the numerically identical structure or the numerically

> identical parts as the life at *t*1. The life just is this storm of parts and relation instances. It isn't as if there is some fundamental thing that has various separable parts and relation instances. A life just is the storm, the collection of parts and relations. The result is that a life does not endure.[53]

So, the event that is the Life of Sally at age 5 involves the same *types* of relations as the event that is the Life of Sally at age 30 (viz., the types of relations in which organic materials stand) as well as the same *types* of parts (viz., organic matter). But the particular bits of matter—that is, the part *tokens*—of the Life of Sally at age 5 are not the same as the part *tokens* as those of the Life of Sally at age 30, and the particular relation *tokens* of the former are not the same as those of the latter. The upshot of this is that the Life of Sally at age 5 is not numerically identical with the Life of Sally at age 30 and so cannot be the ground of Sally's enduring through time. Given the inability of Life itself to endure, physicalism's best hope for compatibility with endurantism—animalism—fails.[54] Once again, then, we do not see how physicalism can account for the numerical identity of those who enjoy pre-resurrection experience of life everlasting and those who enjoy post-resurrection experience of life everlasting.

CONCLUSION

The eternal God who willingly endures through timc wishes to grant life everlasting to the redeemed in Christ. We cannot see how this is compatible with physicalist views of persons and have argued as much. Setting aside the difficulty of construing "the happy and holy communion which the faithful have with God" in merely physical terms, it seems to us that a proper understanding of life everlasting requires strict numerical identity of persons over time. This is integral both to the core notion of *being redeemed* and to explaining the essential continuity between one's pre- and post-resurrection experiences of life everlasting inherent within God's program of redeeming creation. We believe the substance dualist view of human persons makes good sense of these notions. Whatever attractions there may be for "Christian physicalism," it seems to us that physicalism about persons cannot account for the biblical notion of life everlasting.

NOTES

1. Stewart Goetz and Charles Taliaferro, *A Brief History of the Soul* (Malden, MA: Blackwell Publishing, 2011), 131–151.
2. Alvin Plantinga, "What is 'Intervention'?" *Theology and Science* 6 (2008): 369–401.

3. Irenaeus, *Against Heresies*, in *The Apostolic Fathers, Justin Martyr, and Irenaeus*, eds. Alexander Roberts, James Donaldson, and A. Cleveland Coxe, *Ante-Nicene Fathers*, vol. 1 (Peabody, MA: Hendrickson, 1994), 5.1.1. Ben C. Blackwell's *Christosis: Pauline Soteriology in Light of Deification in Irenaeus and Cyril of Alexandria* (Tübingen: Mohr Siebeck, 2011) helpfully discusses this theme in some Greek patristics.

4. John Wesley, "Sermon LXXXII.–*Spiritual Worship,*" in *The Works of the Reverend John Wesley*, ed. John Emory, vol. 2 (New York: B. Waugh and T. Mason), 12; cf. Jerry Walls, *Heaven: The Logic of Eternal Joy* (New York: Oxford University Press, 2002), 51.

5. Richard Bauckham, "Eschatology," in *The Oxford Handbook of Systematic Theology*, eds. John Webster, Kathryn Tanner, and Iain Torrance (Oxford: Oxford University Press, 2007).

6. Douglas Farrow, "Resurrection and Immortality," *Oxford Handbook of Systematic Theology*, 219–222.

7. See, e.g., N. T. Wright, *The Resurrection of the Son of God* (Minneapolis, MN: Fortress Press, 2003), 463.

8. Hans Weber, "Hope and Creation," in *The End of the World and the Ends of God*, eds. John Polkinghorne and Michael Welker (Harrisburg, PA: Trinity Press International, 2000), 193; cf. Jürgen Moltmann, *The Coming of God,* trans. Margaret Kohl (Minneapolis, MN: Fortress Press, 1996), 291f.

9. Edward Epsen, "Eternity is a Present, Time is Its Unwrapping," *The Heythrop Journal* 51 (2010): 417. Cf. Maximus the Confessor, *Chapters on Knowledge*, 1.68–70.

10. R.T. Mullins, *The End of the Timeless God* (Oxford: Oxford University Press, 2016), xi–xii.

11. Miroslav Volf, "'Enter into Joy!' Sin, Death, and the Life of the World to Come," in *The End of the World and the Ends of God*, 270; cf. Brian Hebblethwaite, *Philosophical Theology and Christian Doctrine* (Malden, MA: Blackwell, 2005), 111f.

12. Rory Fox, *Time and Eternity in Mid-Thirteenth-Century Thought* (Oxford: Oxford University Press, 2006), 134ff.

13. Fox, *Time and Eternity*, 226–227.

14. J. M. Child, *The Geometrical Lectures of Isaac Barrow* (London: The Open Court Company, 1916), lecture I.

15. Clarke, *A Demonstration of the Being and Attributes of God and Other Writings,* ed. Ezio Vailati (Cambridge, UK: Cambridge University Press, 1998), 105. Cf. William Uzgalis, ed., *The Correspondence of Samuel Clarke and Anthony Collins, 1707–08* (Ontario: Broadview Press, 2011), 261.

16. Clarke, *Demonstration*, 122–123.

17. John Bigelow, "Presentism and Properties," *Philosophical Perspectives* 10 (1996): 35. See also Dean Zimmerman, "The A-Theory of Time, Presentism, and Open Theism," in ed. Melville Y. Stewart, *Science and Religion in Dialogue* (Malden, MA: Blackwell Publishing Ltd., 2010), 793.

18. Craig Bourne, *A Future for Presentism* (Oxford: Oxford University Press, 2006), 39–69.

19. Theodore Sider, *Four-Dimensionalism* (Oxford: Oxford University Press, 2001).

20. Mullins, *The End of the Timeless God*, 74–98.

21. Augustine, *Confessions*, trans. Henry Chadwick (New York: Oxford University Press, 1998), XI.20.

22. Dean Zimmerman, "Presentism and the Space-Time Manifold," in *The Oxford Handbook of Philosophy of Time*, ed. Craig Callender (Oxford: Oxford University Press, 2011), 163–244. Bradley Monton, "Prolegomena to Any Future Physics-Based Metaphysics," in *Oxford Studies in Philosophy of Religion,* vol. 3, ed. Jonathan Kvanvig (Oxford: Oxford University Press, 2011), 142–165. John Polkinghorne, "The Nature of Time," in *On Space and Time*, eds. Alain Comes, Michael Heller, Shahn Majid, Roger Penrose, John Polkinghorne, and Andrew Taylor (New York: Cambridge Press, 2008), 278–283. Roberto Mangabeira Unger and Lee Smolin, *The Singular Universe the Reality of Time: A Proposal in Natural Philosophy* (Cambridge, UK: Cambridge University Press, 2015).

23. Michael Rea, "Four-Dimensionalism," in *The Oxford Handbook of Metaphysics*, ed. Michael J. Loux and Dean W. Zimmerman (New York: Oxford University Press, 2003), 246–280.

24. Katherine Hawley, *How Things Persist* (Oxford: Oxford University Press, 2001).

25. We are here discussing a version of four-dimensionalism called the stage theory since it has the most ardent contemporary defenders in philosophers like Hawley and Sider. Due to space constraints, we are ignoring another version called perdurance. However, our arguments can easily be amended to count against perdurance.

26. R. Keith Loftin, "On the Metaphysics of Time and Divine Eternality," *Philosophia Christi* 17, no. 1 (2015): 177–187.

27. Peter Lombard, *Sentences* I, trans. Giulio Silano (Ontario: Pontifical Institute of Mediaeval Studies, 2007), distinction XXXVII.7.

28. Boethius, "The Trinity is One God Not Three Gods," in *Theological Tractates*, trans. H. F. Stewart, E. K. Rand, and S. J. Tester (Suffolk: St. Edmundsbury Press, 1973), IV.

29. See James Barr, *Biblical Words for Time* (London: SCM Press, 1962); Henri Blocher, "Yesterday, Today, Forever: Time, Times, Eternity in Biblical Perspective," *Tyndale Bulletin* 52 (2001): 183–202; John Feinberg, *No One Like Him: The Doctrine of God* (Wheaton, IL: Crossway Books, 2001), 255–276; Antje Jackelen, *Time and Eternity: The Question of Time in Church, Science, and Theology* (London: Templeton Foundation Press, 2005), 61–120; G.E. Ladd, "Age, Ages," in *Evangelical Dictionary of Theology*, ed. Walter A. Elwell (Grand Rapids, MI: Baker Books, 1984); Ted Peters, "Eschatology: Eternal Now or Cosmic Future?" *Zygon* 36 (2001): 349–356.

30. Anselm, *Proslogion*, in *Monologion and Proslogion*, trans. Thomas Williams (Indianapolis: Hackett, 1995), XXII. Cf. Sandra Visser and Thomas Williams, *Anselm* (Oxford: Oxford University Press, 2008), 102.

31. Anselm, *Monologion*, XXI.

32. See Robert Pasnau, *Metaphysical Themes: 1274–1689* (Oxford: Oxford University Press, 2011), 374–400.

33. Anselm, *Monologion*, XXI–XXIV. Cf. *Proslogion*, XIX–XXII.

34. Lombard, *Sentences* I, Dist. VIII.3.

35. Anselm, *Proslogion*, XVIII. Cf. *Cur Deus Homo*, VII.

36. R. T. Mullins, "Simply Impossible: A Case Against Divine Simplicity," *Journal of Reformed Theology* 7 (2013): 181–203.

37. R. T. Mullins, "Divine Temporality, the Trinity, and the Charge of Arianism," *Journal of Analytic Theology* 4 (2016): 267–290.

38. R. T. Mullins, *The End of the Timeless God.*

39. George Gasser and Matthias Stefan, eds., *Personal Identity: Complex or Simple?* (Cambridge, UK: Cambridge University Press, 2012).

40. George Gasser and Matthias Stefan, "Introduction," in *Personal Identity*, 3.

41. David B. Hershenov, "Four-Dimensional Animalism," in *Animalism: New Essays on Persons, Animals, and Identity*, eds. Stephan Blatti and Paul F. Snowdon (Oxford: Oxford University Press, 2016), 211–212.

42. Of course, if everything is self-identical, there is a trivial sense in which the worm may have numerical identity. Anything with either more, or less, or different temporal parts would be distinct from that worm. Anything with neither more, nor less, nor different temporal parts would be the same. What is in view, however, is a four-dimensionalism that is (typically) committed to unrestricted composition, on which view any temporal parts whatsoever can make up a worm. The gen-identity relation is a continuity relation that attempts to pick out a particular group of temporal parts.

43. Peter van Inwagen, "What Do We Refer to When We Say 'I'?" in *The Blackwell Guide to Metaphysics*, ed. Richard M. Gale (Oxford: Blackwell Publishers Ltd, 2002), 177.

44. Eric Steinhart, "The Revision Theory of Resurrection," *Religious Studies* 44 (2008): 66–67.

45. Katherin Rogers, "Anselm on Eternity as the Fifth Dimension," *The Saint Anselm Journal* 32 (2006): 3.

46. The physicalist Trenton Merricks affirms the simple view. See his, "There are No Criteria of Identity Over Time," *Noûs* 32 (1998): 106–124.

47. It may be that John Locke is a dualist who affirms the complex view. See E. J. Lowe, *More Kinds of Being*, 2nd ed. (Malden, MA: Blackwell, 2009), 104–116.

48. Andrew M. Bailey, "Animalism," *Philosophy Compass* 10 (2015): 868.

49. Eric T. Olson, *What Are We? A Study in Personal Ontology* (New York: Oxford University Press, 2007), 24.

50. Some of the difficulties facing such a position are laid out in Trenton Merricks, "Composition as Identity, Mereological Essentialism, and Counterpart Theory," *Australasian Journal of Philosophy* 77 (1999): 192–195.

51. Olson, *What Are We?* 28.

52. Olson, 29.

53. Brandon Rickabaugh, 9ff)"An Enduring Problem for Animalism," presented at the *Perspectives on the First-Person Pronoun "I": Looking at Metaphysics, Linguistics and Neuroscience*, at Durham University (Durham, England) May 16 to May 18, 2014. Available online at www.brandonrickabaugh.com.

54. As Rickabaugh goes on to show, taking a Life to be a temporally extended event only leads to further problems (9ff).

SUGGESTED FURTHER READING

Anselm. *Monologion and Proslogion*. Translated by Thomas Williams. Indianapolis: Hackett, 1995.

Augustine. *Confessions*. Translated by Henry Chadwick. New York: Oxford University Press, 1998.

Barr, James. *Biblical Words for Time*. London: SCM Press, 1962.

Blatti, Stephan, and Paul F. Snowdon, eds. *Animalism: New Essays on Persons, Animals, and Identity*. Oxford, UK: Oxford University Press, 2016.

Callender, Craig, ed. *The Oxford Handbook of Philosophy of Time*. Oxford, UK: Oxford University Press, 2011.

Merricks, Trenton. "Composition as Identity, Mereological Essentialism, and Counterpart Theory." *Australasian Journal of Philosophy* 77 (1999): 192–195.

Mullins, R.T. *The End of the Timeless God*. Oxford, UK: Oxford University Press, 2016.

Olson, Eric T. *What Are We? A Study in Personal Ontology*. New York: Oxford University Press, 2007.

Pasnau, Robert. *Metaphysical Themes: 1274–1689*. Oxford, UK: Oxford University Press, 2011.

Walls, Jerry. *Heaven: The Logic of Eternal Joy*. New York: Oxford University Press, 2002.

Chapter 6

Holy Saturday and Christian Theological Anthropology

Jason McMartin

I contend that while the Christian physicalist may be able to provide metaphysically possible accounts of the resurrection of Jesus, theologically adequate models are more difficult to come by. In particular, I argue that physicalist accounts of the intermediate state experienced by the incarnate Christ on Holy Saturday have difficulty aligning with various theological criteria arising from orthodox Christology and soteriology. After outlining these theological principles, I survey three models that physicalists have proposed for understanding postmortem existence: gappy existence, alternate temporality, and immediate resurrection. After explaining these basic models and considering how they might understand Jesus's death, resurrection, and intermediate state, I will consider how well these models align with the theological principles I have described. In the final section, I consider methodological issues pertaining to the entire project.

Christian philosophers have tended to focus on modal issues and on metaphysical models for understanding resurrection. Some have questioned whether physicalists will be able to develop a metaphysically satisfactory story of resurrection or of the intermediate state.[1] In general, Christian physicalists have recognized the challenge to their positions from the resurrection. Some of these also consider how to understand the intermediate state from the standpoint of physicalism. Receiving comparatively little attention, however, has been Christ's death and resurrection. He, too, went through an intermediate state: Holy Saturday in between Good Friday and Easter.

In what follows, I consider how three physicalist models of resurrection would explain the intermediate state of Jesus.[2] I won't consider these models on their own merits, and they may not be equally valid answers to the question at hand. Supposing that metaphysical difficulties such as the preservation of personal identity can be overcome, I intend to explore the extension of these physicalist stories into Christ's intermediate state and their ability to encompass the relevant theological principles. To what extent do physicalist models of resurrection provide a plausible explanation of the biblical evidence and of theological inferences therefrom? Which model of the metaphysics of human personhood best explains the theological data? As we add additional evidence from Scripture and theology, how well do physicalist models fare? I suggest new weaknesses emerge in physicalist accounts of resurrection when we attempt to explain the theological data from Christ's intermediate state. Considering the biblical data concerning Christ's intermediate state contributes to the construction of Christian theological anthropology.

This project may be thought of as an extension of that undertaken by John Cooper in *Body, Soul, and Life Everlasting*.[3] He contends that certain physicalist theories of gappy existence and immediate resurrection are ruled out by Christ's intermediate state. I'll add the theory of alternate temporality and include discussion of a version of immediate resurrection developed since the time of Cooper's writing (fissioning).

I will use "physicalism" to designate the view that a human person cannot exist without a body. The term is contentious, and as I am using it here, likely includes several other views that are not always labeled as physicalist positions. Baker, for instance, uses the term "materialist" and distinguishes between type I and type II materialists.[4] Type I materialists insist that we cannot exist without the particular organic bodies we have. Baker's own type II materialism holds only that we cannot exist without some body or other, since our mental states depend on that body. We can, for example, exist with nonorganic, or with partially organic bodies. As I use the term, physicalism can include both of these forms of materialism. It probably then includes constitutionalists (Corcoran, Baker), four-dimensionalists (Hudson), nonreductive physicalists (Murphy), and animalists (van Inwagen, Merricks). It may also include certain emergentists, such as O'Connor.

THE INTERMEDIATE STATE AND THE THEOLOGY OF HOLY SATURDAY

Various strands of Christian theology have affirmed an intermediate state: a period of time between the death of an individual human person and the future general resurrection. Certain biblical passages suggest life after death and prior to the resurrection of the body. Some theologians deny such a state, perhaps by affirming an extinction-recreation scenario or by postulating an immediate resurrection. With the possible exception of immediate resurrection views, the intermediate state is usually thought of as being an atypical period of human existence. If, for example, Paul refers to an intermediate state in 2 Corinthians 5, then his metaphor of being unclothed suggests an unnatural human state.

Jesus, too, experienced an intermediate state between his death on Good Friday and his resurrection on Easter Sunday. I will avoid becoming entangled in the exegetically and theologically difficult debates concerning *what* Christ was doing on Holy Saturday. I will instead focus on the status of his incarnation and its theological implications on this intervening day, using some minimal theological assertions concerning the work of his passion. The principal concern of this chapter will be the status of the human nature of Christ during the passion. Even apart from making a judgment about the theological significance of Holy Saturday, the death, and resurrection, the intervening time between them will be theologically significant.

One of the primary theological discussions surrounding Holy Saturday concerns the affirmation of the Apostle's Creed that Christ "descended to hell" following burial and prior to resurrection. Several of the positions concerning a supposed descent on Holy Saturday depend quite heavily on the full humanity of Christ. One model suggests that Christ descended to hell on Holy Saturday to provide a full satisfaction of God's wrath and to experience complete torment and abandonment in the place of the rest of humanity.[5] Heppe, for example, states that "like all human souls which separate from their bodies, even Christ's soul had to descend into Hades, because his whole divine human person was punished with real death in order that sin might be atoned for and the covenant of grace consummated."[6] This model of Holy Saturday and the descent into hell requires the full humanity of Christ during his intermediate state. This view also fits with the idea that humans undergo an intermediate state as part of punishment for sin, just as they die resulting from punishment for sin.

THEOLOGICAL PRINCIPLES

Before evaluating physicalist alternatives for understanding Holy Saturday, I will explain a handful of theological principles that will be frequently used as criteria for the discussion. Several theological affirmations govern our understanding of Christ's incarnation and its connection to our human nature, knitting soteriology, Christology, and theological anthropology closely together. In order to redeem us fully, Christ had to be fully human and fully divine. Further, Christ shows us what it is to be fully human as intended by God. Though potentially revisable, these criteria cohere and changes will have far-reaching impact among connected doctrinal affirmations. Alternate accounts of Christ's humanity will have different theological implications.

First, the triune God cannot cease existing. I'll designate this as

Endless Existence (EE): God exists unendlessly.

Therefore, whatever may happen in the activities of the incarnate Christ, the second person of the Trinity can't cease existing. As we'll see, some physicalist accounts of resurrection allow for nonexistence of the human person between the time of death and future resurrection. Perhaps that is possible for human persons, but it is not possible for God.

Second, the Chalcedonian consensus affirms that Christ is one person in two natures, fully divine and fully human, without confusion, without change, without division, and without separation. In all things, he is like us (sin only excepted).

> We, then, following the holy Fathers, all with one consent, teach people to confess one and the same Son, our Lord Jesus Christ, the same perfect in Godhead and also perfect in manhood; truly God and truly man, of a reasonable [rational] soul and body; consubstantial [co-essential] with the Father according to the Godhead, and consubstantial with us according to the Manhood; in all things like unto us, without sin; begotten before all ages of the Father according to the Godhead, and in these latter days, for us and for our salvation, born of the Virgin Mary, the Mother of God, according to the Manhood; one and the same Christ, Son, Lord, only begotten, to be acknowledged in two natures, inconfusedly, unchangeably, indivisibly, inseparably; the distinction of natures being by no means taken away by the union, but rather the property of each nature being preserved, and concurring in one Person (*prosopon*) and one Subsistence (*hypostasis*), not parted or divided into two persons, but one and the same Son, and only begotten God (μονογενῆ Θεόν), the Word, the Lord Jesus Christ; as the prophets from the beginning [have declared] concerning Him, and the Lord Jesus Christ Himself has taught us, and the Creed of the holy Fathers has handed down to us. (Schaff 1919)

Several theologians in recent decades have insisted that human nature ought to be understood by means of Christology, since "only Jesus Christ reveals who and what human persons truly are."[7] This Christological approach to anthropology "has become so pervasive that theologians can speak of a broad consensus regarding the centrality of Jesus Christ in any attempt to understand the nature of human persons."[8] Though theologians disagree concerning the correct way to employ Christology in understanding humanity, Jesus links our inquiries into Christology and theological anthropology. Chalcedon implies that in having human nature fully, Christ is like us. This yields the principle of the theandric union:

Theandric Union (TU): In the Incarnation, Jesus Christ is one person uniting two natures, human and divine.

We share the same human nature as Christ; we are consubstantial. "If Christ is consubstantial with us with respect to his human nature, then Christ's human nature is no different from our human nature in its composition."[9] For this reason, we may inquire concerning the implication the Incarnation has for understanding human constitution and vice versa.

Christ's ascension implies that his human nature, including his embodiment, continues and will continue until his return (Acts 1:11). It would seem that, once incarnate, the second person of the Trinity remains incarnate thereafter. The phrase "without division" would appear to imply the two natures are united and will not be separated. Although I will not develop it here, I believe that the an/enhypostasis distinction used to describe the theology of the fifth council would also imply that the incarnation is inseparable. Once the human nature has been hypostatized, it continues to be impersoned by the logos.

Third, Christ's saving work is closely connected to the nature of Christ's person as expressed in the theandric union principle. It is the work that Christ does in virtue of the person that he is. Here we will focus on the soteriological significance of his full humanity, which opens up the salvific possibility that we may enter into our full humanity in the way we were intended to do. Jesus had to be fully human in order to redeem the entirety of the human person, since that which is not assumed is not healed. Jesus had to endure the consequences of sin, particularly death, as a human in order to effect our salvation. We find our salvation by identifying with him in our deaths and resurrections. Yet we must also distinguish between components of his suffering that function as merely an example for us in our suffering (1 Peter 2:21) and those aspects of suffering we also endure. Most aspects of his passion will not find exact equivalent in our experience. With regard to death, most all humans will die physically, and so Jesus conquers eternal death for us (rather than physical death). In the words of theologians, his substitution is inclusive with regard to physical death (he dies physically and so do we), but exclusive with regard to eternal death and alienation from God (he bears the ultimate consequences of sin so that we don't have to).[10] Perhaps it is also the case that his intermediate state is an instance of what we also will experience in terms of the consequences of sin. Nevertheless, his resurrection is the pattern for ours. This seems to be the meaning of Paul's claim that Jesus's resurrection is the first fruit, the first and best part of the crop that is the promise of the remainder of the crop to come. This is summed up as:

> Soteriological Pattern (SP): Christ is the soteriological pattern for our death and resurrection.

As signified by the sacraments and by the entirety of Christian existence, significant points of continuity may be found between Christ's life, death, and resurrection and our own.

Finally, postmortem existence will have aspects of both continuity and discontinuity with our terrestrial lives. For example, in the resurrection appearances, Christ was both recognizable to the disciples and bore the marks of his suffering, but could also appear in a locked room. The clothing metaphor that Paul uses in 1 Corinthians 15:53–54 suggests that immortality is added to us, rather than getting rid of our earthly mortality. Among various points of continuity, perhaps least controversial would be that we continue to be human in postmortem existence. This is the principle of anthropological continuity:

> Anthropological Continuity (C): Essential properties of human nature endure in postmortem existence.

At the same time, many things will be different in our postmortem existence. In continuity with our earthly lives, we are meant to be embodied beings. That means that if we endure an intermediate state, as I believe we do, it is not comfortable or "natural" for us (2 Cor. 5). Nonessential facets of human life may be and will likely be quite different on the other side of the grave. This yields a discontinuity principle:

> Anthropological Discontinuity (D): In postmortem existence, many nonessential features of human nature will be unlike terrestrial existence.

It is not easy to sort out where continuity ends and discontinuity begins. For example, issues of continuity and discontinuity raise the question concerning the proper theological and biblical way of understanding death. In what ways does death bring an end to usual human existence and in what ways does it provide continuity with usual human existence? I consider the meaning of death in what follows.

THREE THEORIES

I turn now to a description of the three kinds of physicalist theories that have been suggested as explanations for postmortem existence. One version of the third of these has been developed most thoroughly and has received considerable attention. Since these models primarily seek to explain the resurrection, they differ concerning the extent to which they incorporate space for an intermediate state. In this section, I will first explain each physicalist model for understanding resurrection. Then I will apply it to Christ's resurrection. Finally, I will consider theological implications of the physicalist story.

Gappy Existence

One alternative for physicalist resurrection is continued existence with temporal gaps. The main idea on this view is that while I cease to exist at my death, I exist once again at the general resurrection from the dead. That is, in the span of time between my death and the future general resurrection, I do not exist, and death means nonexistence.

Physicalists holding this theory have been primarily concerned to show that it is metaphysically possible to preserve continuity of personal identity when crossing a spatiotemporal gap, since critics have frequently urged that such gaps would make principled differentiation between a resurrected self and a replica of myself impossible. For varying reasons, physicalists have affirmed that such gaps are possible and do not undermine personal identity.[11] Corcoran offers a story for how this might work. He explains, "if God causes that body to exist once, why couldn't God cause it to exist a second time? . . . What makes the first stage of the post-gap body a different stage of the same body that perished is that God makes it so."[12] Later he appears to imply that gappy existence is compatible with belief in an intermediate state, though apparently does not comport well with that position.

Postulation of gappy existence is meant as a way of avoiding the constraint of causal continuity or of an immanent causal connection. (I sense that this model is not overly popular because of these avoidances, which appear to result in some absurd or implausible outcomes.)

Physicalists differ concerning how they propose resumption of life (i.e., resurrection) occurs on the far side of the gap according to their theories of the relationship of mental states to bodies. For nonreductive physicalists, God may reconstitute a suitable collection of material particles into a structure that will subvene the same mental states the deceased had during life, giving rise to the same person. Similarly, constitutionalists allow that God would reassemble material stuff to reconstitute me, and so forth.

Van Inwagen's simulacrum view differs in salient ways from other physicalist accounts on this point.[13] As an animalist, van Inwagen affirms that a human person is a biological organism (an animal) composed of matter that has an internal structure and that has been caught up in a life. Death removes the event that provides the requisite structure of the material parts; decomposition results in the progressive diminishment of relations between the material parts that composed the structure that was my life. Van Inwagen's suggested account of resurrection is that, at death, God creates a simulacrum of the body of the deceased, which he leaves in place of the original body. The original is preserved elsewhere with its proper structure, allowing the material parts of the resurrected body to stand in the correct causal structure to one another and therefore be reconstituted later as the same life.

Given this physicalist story of the resurrection, how should we understand Holy Saturday? First, given EE, the second person of the Trinity does not cease existing on Holy Saturday. (I am not even sure where to go if someone were to deny this.) The divine nature of the incarnate Christ does not cease existing. Presumably, the incarnate Christ continues to have conscious experiences on Holy Saturday independent of the status of his body.[14]

Physicalists may not worry about the temporal gap since the body in the tomb may be plausibly revivified. The material stuff may be caused by God to subvene human conscious states, or to constitute a human person, or to be caught up in the same life.

However, if Holy Saturday is a temporal gap, then it would seem that Christ's human nature does cease existing, since, for the physicalist, to be human is to have a body. Some physicalists may wish to affirm that Christ still had a human nature on Holy Saturday, despite the dead body in the tomb. For example, according to Olson's construal of animalism, death means a life that cannot be restarted. Consider: "As for identity over time, I am inclined to believe that an organism persists if and only if its life continues. This has the surprising consequence that an organism ceases to exist when the event that maintains its internal structure stops and cannot be restarted—that is, when the organism dies."[15] It would follow that no organism is dead whose life can be restarted. Either Christ's human life was restartable or it wasn't. Whether a life is restartable presumably incorporates various medical criteria. Two primary components are clinical death and brain death. Suppose you thought the human life of Jesus restarted on Sunday, then the incarnate Christ still has a human nature on Saturday. Although it avoids the problem of a disincarnated Christ, this response wreaks havoc on our usual understanding of death and resurrection. On this view, Christ was never dead in Olson's sense considered here, though could perhaps have been said to be dead in a colloquial sense. This would imply that Christ was not clinically dead from the time of being taken from the cross until Sunday. Technically, then, he is revived and not resurrected. Theologically and empirically, it is implausible to suppose that Christ did not really die.

If, on the other hand, Christ was dead and ceased to have a human nature, then, given EE, Christ is no longer incarnate on Holy Saturday and no longer human. This would appear to contradict the "without separation" clause of Chalcedon and the enhypostasis of the second person as considered in the fifth council, hence contravening TU. The unity of the person would be in doubt. Setting that aside, how would we understand his resurrection on Sunday? Perhaps the resurrection would best be understood as a second incarnation, and the uniqueness and significance of Christmas would be called into question.[16] Unlike the first incarnation, the logos would assume a human nature in the form of the dead body in the tomb. In what way would it continue the particular instance of the human nature present in the first incarnation? This is a problem that plagues this view in general: what accounts for the continuity across temporal gaps? Since it is the body in the tomb that is incarnated by the logos, perhaps the continuity can be explained by the material stuff being the same body.

In this scenario, the resurrection is more akin to an incarnation, an incarnation into decaying body. Not very many thinkers want to affirm that the very same physical stuff must be employed in the resurrection. For most of us, our original matter will not be available to us. If Christ's resurrection is relevantly similar to human resurrection, then dualism would appear to be implied for the rest of humanity. Otherwise, Christ's resurrection appears to have few similarities to our resurrection, contravening SP.

Additionally, in tension with SP and C, Jesus is not human (or not fully human) on Holy Saturday. Theologically, then, he does not bear the consequence of sin in death as a human, but simply avoids it. Rather than endure death or the intermediate state, he claims divine prerogative for avoiding them.

Alternate Temporality

A second set of physicalist models for resurrection deny or question the existence of the intermediate state by suggesting that it is mistaken to think of a period between death and resurrection. Appealing to God's timelessness, proponents of this view suggest that, at death, a person enters into a timeless state.

> According to the eternalist, it would quite possibly mean experiencing time as God does through the dissolution of the fragmentariness of human existence and an awakening to an experience of life in undivided wholeness. When one dies, perhaps it is not merely the totality of that life, but the whole of history which appears to that person in undivided wholeness. Consequently, to those taken up into the divine eternity, the general resurrection would be just as present as the moment of their death. For them, there would be no interim period before the general resurrection; rather, to die is to be present to the general resurrection, and thus to be ready for participation in resurrected life. Accordingly there would be no need to posit an intermediate state.[17]

Joel Green admits that if we define the intermediate state as the period of terrestrial time that intervenes between death and future resurrection, then it is entailed that Luke 16:19–31 supports its existence. He explains:

> [A]s we have seen, Luke 16:19–31 self-evidently refers to an intermediate state *insofar as "intermediate" refers to the linear marking of time from the*

> *perspective of the rich man's brothers still alive in this world.* Whether the rich man and Lazarus experience their existence beyond death as "intermediate" is an altogether different question, however.[18]

It is this assumption that Green wishes to call into question, and he rightly notes that the passage of time for the earth-bound is not the main issue.

> Whether Luke's parable envisions an intermediate state depends on how one defines "state," whether in temporal or spatial terms, or both. If one presumes time as experienced from an earthly point of view, then it makes sense to speak of an intermediate state—that is a period that passes between the death of the individual and the consummation of all things, as we, Luke's readers experience time. What is not obvious is that Lazarus (for example) experiences the afterlife in this parable as a kind of waiting room between death and final Judgment.[19]

On the alternate temporality view, no disembodied state is required to make sense of the intermediate state, thus providing room for physicalist construals of postmortem existence. Glenn Peoples suggests that divine timelessness also enables the physicalist to affirm that the death of Jesus does not affect the timeless existence of the Son of God (though without linking this explicitly to the resurrection).[20]

As an explanation of postmortem existence, this model does not appear to be common. It does not provide an explanation for bodily continuity, only for why an intermediate state is not needed, which means one of the other two models will likely still be needed. Any duration of time for the deceased prior to the final judgment, even in an alternate temporality, would seem to require either a gap in existence or a physicalist explanation of bodily continuity. The alternate temporality model seems most fruitfully combined with the immediate resurrection model to be considered next.

Whatever its merits for understanding human postmortem survival, the alternate temporality thesis would appear to have less value for explicating Jesus's intermediate state. According to the model, the incarnate logos would cease experiencing terrestrial time on Friday, and begin experiencing it again on Easter Sunday. Neither EE nor TU need be contravened. It is less clear that this view can work well with SP and C. Some differences between Jesus's resurrection and ours are to be expected. Yet, Jesus's resurrection can't be a shift to the future general resurrection, since he experiences normal terrestrial time from Easter onwards. The model may postulate that, upon death, Jesus entered temporal simultaneity with both his resurrection and the future general resurrection. Following his resurrection, he experienced typical terrestrial time. Alternatively, Christ, entered temporal simultaneity at death, and remained in this following resurrection, yet while having typical time-bound interaction with his disciples in the resurrection appearances. I confess difficulty in making sense of these alternatives.

I question whether this view provides an altogether satisfactory account of the person after death. Our temporality comprises a key component of our experience. Can we make sense of wholly material entities that are nontemporal? Supposing we are able, postulation of a timeless postmortem existence would appear to be ad hoc. Physicalism and C seek to preserve significant continuity with present human experience, and temporality is an inescapable feature of our experience.

In the end, it doesn't appear that this view can provide a useful explanation of the status of the incarnate Christ on Holy Saturday. Likely this view collapses into the next model to be considered.

Immediate Resurrection

Physicalists frequently propose an immediate resurrection following death in order to maintain the continuance of the body and hence of the person. Some physicalists even allow for an intermediate state, in addition to an immediate resurrection. Lynne Rudder Baker for example explains:

> I know of no reason—Biblical or philosophical—to suppose that the intermediate state must be a *disembodied* state. For all we know, persons in the intermediate state (assuming that there is one) are constituted by intermediate-state bodies. As we saw, when one is resurrected, one has a "spiritual," or "glorified," or "imperishable" body. If God can so transform or replace our bodies once, he can do it twice.[21]

Baker contends that the intermediate state can be compatible with a physicalist account of the person if two resurrections are postulated. The first resurrection is from the earthly body to the intermediate body, and the second is from the intermediate body to the glorified body. This hypothesis makes physicalism consistent with an intermediate state, though some have questioned the motivation for postulating this second resurrection. For example, in considering the compatibility of animalism with purgatory, David Hershenov opines, "if you will be in Purgatory as a material being, then it is hard to envision what would be the point of the later resurrection promised upon Jesus' return."[22] Perhaps the key motivation should be SP: our bodies experience two resurrections because according to the model, Christ's did. In any case, this model postulates that all postmortem existence is embodied in one way or another.

Zimmerman's fissioning model has risen to prominence among models of resurrection.[23] On this view, at death, each of my individual particles (or simples, in Corcoran's language) fissions, creating two complete bodies, both of which are immanently causally connected to my life. One of these fissioned collections of material stuff is the dead body that is buried in the ground. The other fissioned collection is preserved by God in a different time and place. (I will continue to use the term fissioning, even though in the more recent development, Zimmerman calls the process budding.) The second fission product continues the life of the premortem person. The fissioning view avoids a gappy existence view (or at least one without immanent causal connections) and could be used to support the idea of an immediate resurrection. As Baker suggests, immediate resurrection may be combined with an intermediate state, involving two resurrections.[24] Likewise, the fissioning view may be used to postulate the mechanism of at least the first resurrection, and perhaps of the second also.

How should we think of the fissioning theory in relationship to Christ's resurrection? According to Zimmerman's fissioning account, Jesus could have remained fully human by having a body after his death. At death, Jesus would have fissioned, leaving his original body to be taken from the cross and then buried in the ground. Jesus continues to be fully human and embodied on Holy Saturday, whatever activities in which he may have been engaged. There are now two bodies, or at least, two collections of matter. The second of these belongs to the living Christ. The first of these is carried down from the cross, prepared, and buried in the tomb owned by Joseph of Arimathea.

An initial concern for the story thus far, is that the resurrection of Christ is marked by an empty tomb. The women find the tomb empty (Luke 24:2–3). So what happens to the first body? Perhaps Jesus's original body disappears by being stolen. More promisingly for the model, God could have removed the first fissioned body, perhaps by dissipating it rapidly once the tomb was sealed. These scenarios generate other problems, such as attributing deceit to God (as has been pressed against van Inwagen's account) or generating massive confusion to people who saw both the resurrected Lord and the stolen body (that could conceivably be in the same room!). The theory also has the potentially incongruous consequence that, contrary to the creed, Jesus was not buried.

Another route for the physicalist would be to hypothesize that Jesus returned from his second, fissioned body to his original body. In order to preserve the immanent causal connection among the particles composing Christ's body, we might postulate a new process: fusioning. The fissioned, closest continuer person of Christ fused with the body in the tomb. Unlike the dissipation model, the fusion model holds that the body of the post-resurrected Christ was prepared by his disciples and buried.

Neither fusion nor dissipation appear to contravene TU, but they comport less well with SP and C. For each of these, the theologically significant event for humans happened immediately upon Jesus's death, prior to his burial, and not on Easter. Given dissipation, the resurrection happened at the moment of Jesus's death; Easter only reveals this state of affairs. On the fusion model, there are two kinds of resurrection: fissioning and fusioning. Christ experienced fusion-resurrection on Easter. Fission-resurrection provides the pattern for most of humanity, since it is the way humans will survive their deaths and most humans won't be returned to the corpses of their bodies. Should we then change our liturgical orientation to Good Friday and Easter? According to these two versions of the immediate resurrection model, the hope-conferring event of the Christian faith happens at Jesus's death and is only marked by the later discovery of the empty tomb and in the post-resurrection appearances. The resulting flattened narrative alters the Christian narrative of identification with Christ's soteriological pattern.

Both the dissipation and fusion models affect our account of postmortem bodily continuity as well. When Jesus appears to the disciples after the resurrection, he provides them with bodily evidence for his identity. "See my hands and my feet, that it is I myself. Touch me, and see. For a spirit does not have flesh and bones as you see that I have" (Luke 24:39). According to the fusion model, this statement by Jesus obscures a counterintuitive truth: the body he presents to them both is and isn't the body taken from the cross and the body from the tomb. On the dissipation model, the post-resurrection body was not taken from the cross and buried. On both, the original body of Jesus did not accompany Christ during the events of Holy Saturday as did the second fissioned body. This means that the post-Easter resurrected human nature of Jesus either didn't endure death by going through Holy Saturday or both did and didn't as a fusion of the two bodies. Similar problems may plague dualist conceptions of bodily continuity in resurrection, but the challenges of physicalist models appear to be intensified in relation to the unique features of Christ's resurrection.

EVALUATING COMPETING EXPLANATIONS

In the previous section, I explored theological implications of varying physicalist models of Christ's intermediate state. Although extensive comparison with dualist models would be beyond the scope of this chapter, I raise, in this section, three areas for such an evaluation.

First, all models of postmortem existence intersect with questions concerning the nature of death. Although Scripture views death as a defeated foe, it is nonetheless characterized in an overwhelmingly negative fashion. Death is a curse, a punishment, and an enemy. Physicalists have pressed the objection that dualism fails to provide a compelling reason for a negative evaluation of death. For example, Merricks explains that, "it is not clear that the dualist can agree that death is bad. When the Christian dies, according to the dualist, he or she goes immediately to a much better place . . . death, it would seem, is even better than quitting your job and moving to a beachfront villa in Hawaii . . . if dualism is true, it is hard to see how death is an enemy."[25] It is true that some dualists have viewed immediate postmortem existence as a state of blessedness, following the seeming implication of St. Paul in Philippians 1:23. On the other hand, Merricks's objection would seem to apply equally to physicalist accounts of resurrection. For each physicalist theory considered here, after closing my eyes in death one moment, I would awake the next moment to something better than a vacation.[26] Perhaps the only ill effect I might experience would be something like jet lag and a general disorientation at having awoken elsewhere or elsewhen.[27]

Theologians have commonly defined death as separation rather than as extinction in order to preserve the similarity between physical death and eternal death. For example, Berkhof, after surveying the biblical evidence, states the following:

> In view of all this it may be said that, according to Scripture, physical death is a termination of physical life by the separation of body and soul. It is never an annihilation . . . Death is not a cessation of existence, but a severance of the natural relations of life. Life and death are not opposed to each other as existence and non-existence, but are opposites only as different modes of existence . . . Death means a break in the natural relations of life.[28]

If disembodiment is unnatural, the negative facets of death can be encompassed by a dualist reading of the intermediate state. The goods of embodiment are lost during the intermediate state.[29] Death is a state of existence to be endured and can be gratefully set aside in the resurrection. Further, this understanding allows for a robust facet of the discontinuity (D) between terrestrial and postmortem existence not available to the physicalist.

Second, Christian dualists and physicalists should make the manner of their incorporation of the biblical material explicit. It is common to claim that resurrection and the afterlife are mysterious and that we ought not expect to gain a detailed understanding concerning how it works. Accordingly, a physicalist may deny the need to account for all of the details brought into consideration above. While it is certainly true that we ought not expect all of our questions to be answered, we also don't want to punt to mystery on first down! It is not at all clear when we have an epistemological mystery before us (something we don't know) or a theological mystery (something we can't know apart from revelation, or perhaps at all). We should be hesitant about claiming theological mystery too early.

Some have claimed that the biblical texts provide an ambiguous or inconsistent picture of resurrection and the state of life after death and hence prove of little use in developing theories of human constitution. Noting apparent contradictions among the texts describing Jesus's post-resurrection appearances, Nancey Murphy contends that

> the church, in canonizing a collection of documents with genuinely inconsistent accounts of the resurrected body, is telling us something very important about resurrection—namely, that the language of the present aeon is incapable of describing a resurrected body. No ordinary description is possible. Rather, we must be content with a variety of contrasting verbal pictures of Jesus.[30]

She goes on to say that

> The futuristic pictures that characterize Christian belief, such as life after death and the Last Judgment, are not based on ordinary sorts of evidence, and while these pictures need to be connected aright to the rest of the Christian faith, we need not be able to specify all of the spatial, temporal, and causal connections. That is, in order to believe in eternal life we need not be able to fit it into a chronology of historical events nor locate heaven with respect to earth, sun, or stars.[31]

In response, we may note first that the accounts of Scripture are not the same as the teaching of Scripture. Murphy contends that contradictory accounts aim to teach the mystery and unknowability of matters having to do with the resurrection. Murphy then absolves herself of the need to harmonize the statements of the biblical texts, a tactic followed by others who discuss the relevance of the biblical evidence to human constitution and postmortem existence. Do we need to develop a coherent and harmonized account of all the various aspects of biblical teaching? This depends on our theology of Scripture and on our conception of systematic theology. While the scope of salient evidence may be debated, central Christian doctrines ought to constrain development of our models. The theological principles I have considered here should have broad agreement.

Third, as with all models, distinguishing among useful explanations and ad hoc additions proves difficult. As characterized by Zimmerman and others, the fissioning of particles upon death would be a miracle conferred upon those material parts by God. By contrast, Timothy O'Connor (whose version of emergentism would qualify as a physicalist view as I have defined it) posits fissioning as a latent disposition of all matter. For him, matter possesses a wide variety of dispositions that can be realized in the correct circumstances, among these are the disposition to generate an individual from a suitably complex physical arrangement. O'Connor and Jacobs have argued that their account allows for resurrection, and seemingly that the individual does not require a specific physical substrate for the sake of survival. Consider, for example, their discussion of the resurrection of the emergent individual, Augustine:

> We need only suppose that the features of the constituents of Augustine's body—and as these are no different in kind from the constituents of any material thing, of all material things—and the emergent level aspects of Augustine jointly have a hitherto entirely latent tendency to jointly cause the composing simples to fission in the requisite context, which is providentially connected solely to situations of imminent demise.[32]

They go on to make a more general point about the various dispositions that matter has in terms of emergence.

> Which emergent features, if any, are latent in the fundamental constituents of our universe cannot reasonably be assigned any particular a priori probability. They are discovered empirically, having to be accepted, in the phrase of the early 20th century emergentist, Samuel Alexander, with "the natural piety of the investigator." Given that the posited consequence is ex hypothesi not observable to us in this life, who can say? . . . Experience teaches that the simples that compose us and all other material things have latent dispositions such that, when organically arranged in the right sorts of ways—in the first instance, into cells, then into more complex structures such as functioning nervous systems—they collectively cause and sustain emergent mental phenomena. It may be that those latent dispositions are sufficiently robust that when matter is arranged in functionally equivalent ways from the level of molecular biology on up—with non-organic components that are differently constituted from but functionally equivalent to ordinary cells—we'd get the same emergent phenomena. Maybe. And if so, our view can cheerfully accept it.[33]

The discovery of these dispositions is an empirical matter. Whether latent disposition or miraculous intervention, the processes considered here can only be postulated and not observed. For example, one model considered here posits two processes: fissioning and fusioning. Whether these processes miraculously appear or are latent dispositions, how should we decide when these hypotheses are ad hoc rather than just undiscovered? The theological evidence cannot indicate that such dispositions (or miracles) do not exist. The dialectic of this chapter suggests that physicalist models require increasingly complex mechanisms to explain the theological principles connected to Christ's intermediate state on Holy Saturday.

CONCLUSION

How do the prospects for physicalist models compare to those for dualist accounts of postmortem existence and the status of the Incarnation on Holy Saturday? While admitting that the problem of resurrection is formidable for the Christian physicalist, Corcoran claims that dualists

> are no better off when it comes to making sense of the afterlife than their materialist siblings. For it is plausible to believe that a Christian Dualist, whether he or she realizes it or not, faces one of the same challenges as the constitutionalist: that of accounting for how a body that apparently falls apart and ceases to exist can nevertheless put in an appearance in the heavenly city.[34]

Dualists face some of the same problems as physicalists, but not all of the same ones. Arguably, Corcoran has attempted to foist his problem onto the dualist by equivocating on the term body. For the dualist, something is my body if it is ensouled by me. Bodily continuity, then, does not depend on continuity of the same material stuff. Although I am doubtful that Corcoran's claim on this point is correct, it is likely still worthwhile to consider briefly the prospects for the dualist position in this discussion. The challenge to the dualist position arises primarily at the point of the resurrection itself. The empty tomb implies that Jesus returns to his body. Up until this point, it is similar to Lazarus's resurrection, though without the second death that Lazarus presumably must endure. The key point of discontinuity with our resurrection is that Christ's resurrection body is a transformation of the matter that comprised his original body, whereas for most believers through history, their resurrected bodies will not be the same matter as comprised their original bodies. Resurrection is not reassembly, for all of the reasons that are pressed against that position.

Perhaps this is not overly alarming, since Paul's view is that those who are still alive at Christ's return will also be changed. Presumably, their bodies will be changed into the same sort of bodies that their resurrected brethren possess. So, there are multiple ways for resurrection to occur: return to ensoul the matter that comprised one's original body (as in Christ and perhaps others), ensouling new material stuff after death, or ensouling new material stuff without dying. These varying modes of resurrection may appear to contravene SP and C, which affirm continuity between our deaths and resurrections and Christ's. However, the dualist has a relatively straightforward way of explaining the continuity of these cases in the midst of differences related to mode and matter to be resurrected. While physicalist models can provide an account that satisfies both continuous (C) and discontinuous (D) facets of postmortem existence, it proves more difficult to do so while also affirming the union of person and full humanity of the incarnate Christ (TU) on Holy Saturday that provides a soteriological pattern (SP) for our own deaths and resurrections.[35]

NOTES

1. Jonathan J. Loose, "Constitution and the Falling Elevator: The Continuing Incompatibility of Materialism and Resurrection Belief," *Philosophia Christi* 14, no. 2 (2012): 439–450; and "The Metaphysics of Constitution and Accounts of the Resurrection," *Philosophy Compass* 8, no. 9 (2013): 857–865.

2. I don't include discussion of hylomorphic theories here. Insofar as those theories resemble physicalist ones, the discussion will be relevant.

3. John W. Cooper, *Body, Soul, and Life Everlasting: Biblical Anthropology and the Monism-Dualism Debate* (Grand Rapids, MI: Eerdmans, 1989).

4. Lynne Rudder Baker, "Christian Materialism in a Scientific Age," *International Journal for Philosophy of Religion* 70, no. 1 (2011): 47–48.

5. David Lauber, *Barth on the Descent into Hell: God, Atonement, and the Christian Life* (Farnham, UK: Ashgate 2004), 1–41. Lyra Pitstick, *Christ's Descent into Hell: John Paul II, Joseph Ratzinger, and Hans Urs von Balthasar on the Theology of Holy Saturday* (Grand Rapids, MI: Eerdmans, 2016), 1–6.

6. As quoted in Thomas C. Oden, *The Word of Life* (San Francisco: Harper & Row, 1989), 442.

7. Marc Cortez, *Embodied Souls, Ensouled Bodies: An Exercise in Christological Anthropology and its Significance for the Mind/Body Debate* (London: T&T Clark, 2008), 4.

8. Cortez, 4.

9. Oliver Crisp, "Materialist Christology," in *God Incarnate: Explorations in Christology* (London: T&T Clark, 2009), 141.

10. Stanley J. Grenz, *Theology for the Community of God* (Nashville, TN: Broadman & Holman, 1994), 353.

11. Trenton Merricks, "How to Live Forever without Saving your Soul: Physicalism and Immortality," in *Soul, Body, and Survival: Essays on the Metaphysics of Human Persons*, ed. Kevin Corcoran (Ithaca, NY: Cornell University Press, 2001), 183–200.

12. Kevin Corcoran, *Rethinking Human Nature: A Christian Materialist Alternative to the Soul* (Grand Rapids, MI: Baker Academic, 2006), 131.

13. Peter Van Inwagen, "The possibility of resurrection," *International Journal for Philosophy of Religion* 9, no. 2 (1978): 119–121.

14. That is, without getting embroiled in debates between social and nonsocial Trinitarians concerning the manner in which we should think of the second person as a "person."

15. Eric T. Olson, *What are We? A Study in Personal Ontology* (Oxford: Oxford University Press, 2007), 29.

16. Thomas H. McCall considers this challenge to physicalist Christologies in *An Invitation to Analytic Theology* (Downers Grove, IL: InterVarsity Press, 2015), 119–120.

17. Charles E. Gutenson, "Time, Eternity, and Personal Identity: The Implications of Trinitarian Theology," in *What About the Soul? Neuroscience and Christian Anthropology*, ed. Joel B. Green (Nashville, TN: Abingdon Press, 2004), 122.

18. Joel B. Green, *Body, Soul, and Human Life: The Nature of Humanity in the Bible* (Grand Rapids, MI: Baker Academic, 2008), 165.

19. Green, *Body, Soul, and Human Life,* 160.

20. Glenn Andrew Peoples, "The Mortal God: Materialism and Christology," in *The Ashgate Research Companion to Theological Anthropology,* eds. Joshua R. Farris and Charles Taliaferro (Farnham, UK: Ashgate, 2015), 339–341.

21. Baker, "Christian Materialism in a Scientific Age," 55.

22. David B. Hershenov, "Soulless Organisms?" *American Catholic Philosophical Quarterly* 85, no. 3 (2011): 480.

23. Corcoran, *Rethinking Human Nature.* Garrett J. DeWeese, "Is There Hope for Christian Physicalists?" (unpublished manuscript); Dean W. Zimmerman, "The Compatibility of Materialism and Survival," *Faith and Philosophy* 16, no. 2 (1999): 194–212; Dean W. Zimmerman, "Bodily Resurrection: The Falling Elevator Model Revisited," in *Personal Identity and Resurrection: How Do We Survive Our Deaths?* ed. Georg Gasser (Farnham, UK: Ashgate, 2010), 33–50.

24. Baker, "Christian Materialism in a Scientific Age," 55.

25. Trenton Merricks, "The Resurrection of the Body and the Life Everlasting," in *Reason for the Hope Within*, ed. Michael J. Murray (Grand Rapids, MI: Eerdmans: 1999), 284–285.

26. Setting purgatory aside for the time being.

27. My thanks to Jonathan Loose for suggesting this metaphor to me.

28. Louis Berkhof, *Systematic Theology* (Grand Rapids, MI: Eerdmans, 1996), 668.

29. Taliaferro and Goetz, "The Prospect of Christian Materialism," 312–313.

30. Nancey Murphy, "The Resurrection Body and Personal Identity: Possibilities and Limits of Eschatological Knowledge," in *Resurrection: Theological and Scientific Assessments*, eds. Ted Peters, Robert J. Russell, and Michael Welker (Grand Rapids, MI: Eerdmans, 2002), 204–205.

31. Ibid., 205.

32. Timothy O'Connor and Jonathan D. Jacobs, "Emergent Individuals and the Resurrection," *European Journal for Philosophy of Religion* 2, no. 2 (2010): 79.

33. Murphy, "The Resurrection Body and Personal Identity," 79–80.

34. Corcoran, *Rethinking Human Nature,* 120.

35. I am grateful to those who have helped to shape my thinking on this topic thus far. Tom Crisp and Gregg Ten Elshof gave initial feedback on the shape the argument when I began research on it as a fellow at the Biola University Center for Christian Thought. Conversations with various colleagues have guided my thinking; among them: Garry DeWeese, Doug Huffman, and attendees at a presentation of an earlier version at the Evangelical Philosophical Society meeting in 2014, especially Keith Hess and Jonathan Loose. I am also grateful for helpful evaluation from the

participants in the Interim State Writing Workshop in McCall, Idaho in July 2015. In particular, I'm grateful for the commentary of Thom Atkinson. I am also thankful to the editors of this volume.

SUGGESTED FURTHER READING

Baker, Lynne Rudder. "Christian Materialism in a Scientific Age." *International Journal for Philosophy of Religion* 70, no. 1 (2011): 47–48.

Cooper, John W. *Body, Soul, and Life Everlasting: Biblical Anthropology and the Monism-Dualism Debate.* Grand Rapids, MI: Eerdmans, 1989.

Cortez, Marc. *Embodied Souls, Ensouled Bodies: An Exercise in Christological Anthropology and its Significance for the Mind/Body Debate.* London: T&T Clark, 2008.

Crisp, Oliver. *God Incarnate: Explorations in Christology.* London: T&T Clark, 2009.

Loose, Jonathan J. "Constitution and the Falling Elevator: The Continuing Incompatibility of Materialism and Resurrection Belief." *Philosophia Christi* 14, no. 2 (2012): 439–450.

———. "The Metaphysics of Constitution and Accounts of the Resurrection." *Philosophy Compass* 8, no. 9 (2013): 857–865.

Merricks, Trenton. "The Resurrection of the Body and the Life Everlasting." In *Reason for the Hope Within*, edited by Michael J. Murray. Grand Rapids, MI: Eerdmans: 1999.

Olson, Eric T. *What are We? A Study in Personal Ontology.* Oxford, UK: Oxford University Press, 2007.

Schaff, Philip. *The Creeds of Christendom.* New York: Harper and Brothers, 1919, vol. 2, p. 62.

Van Inwagen, Peter. "The Possibility of Resurrection." *International Journal for Philosophy of Religion* 9, no. 2 (1978): 119–121.

Chapter 7

Physicalism, the Incarnation, and Holy Saturday

A Conversation with Karl Barth

Marc Cortez

Christians have long maintained the conviction that Jesus should play some fundamental role in establishing what we think about the nature of humanity. As David Kelsey notes, "the way Christians understand these matters is shaped in some way by their beliefs about Jesus Christ and God's relation to him. That is ultimately what qualifies theological answers to proposed anthropological questions as authentically Christian theological anthropology."[1] Although he goes on to acknowledge that phrases like "the way Christians understand these matters" and "in some way" indicate the considerable diversity that exists between various construals of the Christology-anthropology relationship, it remains the case that most Christians would agree that our beliefs about Jesus should inform our beliefs about humanity in *some* way.[2] Indeed, we could probably go further and maintain that most orthodox Christians would agree with the following claim of a christological anthropology (CA):

> CA: We should reconsider any anthropological proposal deemed incompatible with essential beliefs about the person and work of Jesus Christ.

Of course, this raises its own questions about what qualifies as "essential beliefs," what it means for two such claims to be "incompatible," and who gets to make that determination. Nonetheless, such a statement can still serve as a worthwhile point of reflection for considering the relationship between Christology and anthropology. If something like CA is correct, then we at least have good grounds for interrogating an anthropological claim if it appears to be in conflict with one or more essential christological beliefs.

One of the truths typically deemed as essential for any orthodox view of the incarnation is the idea that Jesus remains fully and truly human even after his death, resurrection, and ascension. As the author of Hebrews maintains, the eternal priesthood of Christ requires his eternal humanity and thus an eternal incarnation (Heb. 5:1–6; 7:17–21). Let us refer to this as the permanent incarnation (PI) principle.

> PI: The incarnation is a permanent reality such that Jesus remains fully and truly human from the moment of the incarnation into eternity.

In addition to these long-standing theological intuitions, modern theological anthropology is increasingly shaped by yet a third conviction: Christian physicalism (CP). Many contemporary theologians reject the various kinds of substance dualism that have dominated Christian reflection about humanity, focusing instead on the idea that humans are entirely physical beings. Without getting overly technical, we can define this third principle as follows:

> CP: The human person is entirely comprised of, though not necessarily reducible to, "those entities and processes that are studied by the physical sciences, either as those sciences are currently understood, or in some future form that will not be radically different from their present state."[3]

The focus of this chapter is on the intersection of these three theological convictions. Given the widespread affirmation of CA in the history of theology, it should come as a bit of a surprise to discover that the modern trend toward more physicalist ways of understanding human ontology have not yet been characterized by any extensive engagement with Christology or the implications of the incarnation for understanding what it means to say that humans are embodied beings. Yet that has largely been the case. Over the last several decades, during which Christian physicalists have refined their biblical and theological arguments considerably, they have dealt extensively with the implications of physicalism for things like spiritual growth, worship, free will, the uniqueness and dignity of human persons, the resurrection, and a number of other theological issues. Until recently, though, few have given any significant attention to Christology in their discussions, often making no more than a passing comment about the "mystery" of the incarnation with the implication that it is inappropriate to try and solve one mystery (the mind-body relationship) by appealing to another (the hypostatic union).

This has begun to change recently with a number of works focusing specifically on what it would mean to develop an explicitly physicalist account of the incarnation.[4] Even in these works, however, people rarely engage extensively with the relationship between the permanence of the incarnation and Christian physicalism, particular as it relates to Holy Saturday (that is, the period between Christ's death on the cross and his resurrection). On a physicalist account of the resurrection, it would seem that we need to affirm that the death of Jesus involved the complete cessation of his human life. If so, how should we understand the state of the incarnation between Good Friday and Easter? If human persons are comprised entirely of physical entities and processes, and if the life of the human person consequently comes to an end at biological death, would this not mean that the human life of Jesus terminated on the cross? Even if we maintain that this human life resumed when Jesus rose from the grave on Sunday, we still have to wrestle with the status of his humanity in the interim. At first glance, CP seems to suggest that since the human life of Jesus ended on Friday and only resumed on Sunday, we have no real incarnation on Saturday. This may seem like a relatively short period of time, but it suggests that physicalists need to reject PI and affirm a kind of gappy incarnation. Indeed, we might wonder if on such an account it becomes necessary to talk about a kind of *re*incarnation since the eternal Son appears to divest himself of humanity at one point and then incarnate himself again in humanity shortly thereafter.

This brief sketch at least raises prima facie problems for physicalist views of the human person, but I do not want to pursue here the broader question of whether physicalism in general has the resources to handle this challenge.[5] What I want to do instead is consider one influential proposal that appears to offer a way of navigating the tension between these three theological principles. In the first half of this essay, we will see that Karl Barth offers a vision of the human person that is firmly committed to a christological anthropology (CA), the permanence of the incarnation (PI), and Christian physicalism (CP). Barth clearly affirms the first two, and although he is less clear with respect to the third, we will see that there are good reasons for associating Barth with this principle, as well. Given the influence of Barth's theology in general and his theological anthropology in particular, he can thus serve as an interesting test case for exploring how a theologian might resolve the apparent tension between these principles. Having laid that groundwork, the second half of the chapter will assess the adequacy of Barth's approach. Here, I will argue that although Barth offers a way of affirming all three of these propositions coherently, his approach raises a number of other important concerns that undermine the overall viability of this approach to reconciling PI and CP.

KARL BARTH AND THE THREE THEOLOGICAL PRINCIPLES

Let us begin developing our Barthian response to the problem of Holy Saturday by establishing that Barth does indeed affirm all three of the relevant principles. His commitment to CA is beyond doubt.[6] As he boldly declares, "the nature of the man Jesus alone is the key to the problem of human nature."[7]

Indeed, Barth pursues the logic of CA with such rigor that many have worried that his theology exhibits a tendency toward reducing all theological truths, anthropology included, to Christology. We do not need to concern ourselves with responding to that worry here, remaining content with noting that Barth would clearly be comfortable with a logic that requires anthropological truths to be guided by essential christological truths.

Establishing his commitment to PI can also be done relatively quickly. According to Barth, the incarnate Christ is "Lord of Time" in the past, the present, *and* the future,[8] routinely emphasizing the importance of Hebrews 13:8: "Jesus Christ is the same yesterday, today, and forever." As the God-man, Jesus is the firm and certain ground upon which all hope for the future resides. Consequently, his incarnate being exists εἰς τοὺς αἰῶνας.[9]

A more challenging task arises with respect to CP. At first glance, it seems relatively easy to identify Barth as a physicalist since he explicitly disavows dualism of any kind, preferring instead to describe the human in terms of a "concrete monism" in which body and soul are viewed as an indissoluble union.[10] However, the picture becomes somewhat more complicated when we appreciate that he also affirmed a "concrete and Christian dualism,"[11] eschewing any kind of reductive physicalism that would view the human person as no more than the sum total of physical operations. Many have thus concluded that we should view Barth as some kind of holistic dualist, one who recognizes the duality of human ontology while still affirming that body and soul are both necessary to constitute a rightly functioning human person.[12]

As I have explained elsewhere, part of the difficulty in identifying Barth's ontology is that he focuses primarily on identifying the christological shape of human existence and is relatively unconcerned with supporting any particular theory regarding the body-soul relationship.[13] On the basis of this christological starting point, he concludes that the human person is an essentially unified being, that we must nonetheless recognize a body-soul, and that there is an ordered relationship in that duality such that the soul is the leading aspect of the human person. The first point supports the conclusion that Barth was a physicalist, but it is quite possible to hear the latter two as more friendly to dualism.

Nonetheless, we have at least two reasons for thinking it best to identify Barth as some kind of physicalist. Despite using the language of "duality" to talk about the body and the soul, Barth does not envision these as discrete substances, explicitly disavowing any form of substance dualism.[14] Instead, he prefers to label body and soul as the two "moments" of the one human person, suggesting something closer to the idea of "aspects" common in various kinds of Christian physicalism.[15] Given this more aspectual understanding of the body-soul relationship, Barth also does not appear to support the conceivable separation of the two.[16] Since dualists envision body and soul as discrete substances, most maintain at least the possibility that they could be separated, even if they deny that this ever actually happens in practice. Barth, on the other hand, presents the body-soul relationship in such a way that neither could possibly exist without the other. The soul is the active principle that leads and the body is the passive potentiality that follows. Without the other, each would be a mere abstraction. Barth thus affirms clearly that the human person ceases to exist at death, which he describes as a state of "non-being."[17] Death is "the limit of our existence in time"[18] and "the end of all human and creaturely life and creativity and work."[19] He does use occasional rhetoric that suggests the possibility of the soul continuing to exist in isolation from the body, referring to "a bodiless soul and a soulless body"[20] that are now "alienated" from one another.[21] Yet the likeliest explanation for such rhetoric is that Barth wants to emphasize that death involves the complete destruction of the human person and the fact that in death the person is utterly incapable of performing those actions for which they were created by God.[22]

In light of all this, we have good reasons for thinking that Barth would have affirmed all three of the theological principles relevant to this discussion. In this next section, we will consider these principles in relationship to Holy Saturday, seeking to determine if Barth can maintain all three of these principles in a coherent fashion.

A BARTHIAN RESPONSE TO THE PROBLEM OF HOLY SATURDAY

As I mentioned in the introduction, the problem of Holy Saturday arises from the juxtaposition of our three principles. At first glance, CP and PI seem to be at odds with one another in light of Holy Saturday. Since CA precludes simply rejecting PI, we seem to have only two options: (1) conclude that PI is not an essential christological truth, or (2) reject/revise CP. Since Barth is committed to the essentiality of PI, the latter option appears to be the only one available to him. Yet, as we have seen, Barth affirms all three. Although Barth himself does not spend significant time reflecting on the significance of Holy Saturday, I will argue in this section that his theology offers enough resources for constructing a Barthian response to the problem of Holy Saturday that maintains a coherent commitment to all three of these principles by drawing on Barth's distinctive account of creaturely finitude, death, and the nature of the eschatological state.

As I mentioned, Barth has little to say about Holy Saturday.[23] This is partly because the Bible simply does not say much about the nature of Jesus's existence during this period. The relevant texts are brief and rife with interpretive challenges (for example, Eph. 4:9; 1 Pet. 3:19). One could argue, then, that the safest approach would be to avoid making much of Holy Saturday in any theological discussion, which is the option primarily deployed by Barth. Fortunately, though, we do not need to speculate about precise details of Jesus's existence on Holy Saturday for this event to offer resources for thinking about human ontology. Barth speaks extensively about the nature of Jesus's death and resurrection in general, dealing with his resurrected life at various points in the *Church Dogmatics*.[24] By connecting some dots between this material and Holy Saturday, we can begin to construct a broadly Barthian vision of what it means to say that Jesus remained the incarnate God-man between Good Friday and Easter.

Consistent with a commitment to physicalism, Barth clearly affirms that Jesus's biological death on the cross involved the cessation of his human life. Jesus "ceased to be" after his physical death such that his existence was "terminated by death like that of every other man."[25] Stated even more bluntly, "to be dead means not to be."[26] Barth thus leaves no room for even a modified or holistic form of dualism in which Jesus continues to exist on Holy Saturday in virtue of some kind of disembodied soul. Death simply is the end of Jesus's human life. For Barth, then, whatever we think about Jesus's existence after his biological death, we cannot think that he "was given further time beyond the unique time of his given life on earth back then."[27]

This will sound to some as though Barth denied the reality of PI after all, despite his claims to the contrary. If Jesus's death is the termination of his human life, how can we affirm that the eternal Son remains human in any meaningful sense. Here we need to wade briefly into Barth's theology of death. According to Barth, death is the necessary terminus of any finite, creaturely existence, as established by the fact that even Jesus died.[28] Creaturely finitude requires that we have both a beginning and an end. Consequently, Barth rejects any attempt to view the resurrected state of the human person as a mere continuation of our creaturely histories, as though eternity involved "an unlimited and unending time."[29] Instead, he contends that the entire history of the human person—that which establishes our identities as the creatures we are—ends with our biological deaths. Once we have died, there is nothing more to be added to our stories.

If this is all Barth had to say on the matter, we would have to conclude that he rejected PI. However, he clearly affirms the importance of the resurrection, viewing this as the event in which human persons are taken up and "eternally preserved" in the eternal life of God.[30] Since we are entirely physical beings, we cannot appeal to some kind of immaterial soul that possesses immortality intrinsically to ground our postmortem hopes. Instead, in death the human person must "throw himself upon God's free grace."[31] Although Barth is not clear on the matter, he seems to envision our eschatological state as being secure in the fact that God eternally knows us as persons established by the history of the lives they lived between birth and death. Barth even makes room for the resurrection of the body in this account, suggesting that since my body is intrinsic to my identity as the historical person I am, it must also be a part of the "I" who is made secure in the eternal life of God. To say that Jesus was raised from the dead, then, is to say that the full reality of who he was as a human person, including his human body, has been secured in the eternal life of God. According to Barth, then, the resurrection appearances of Jesus should not be viewed as additional events in the history of Jesus, which would suggest that his human history was *not* brought to an end at his death. Instead, Barth describes these post-resurrection events as the revelation of the life and identity of Jesus established by his life and death.[32]

Using this robustly christological account of creaturely finitude, death, and resurrection, then, we can begin to see its implications for how Barth might have responded to the problem of Holy Saturday. Consistent with CP, Barth maintains that death involves the end of Jesus's human life, going further than many Christian physicalists by contending that the resurrection of the body does *not* involve the continuation of embodied life as a succession of temporal moments.[33] At death, Jesus would have entered immediately into the resurrected state in which his historic, human identity is forever secured in the eternal life of the triune God. Consequently, Barth does not need to think that the incarnation somehow came to an end on the cross, even for one day. Instead, the identity of the Son is forever shaped by this particular, embodied human history.

SOME LINGERING CONCERNS

Barth thus seems able to maintain coherently all three of our theological propositions. Throughout his discussion, he remains committed to grounding his anthropological considerations in the essential truths of Christology (CA), to affirming that the incarnation is an eternal reality (PI), and to maintaining that Jesus should be understood (in his humanity) as a physical creature, making no appeal to immaterial substances as part of human ontology (CP). Nonetheless, his account is not without difficulties, particularly as it relates to the viability of his way of understanding the eschatological state of the human person. Although discussing Barth's view of the resurrection and any attendant problems in detail would take us too far astray, we can note several potential difficulties.

First, to the extent that Barth emphasizes that the resurrected state does not involve any continuation of our creaturely histories, he runs into a problem with his own christological starting point. The New Testament narratives seem to present a rather clear picture in which Jesus has postmortem experiences (resurrection, fellowship with the disciples, ascension, etc.) that contribute to his human history and identity. Although Barth does not want to downplay the historicity of these events, his view of death requires him to interpret them more as instances in which the already complete historical identity of Jesus is *revealed* in time rather than seeing them as making any real contribution to that identity. According to Nathan Hitchcock, this runs the risk of downplaying the real significance of these events, something that Barth himself would not want to do.

> The startling consequence of Barth's understanding of the incarnation is that Christ's resurrection, ascension, heavenly session, and return are not exaltation for Him. In fact they do not add a single iota to His person or reconciling work. The "afterlife" of Jesus generates nothing new on the ontological level, since the history between conception and cross constitutes His full identity.[34]

However, if Barth were to jettison the idea that death brings human histories to their ultimate terminus, then his approach would again be susceptible to the "gap" worry generated by Holy Saturday.

A similar problem arises with respect to the idea of an immediate resurrection proposed in the prior section. To avoid the gap problem, our Barthian solution suggested that we should view death as a doorway through which Jesus enters immediately into the resurrected state. This has the distinct advantage of maintaining the close link between the cross and the resurrection that is so important throughout the New Testament. However, such an approach again risks missing the significance of the fact that the gospel writers present Friday and Sunday as distinct events in Jesus's history, theologically inseparable but temporally distinguishable. By collapsing death and resurrection into a single event, some worry that Barth's theology misses the importance of the resurrection for understanding Christ's atoning work. Colin Gunton thus warns of a "partial failure" in Barth's theology because Christ's existence "acquires a certain static quality at His death," in which subsequent events do not appear to add anything of theological significance to that which has already been done. Hitchcock offers similar concerns: Barth teaches that Jesus's human-temporal identity is exhausted in His death, and that this Jesus-history, compiled and immutable, is eternalized in the resurrection. But if Barth means to say that the risen Jesus adds nothing to His history in the resurrection, then it makes little sense to say He has a *continuing* history.[35] Given that Barth emphasizes the centrality of the resurrection, maintaining that the bodily resurrection of Christ is the starting point for Christian theology, he would certainly object to any suggestion that his view of eternity undermines the significance of the resurrection. Yet it still seems that there are some important questions that need to be answered here before we can appeal to any notion of an immediate resurrection as a way of dealing with the issue of Holy Saturday.

Another problem arises when we consider the fact that in addition to his emphasis on death as the completion of our creaturely histories, Barth routinely speaks of the eschatological state as one in which humans seem to have at least some kind of ongoing experiences (for example, joy, hope, fellowship). As Berkouwer rightly noted in his early analysis of Barth's eschatology, there is a tension in Barth's description of the resurrection between the idea that eternity does not involve "a history without an end"[36] and an equally strong emphasis on eternity as a source of hope for the human person as they somehow "participate" in God's eternal life.[37] Berkouwer thus comments, "this exposition has brought us face to face with the central problem of Barth's eschatology. Continually Barth repeats the polemic against the idea of 'continuation,' *and* emphasizes man's existence on this side of death."[38] Yet it remains entirely unclear how human persons can have the requisite kinds of experiences without these experiences constituting a continuation of our creaturely histories. And if our creaturely histories continue *in any way* after our biological deaths, then the question of Holy Saturday arises once again.

We could try to resolve this tension by appealing to Barth's distinctive account of the relationship between time and eternity.[39] According to Barth, we should not view these as standing in opposition to one another. Instead, he contends that God's eternity encompasses our temporality, taking the latter to himself in the incarnation. Thus, as he declares, "eternity itself is not timeless."[40] If resurrected humanity somehow participates in the life of God, then maybe we can view this as a state in which real experiences are possible despite the absence of any succession of before and after. Yet it is not clear that such a move will work for Barth. Even on this account, human persons still seem to be having eschatological experiences that transcend that which they experienced during their earthly histories. If so, then their histories are *not* ended by death. And if human persons continue to have histories after their biological deaths, even transcendently eternalized histories, then we seem to have returned to the challenge of Holy Saturday.

Berkouwer offers a likelier possibility when he suggests that Barth views the resurrection as a state in which humans are forever known by God. "There is not a continuation beyond this life, but there is a *standing* in the attention of God through eternity."[41] In other words, what Barth has in mind when he says that we continue to exist despite our creaturely histories coming to an end is that in the resurrection the entirety of our creaturely existence and identity is eternally known and cherished by God.[42] Ultimately, all of creation will be "eternally preserved" in the sense that it will always be "open and present" to God.[43] Barth can thus affirm that "our future non-existence cannot be our complete negation."[44] Such a view immediately raises questions about whether this is truly an adequate understanding of the resurrected life of human persons. Barth should be commended for his concern to center our understanding of the eschatological state around God himself. Answering the question of what resurrected humanity will be like, he answers, "come what may, we shall be what we shall be under and with God."[45] Barth thus denies that the human person has any "beyond" of his own, maintaining instead that "God is his beyond."[46] As important as this might be, however, Barth's approach still results in an overly idealized vision of resurrected humanity. In other words, it is difficult to avoid the conclusion that we exist as anything more than static portraits of our earthly histories retained, and yes loved, in the eternal mind of God.[47]

This worry about the idealization of the resurrection becomes even sharper when we consider Barth's account of the resurrection body. Barth famously emphasized the importance of bodily resurrection in debates with many of his contemporaries, affirming with Paul that the bodily resurrection lies at the center of the Christian faith.[48] However, when viewed in light of his comments about the nature of human life in the eschaton, it begins to sound as though the resurrection of the body is merely the fact that we are "eternally present" to God as the embodied beings that we have been in our earthly lives. In other words, since the body is intrinsic to my identity as a human person, it must also be intrinsic to the person that God cherishes as "me" throughout eternity. Yet this requires us to surrender widely held beliefs about the resurrection as involving the creation of new, albeit transcendent, bodies that constitute a fundamental aspect of our continued creaturely lives.

Some might prefer to flip this argument around and contend that since Barth consistently affirms the resurrection of the body, he must not envision the eschatological state in the idealistic terms described above. As Hitchcock notes, the resurrection of the body often serves for Barth as a way "to hold open the space between God and redeemed humanity."[49] Consequently, maybe we should take his emphasis on the resurrection of the body at face value and use this as the basis for maintaining some kind of continued creaturely existence in the eschaton despite Barth's rhetoric to the contrary. However, this way of resolving the difficulty just returns us again to the challenge of Holy Saturday. If humans have creaturely histories after death, then we still have to account for how Holy Saturday fits into that history. If Barth tries to avoid this by maintaining that we have physical bodies in the eschaton while still denying that this means we must have continuing histories, we can rightly ask about what it means to say that these bodies are "physical" if they are not characterized by the sequentiality that characterizes all physical realities. I suppose Barth could appeal to the fact that we are continually discovering that the physical universe is far stranger than we realize, using this as a basis for describing the body of the resurrection as physical even while denying its temporality, and probably things like its spatiality and divisibility as well since those characteristics are difficult to separate from at least some form of sequentiality. It seems reasonable to ask, though, whether such a view really qualifies as "physical." A "body" that is nontemporal, nonspatial, and non-divisible begins to sound much more like an immaterial soul than a material body. Indeed, such a view would press toward conclusions that even many dualists would reject as a proper view of an immaterial soul (e.g., nontemporality). Consequently, it is not clear that Barth can solve the problem of what sounds like an overly idealized picture of the resurrection body by appealing to some kind of radically transformed body without raising worries about whether this is still a form of physicalism. It seems legitimate to ask at times whether Barth is claiming to be a physicalist while using dualist-sounding rhetoric to avoid some of the problems, especially those related to eschatology, generated by physicalism.

CONCLUSION

We began this discussion by looking at three theological principles held by many modern theologians that seem to produce some interesting challenges when applied to Holy Saturday. According to the first, essential christological truths should play a fundamental rule in determining what we think about humanity (CA). Second, that Jesus Christ remains fully and truly human even after his death and resurrection is an essential christological truth (PI). Third, human persons are complex physical beings such that we do not need to appeal to any kind of immaterial soul to understand human ontology (CP). When we consider the state of the incarnation on Holy Saturday, the second and third principles generate an interesting tension. How can Jesus still be fully and truly human between his death and resurrection if his human life comes to a complete end with his death on the cross? Even if we appeal to his resurrection on Sunday to ground the eternal significance of the incarnation, we still seem to have an odd "gap" in our understanding of the incarnation that needs to be addressed. And in light of the first principle, we cannot address the problem by jettisoning the permanence of the incarnation unless we think we can make an argument that theologians have incorrectly identified this as an essential christological truth.

We then explored Karl Barth's theology as offering an interesting case study in how someone might try to affirm all three of these theological principles coherently. This was not to suggest that Barth's theology presents the only way of addressing this difficulty, or even the best one. The widespread influence of his theology combined with the fact that he deals so extensively with many of the issues involved in this discussion, though, make him an excellent dialog partner.

In the end, I argued that Barth's distinctive view of human death and the resurrected state provides resources for affirming all three of these theological principles coherently. In his humanity, Jesus is an entirely physical being. Consequently, his human life comes to a complete end with his death on the cross. Nonetheless, he does not cease to exist entirely because death is merely the doorway into the resurrected state in which human persons stand before God and are preserved and cherished by him forever. If we apply this framework to Holy Saturday, then, we can say that the incarnation does not end, even for a moment, with Jesus's death. Jesus's own humanity enters immediately into the presence of God and participates in the life of God forever.

Despite the fact that such an account can affirm all three of our theological principles coherently, however, I also argued that Barth's distinctive way of approaching this issue has its own drawbacks, notably relating to his way of understanding death and the afterlife. Viewed from one perspective, Barth's understanding of the human person requires him to say that our creaturely histories come to a complete end at death, thus viewing the afterlife in largely idealistic terms as we are "preserved" in God's love. Such an account quickly runs into difficulties with affirming the embodied nature of resurrected life and long-held convictions about the eschaton involving at least some kind of continued history for God's creatures. More importantly for the purposes of this chapter, such a conclusion also runs into difficulties with the fact that Jesus himself seems to have a continued history after his death—including his resurrection, the post-resurrection appearances, and the ascension. Despite his affirmation of CA, Barth unfortunately does not allow these events to play an adequate role in guiding his understanding of humanity in general. If we take a different perspective on Barth's view of the afterlife and emphasize instead his language about a bodily resurrection that involves at least some kind of continued experiences for the human person, then we do not seem to have made much progress in dealing with the question of Holy Saturday.

In the end, to the extent that Barth offers an interestingly different way for a Christian physicalist to deal with Holy Saturday, his account runs into substantial difficulties in other areas. However, if we revise his account to deal with those difficulties, we lose the distinctive elements that offered unique resources for dealing with the continuity of the incarnation after Jesus's death. Consequently, Christian physcialists may need to look elsewhere to address this interesting question.

NOTES

1. David H. Kelsey, *Eccentric Existence: A Theological Anthropology* (Louisville, KY: Westminster John Knox Press, 2009), 8–9.

2. See especially Marc Cortez, *Christological Anthropology in Historical Perspective: Ancient and Contemporary Approaches to Theological Anthropology* (Grand Rapids, MI: Zondervan, 2016).

3. Marc Cortez, *Embodied Souls, Ensouled Bodies: An Exercise in Christological Anthropology and Its Significance for the Mind/Body Debate* (New York: T&T Clark, 2008), 117.

4. For example, Trenton Merricks, "The Word Made Flesh: Dualism, Physicalism, and the Incarnation," in *Persons: Human and Divine*, eds. Peter van Inwagen and Dean W. Zimmerman (New York: Oxford University Press, 2007), 281–300; Oliver Crisp, *God Incarnate: Explorations in Christology* (New York: T & T Clark, 2009), 137–154; Glenn Andrew Peoples, "The Mortal God: Materialism and Christology," in *The Ashgate Research Companion to Theological Anthropology*, eds. Joshua R. Farris and Charles Taliaferro (Burlington, VT: Ashgate, 2015), 331–344; Kevin W. Sharpe, "The Incarnation, Soul-Free: Physicalism, Kind Membership, and the Incarnation," *Religious Studies* 53, no. 1 (2017): 117–131.

5. See especially the chapter in this volume by Jason McMartin.

6. See Marc Cortez, "What Does It Mean to Call Karl Barth a 'Christocentric' Theologian?," *Scottish Journal of Theology* 60, no. 2 (2007): 127–143 and Marc Cortez, "The Madness in Our Method: Christology as the Necessary Starting Point for Theological Anthropology," in *The Ashgate Research Companion to Theological Anthropology*, ed. Joshua R. Farris and Charles Taliaferro (Burlington, VT: Ashgate, 2015), 15–26.

7. Karl Barth, *Church Dogmatics*, 13 vols., eds. G. W. Bromiley and T. F. Torrance (Edinburgh: T&T Clark, 1956–75), III/2, 43. Hereafter referred to as *CD*.

8. *CD* III/2, 437–511.

9. *CD* III/2, 485.

10. *CD* III/2, 393.

11. *CD* III/2, 394.

12. For example, Robert E. Willis, *The Ethics of Karl Barth* (Leiden: Brill, 1971), 236; Paul W. Newman, "Humanity with Spirit," *Scottish Journal of Theology* 34, no. 5 (1981): 423; Jürgen Moltmann, *God in Creation: An Ecological Doctrine of Creation*, ed. and trans. by Margaret Kohl (London: SCM, 1985), 252; Daniel J. Price, *Karl Barth's Anthropology in Light of Modern Thought* (Grand Rapids, MI: William B. Eerdmans, 2002), 254–255.

13. Cortez, *Embodied Souls, Ensouled Bodies*.

14. For example, *CD* III/2, 380–381. Barth applies this critique to various forms of substance dualism, but many of his worries fail to capture more recent forms of dualism with their emphasis on the holistic nature of the human person.

15. *CD* III/2, 419.

16. See especially *CD* III/2, 418–436.

17. *CD* III/2, 595–595.

18. *CD* III/4, 588.

19. *CD* IV/2, 295.
20. *CD*, III/2, 355.
21. *CD* III/2, 425.
22. For other possible interpretations of Barth's language regarding the intermediate state, see Cortez, *Embodied Souls, Ensouled Bodies*, 89–92. None of this resolves the question of precisely what *kind* of physicalism best accounts for Barth's position (for example, dual-aspect monism, emergent monism, nonreductive physicalism). Since the distinctions between those broad types of physicalism do not have any direct bearing on the arguments we will be considering here, though, we can set that question aside for now.
23. For a more extended analysis of Barth on this point, see David Lauber, *Barth on the Descent into Hell: God, Atonement, and the Christian Life* (Farnham, UK: Ashgate, 2004).
24. For more on Barth's theology of death, see Marc Cortez, *ReSourcing Theological Anthropology: A Constructive Account of Humanity in the Light of Christ* (Grand Rapids, MI: Zondervan, 2018), chapter 8.
25. *CD* IV/3.1, 312.
26. *CD* IV/1, 301.
27. *CD* III/2, 477; cf. I/2, 53.
28. See especially *CD* III/2, 440, 596.
29. *CD* III/4, 572–573.
30. *CD* III/3, 89.
31. *CD* III/2, 569.
32. For a nice discussion of this, see Nathan Hitchcock, *Karl Barth and the Resurrection of the Flesh: The Loss of the Body in Participatory Eschatology* (Eugene, OR: Pickwick, 2013), 109–146.
33. As a result, Barth's approach has the added benefit of avoiding some of the difficult questions that surround how a physicalist can maintain continuity of personal identity through such a radical change as death and resurrection. For Barth, my eschatological identity just is the eternal repetition of my historical, embodied identity.
34. Hitchcock, *Karl Barth and the Resurrection of the Flesh*, 121.
35. Hitchcock, *Karl Barth and the Resurrection of the Flesh*, 93–94.
36. *CD*, III/3, 233.
37. G. C. Berkouwer, *The Triumph of Grace in the Theology of Karl Barth* (London: Paternoster, 1956), 151–165.
38. Berkouwer, *The Triumph of Grace in the Theology of Karl Barth*, 160.
39. See George Hunsinger, "Jesus as the Lord of Time According to Karl Barth," *Zeitschrift Für Dialektische Theologie* (2010): 113–127.
40. *CD* III/2, 526.
41. Berkouwer, *The Triumph of Grace in the Theology of Karl Barth*, 163–164.
42. It is possible, then, that Barth had in mind something similar to what is now described as a four-dimensional view of identity (see Theodore Sider, *Four-Dimensionalism* [Oxford: Clarendon Press, 2001]; Michael C. Rea, "Four-Dimensionalism," in *The Oxford Handbook of Metaphysics*, eds. Michael J. Loux and Dean Zimmerman [New York: Oxford University Press, 2003], 246–280).
43. *CD* III/3, 89.

44. *CD* III/2, 611.

45. *CD* III/2, 545.

46. *CD* III/2, 632. Schurr points out that Barth routinely used language that suggests continued personal existence, so we must take that into account. Yet he maintains nonetheless that "Barth does not seem to have provided the categories within which the identity of subjective continuity beyond death could be affirmed, but he nevertheless insists on it, and with his present emphasis on durable eternity could allow for it" (George M. Schurr, "Brunner and Barth on Life after Death," *Journal of Religious Thought* 24, no. 2 [1967]: 102).

47. By raising the concern about an "idealized" view of the body in the resurrection, I am not suggesting that Barth qualifies as an idealist in the more technical sense (for example, Spinozan or Berkleyan idealism). Indeed, Barth is equally clear in his rejection of all forms of such "monistic spiritualism" (*CD* III/2, 390). Instead, I intend the adjective to refer only to the worry that the resurrected body only endures as some kind of eternal "idea"—that is, an object eternally known and loved by God.

48. See esp. Karl Barth, *The Resurrection of the Dead*, ed. R. Dale Dawson, trans. H. J. Stenning (Eugene, OR: Wipf & Stock, 2015).

49. Hitchcock, *Karl Barth and the Resurrection of the Flesh*, 54.

SUGGESTED FURTHER READING

Barth, Karl. *Church Dogmatics.* Translated by Geoffrey W. Bromiley and Thomas F. Torrance. Vol. III, no. 2. Edinburgh: T. & T. Clark, 1960.

Barth, Karl. *The Resurrection of the Dead*, edited by R. Dale Dawson. Translated by H. J. Stenning. Eugene, OR: Wipf & Stock, 2015.

Cortez, Marc. *Embodied Souls, Ensouled Bodies: An Exercise in Christological Anthropology and Its Significance for the Mind/Body Debate*. New York: T&T Clark, 2008.

Dawson, R. Dale. *The Resurrection in Karl Barth.* Farnham, UK: Ashgate, 2007.

Fergusson, David. "Barth's Resurrection of the Dead: Further Reflections." *Scottish Journal of Theology* 56, no. 2 (2003): 65–72.

Hitchcock, Nathan. *Karl Barth and the Resurrection of the Flesh: The Loss of the Body in Participatory Eschatology*. Eugene, OR: Pickwick, 2013.

Lauber, David. *Barth on the Descent into Hell: God, Atonement, and the Christian Life*. Farnham, UK: Ashgate, 2004.

Lewis, Alan E. *Between Cross and Resurrection: A Theology of Holy Saturday.* Grand Rapids, MI: Eerdmans, 2003.

Merricks, Trenton. "The Word Made Flesh: Dualism, Physicalism, and the Incarnation." In *Persons: Human and Divine*, edited by Peter van Inwagen and Dean W. Zimmerman, 281–300. New York: Oxford University Press, 2007.

Peoples, Glenn Andrew. "The Mortal God: Materialism and Christology." In *The Ashgate Research Companion to Theological Anthropology*, edited by Joshua Farris and Charles Taliaferro, 331–344. Farnham, UK: Ashgate, 2015.

Chapter 8

Physicalist Christology and the Two Sons Worry

R. T. Mullins

Christian physicalists like Trenton Merricks have suggested that physicalism helps solve various problems with the doctrine of the incarnation.[1] Though the suggestions are underdeveloped, one claim seems to be that physicalism is better suited to avoid the heresy of Nestorianism than substance dualism.[2] In this chapter, I shall argue that physicalism does not help the Christian theologian avoid Nestorianism. I shall do this by examining an important distinction developed after the Council of Chalcedon known as the anhypostasia/enhypostasia distinction. This distinction is part of the neo-Chalcedonian Christology that is endorsed at the Fifth Ecumenical Council. Though this distinction is widely held by theologians to be part of the orthodox Christological deposit, it has been grossly overlooked by Christian philosophers. In the first part of this chapter, I shall briefly articulate the main differences between substance dualism and physicalism. In the second part, I shall outline the desiderata that must be satisfied by any ecumenically orthodox account of the incarnation. In the third part, I will explain why a physicalist might think that her anthropology better avoids Nestorianism than dualism. In the fourth part, I shall explain the content and purpose of the anhypostasia/enhypostasia distinction. In particular, I shall explain how it is used to refute Nestorianism. In the fifth part, I shall argue that physicalism does not in fact avoid Nestorianism because it easily violates the anhypostasia/enhypostasia distinction.

PHYSICALISM AND DUALISM

Physicalism is the view that a human person is identical to, or consists only of, a physical substance. Physicalists are divided over which physical substance a human person is in fact identical to, but the two most common claims are that a human person is either identical to a human brain, or identical to a human body.[3]

A further division among physicalists is over mental properties. A reductive physicalist will say that mental properties are reducible to physical properties, whereas a nonreductivist will maintain that mental properties are irreducible.[4] I shall narrow my focus on nonreductive physicalism since it is the most widely held view today among physicalists.

Substance dualism is the view that a human person is comprised of two substances: an immaterial substance and a physical substance. On the version of dualism that I defend, a person is identical to an immaterial substance, often called a mind or a soul. A person is a soul that has the capacity to think and perform free actions.[5] A person is a center of consciousness that can exemplify a variety of mental properties like thought, emotion, self-reflection, self-awareness,etc. Another way to put this is that a person has a first-person perspective. Further, I maintain that in order for a person to be *human*, that person must be, or previously have been, appropriately related to a human body.[6] This appropriate relation is called embodiment. Through embodiment, a soul comes to have physical properties.

Trenton Merricks maintains that only physicalism can make good on the claim that a human person has physical properties. According to Merricks, only physical objects can have physical properties.[7] He speaks as if the substance dualist will agree with this statement, but I don't think that the substance dualist should agree with this statement. According to the dualist E. J. Lowe, a person can have physical properties in a derivative way in virtue of having a physical body.[8] Perhaps a physicalist like Merricks should say that only physical objects can have physical properties in a fundamental or nonderivative way. I believe that a dualist can agree to this. On the version of dualism that I defend, souls can have physical properties in a derivative way in virtue of being embodied.

I maintain that souls also can have spatial location. One might worry that this confuses immaterial and material objects. This worry is due to the fact that one way of demarcating immaterial and material objects is spatial location. This is a demarcation suggested by René Descartes. Not all dualists agree with this demarcation. In fact, there is a long tradition of rejecting this demarcation. For example, the Cambridge Platonists found Descartes's suggestion utterly shocking since it entailed that immaterial substances like souls and God literally had no location, thus giving Descartes's view the derogatory term "nullibism."[9]

Contemporary dualists like Charles Taliaffero, Stewart Goetz, and Richard Swinburne maintain that a human soul is located where its body is.[10] How, then, can an immaterial and material object be distinguished? One way to distinguish the two is suggested by the medieval philosopher Nicole Oresme. According to Oresme, a soul and a body both have spatial location. A physical object, however, fills space in a way that is impenetrable such that other physical objects cannot be located in the same spatial location. Say that an atom is located in point p_1 of space. As long as that atom is located at that spatial point, no other atom can be located there. According Oresme, souls are different. Souls can occupy a point in space without filling it in this impenetrable way. Several immaterial objects could be wholly located at the same point in space. For instance, God can be wholly located at the same point in space as you or I. Oresme also offers the example of Jesus's encounter with a demon-possessed man. The man in question is said to be possessed by a legion of demons. Oresme thinks that all of these demons are located in the same spatial region.[11]

Embodiment: The Physicalist Story

Merricks maintains that whatever is involved in the incarnation of the Son of God, it must involve embodiment.[12] I believe that the dualist should agree with Merricks on this point. I cannot understand how God the Son could become human without being embodied. Of course, a question naturally arises at this point. What does it mean to be embodied? There are quite a few discussions of embodiment in the literature, but there appear to be two basic accounts.[13] The first is physical realization. This assumes a physicalist anthropology of human persons. This view holds that "a person *P* is embodied in body *B* if and only if all the (intrinsic) states of *P* are wholly realized by (intrinsic) states of *B*."[14] One way to put this is that all of *P*'s mental states supervene upon the brain states of *B*.[15] Another way to put this is that a person is identical to a body, though a person can have mental states that are irreducible to the physical. This is the sort of view that a physicalist will wish to endorse in her Christology.

However, Brian Leftow and Robin Le Poidevin note that such a thing is impossible because the Son—an immaterial thing—cannot become wholly material.[16] My inclination is to agree with Leftow and Le Poidevin here, but I will not press the point too much. Like most critics of physicalist Christology, I find it utterly mysterious as to how an immaterial person like the Son can become identical to a physical body. Merricks offers some remarks about the possibility of this, but I find them to be unilluminating.[17]

However, I do not wish to push this point too much since the majority of criticisms of physicalist Christology focus on the impossibility of an immaterial person becoming wholly material. My incarnation-based objections against physicalist Christology will focus on different issues that have yet to be explored.

Embodiment: The Dualist Story

The previous account of embodiment will not help the dualist understand the doctrine of the incarnation. There is a second broad account of embodiment that is more congenial with immaterial minds and cuts things in terms of a causal connection between the mind and the body. A mind is fully embodied in a physical body if and only if the following five conditions are met. The first condition is that the disturbances of the physical body can cause pain in the mind. Also, the various goings-on in the body can cause pleasure in the mind. If the body stubs a toe, the mind will feel pain. If the body is hugged in the right way, the mind will feel pleasure. Second, the mind can feel the inside of the body. An example would be the feeling of an empty stomach. Third, the mind can move the body through a basic action. A basic action is when an agent can perform an act without having to perform some other action in order to accomplish the first act. For instance, I move my arm by a basic act. I do not move the cup of water on my desk by a basic act. Fourth, the mind can look out from the world from where the body is. The body is the mind's locus of perception of the world. The mind acquires perceptual knowledge as mediated through the body. Fifth, the thoughts and feelings of the mind can be affected by the things that go on in the body.[18]

Merricks argues that this dualistic account of embodiment runs into a series of problems. In particular, Merricks argues that the dualist cannot explain how the Son is uniquely embodied in a particular body. Further, he says that the dualist cannot explain how the Father and Holy Spirit are not embodied in the same body as the Son.

For his argument, Merricks boils down these conditions to two main points: direct causal control, and immediate epistemic access to, a body. Given God's omnipotence, God can directly control every single body that exists. Given God's omniscience, Merricks thinks that God has immediate epistemic access to every single body that exists. If to be embodied is to have direct control and immediate epistemic access to a body, then God surely must be embodied in every physical object.[19] So, according to Merricks, the dualist cannot explain the unique embodiment of the Son in His particular human nature. Nor can the dualist explain how the Father and Holy Spirit are not embodied in the Son's human nature, as well. This is a serious problem, but I believe that several of the conditions for dualistic embodiment can be nuanced in order to avoid this problem.

First, it seems to me that Merricks has overlooked a particular assumption at play in these conditions. Embodiment is not simply direct causal control and immediate epistemic access to a body. As Luke Van Horn points out, embodiment involves the *body* having a causal disposition to cause various states in the mind.[20] We have no good reason to think that every creaturely body in the universe has the causal disposition to cause certain mental states in God the Son in the way that is captured by embodiment. For example, when I stub my toe, it causes me direct and immediate pain. When I see you stub your toe, I might have sympathy pain, but this is quite different. My sympathy pain is not immediately caused by your body. Instead, my sympathy pain is derivative. Sympathy, or empathy, involves a person imaginatively constructing a copy of another person's conscious states. So the empath's sympathy pain is derivative, and not direct nor immediate. The empath's pain is not directly and immediately caused by the other person's body. I believe that we have good reasons for affirming that God has this kind of empathetic knowledge, though a full discussion of this would take us off topic.[21] I see no good reason to believe that every creaturely body in the universe has the dispositional power to cause direct and immediate states in the mind of God in the way captured by embodiment.

Second, Merricks makes an interesting suggestion to which I think the dualist should pay attention. Merricks suggests that whatever embodiment is for the dualist, it must be the precondition for having the sort of direct causal control and immediate epistemic access captured in the five conditions noted above. I think this is right. The causal dispositions that Van Horn identifies go some way toward capturing this, but more might be needed.

Recall that condition four says that the locus of a mind is from the particular body in which it is embodied. I feel that this condition is important for several reasons. First, imagine the case of an individual who is paralyzed by a car accident. This person now no longer has direct control over most of her body. Yet she still has her locus on the world bound to this particular body. What explains this boundedness? Why exactly is a soul bound to one body and not another? Why is it that my soul has direct causal control and immediate epistemic access to my body and not yours?

A dualist can posit that there are relatively simple laws of nature called psychophysical laws.[22] Psychophysical laws create a binding relation between a soul and a body such that a) the soul is causally disposed to causally interact with a particular body, and b) a particular body is causally disposed to causally interact with a particular soul. This binding relation is what gives rise to the other conditions discussed above for embodiment.

Someone like Merricks might point out that God could bind Himself to everybody, even though God does not do so. One might develop an objection that says that the mere possibility of being so bound is sufficient for embodiment. However, this is mistaken. The mere possibility of being bound to a body is not enough to be embodied. One must actually be bound through the appropriate psychophysical laws to a particular body in order for embodiment to obtain.[23]

One might worry that positing this binding relation is ad hoc. I beg to differ. This binding relation looks similar to the early Christian usage of the term *perichoresis*. What I am asserting is that there is prima facie motivation to posit a perichoretic binding relation since it has a great deal of precedent within the Christian tradition. I don't claim that the mere existence of a widespread acceptance of this in the tradition provides the dualist with *ultima facie* motivation for affirming perichoresis. All that I need for the moment, however, is the prima facie motivation in order to ward off the charge of being too ad hoc.

What exactly is perichoresis? This is a tricky term because Christian theology uses this term in different ways in the doctrines of the Trinity and the incarnation.[24] Quite literally, perichoresis means "interpenetration."[25] In Neoplatonic thought, perichoresis was used to describe the relationship between a soul and body.[26] Many early Church Fathers would appeal to the soul's relation to the body as a way of describing the incarnation of the Son. The idea from the early fathers is that the Son stands in a perichoretic relation to His humanity, but does not stand in a perichoretic relation to the rest of humanity. The dualist can appeal to the Church tradition on this point to avoid the accusation that her view is ad hoc.

CHRISTOLOGICAL DESIDERATA

Now that we have the basic dualist and physicalist story before us, we can delve deeper into the doctrine of the incarnation. I shall develop some of the basic Christological desiderata derived from the seven ecumenical councils: Nicaea I, Constantinople I, Ephesus, Chalcedon, Constantinople II, Constantinople III, and Nicaea II. Due to space constraints, I cannot outline all of the Christological desiderata. I shall limit myself to certain core desiderata that are relevant for the discussion at hand. For interested readers, I have elsewhere offered a more thorough discussion of the Christological desiderata.[27]

On the standard story, the most significant aspects of ecumenical Christology come from the Council of Chalcedon in 451. What does Chalcedonian Christology look like? Oliver Crisp summarizes five relevant desiderata from the Chalcedonian creed:

1. Christ is of one substance (homoousious) with the Father.
2. Christ is eternally begotten of the Father according to his divinity and temporally begotten of the Virgin Mary according to his humanity.
3. Christ is one theanthropic (divine-human) person (hypostasis) subsisting in two natures (*phuseis*), which are held together in a personal union.

4. Christ's two natures remain intact in the personal union, without being confused or mingled together to form some sort of hybrid entity or *tertium quid.*
5. Christ's two natures are a fully divine nature and a fully human nature, respectively, his human nature consisting of a human body and a "rational" soul.[28]

For the purposes of this chapter, I shall focus my critique of physicalist Christology on its inability to offer an adequate account of (3). In particular, I shall argue that a physicalist Christology is subject to the charge of Nestorianism. Nestorianism is a heresy condemned at Chalcedon, deriving its name from the early Church theologian Nestorius. In contemporary parlance, Nestorianism is a term applied to any view that entails that there are two persons in Christ, instead of one person.

Christological Models

There are various models of the incarnation at large today, many of which claiming to be consistent with ecumenical Christology.[29] The dominant strands are called composite Christologies because they identify different "parts" that constitute Jesus Christ.[30] (The language of "parts" is being used here rather loosely. It does not assume a classical mereology.[31]) All of the Christologies under consideration here claim that God the Son is one part of Jesus Christ, and that a human body is another part. However, composite Christologies can involve two, three, or four parts of the composite Christ depending on one's philosophical anthropology. The traditional Christian view is sometimes called the *two-minds view*, and it is a three-part Christology since it posits that Jesus Christ is a divine mind (God the Son), a human mind (a concrete soul), and a human body.[32] Someone who is a substance dualist may find this attractive. However, she may also find a two-part Christology equally attractive. A two-part Christology could involve God the Son and a human body. This is where the divine mind constitutes a human person by being connected to a human body in the appropriate way.[33] A trichotomist will most likely have a four-part Christology since she holds that human persons are comprised of a body, soul, and spirit. In this instance, God the Son would take on a human body, a human soul, and a human spirit. However, it is not necessary for a trichotomist to hold to a four-part Christology.

The early Church theologian Apollinaris was a trichotomist who believed that human persons are comprised of a human body, a rational soul, and an animal soul. He had a three-part Christology since, on his view, the Son already is a rational soul. According to Apollinaris, if the Son assumed another rational soul, that would involve the Son assuming another person.

That would be the heresy we are calling Nestorianism, and Apollinaris sought to avoid this. For Apollinaris, the Son counts as fully human because the Son is a rational soul with an animal soul and a human body. Apollinaris says, "If, then, a human being is made up of three parts, the Lord is also a human being, for the Lord surely is made up of three parts: spirit and soul and body."[34]

Closely related to this discussion of philosophical anthropology is the question of the divine and human will. Monothelites hold that the Son only has one will, whereas dyothelites hold that the Son has two wills: a human and a divine will.[35] A monothelite will say that only persons have a will, whereas a dyothelite will maintain that natures have a will.[36] A dyothelite will say that since Christ took on a human nature, He must have taken on a human will, as well. What might this look like? Say one has a three-part Christology and is a dyothelite. On this view, God the Son—a divine mind—assumes a human mind, a human body, and a human will.

A three-part dyothelite Christology becomes the majority view after the seventh century for all of those who adhere to the seven ecumenical councils.[37] On this version of the three-part dyothelite Christology, Jesus Christ is composed of God the Son, a rational soul with a numerically distinct will, and a human body.[38]

A WORRY ABOUT NESTORIANISM

Merricks complains that it is unclear how the standard three-part dyothelite Christology can avoid the entailment that there are two persons in Jesus Christ. He points out that the individual human nature that God the Son assumes is intrinsically like a complete human person. Thus, he finds it difficult to see how this complete human nature is not in fact a complete human person. Merricks notes that if this complete human nature had not been assumed by the Son, this complete human nature would have been a human person. This, says Merricks, looks like Nestorianism.[39] What Merricks does not seem to realize is that he has articulated a deep worry that the theologians sought to address prior to, and in the aftermath of, the Council of Chalcedon. I shall call it the Two Sons Worry in order to avoid the anachronism of identifying various and diverse views as Nestorian.

The Two Sons Worry is an incredibly popular argument in the early Church, and it comes in several forms. One form of the argument from Eunomius goes as follows: a human person is a soul and a body. If the Son assumed a soul and a body, the Son assumed a human person. So there are two sons (two persons) in the incarnation.[40] That is heresy.

It isn't hard for a physicalist to construct a Two Sons Worry argument against the dualist who affirms the standard three-part dyothelite Christology.

On this Christology, God the Son is a divine mind with a will. God the Son assumes a human soul and body. This human soul has its own distinct will. Earlier it was noted that a dualist affirms that a person is an immaterial substance that has the capacity to think and perform free actions. On the standard three-part dyothelite Christology, there are two immaterial substances in Christ that have the capacity to think and perform free actions. Given the dualist's own understanding of personhood, this Christology seems to entail that there are two persons in Christ—that is, two sons plain and simple.

Given this, it makes sense why a physicalist might think that her view fits better with ecumenical Christology. On a physicalist Christology, the Son does not assume a human soul. Instead, the Son becomes identical to a human body. So there should be no worry of having too many thinkers in Christ given this physicalist anthropology. In other words, it seems like a physicalist Christology is the best way to affirm that there is only one person in Christ.

However, I don't find this physicalist move to be obvious. The physicalist does have some explaining to do. She has not yet explained how God the Son is the only person in the incarnation. By itself, the body of Jesus would have all of the mental properties needed to be a complete human person. The body would be a complete human person without God the Son given a physicalist anthropology. What about the incarnation prevents this from being the case? I will have more to say on this later.

In the fifth section, I shall argue that the physicalist does not escape the Two Sons Worry quite so easily. Before doing so, I must articulate an important ecumenical constraint on Christological theorizing. As I shall argue, this constraint prevents the physicalist from claiming victory over the dualist with regards to avoiding the Two Sons Worry.

THE NEO-CHALCEDONIAN CHRISTOLOGY OF CONSTANTINOPLE II

In this section, I shall articulate the anhypostasia/enhypostasia distinction developed in the aftermath of the Council of Chalcedon. One of the motivations for this distinction is to avoid the Two Sons Worry by placing a constraint on Christological theorizing. After articulating this distinction, I shall argue that Merricks's physicalist anthropology does not avoid the Two Sons Worry.

Merricks, like most Christian philosophers, focuses his Christology almost exclusively on the Council of Chalcedon. He shows no awareness of the Christological developments after Chalcedon. This is understandable. As Andrew Louth points out, "Older treatments of the history of Christian doctrine have usually presented the Definition of Chalcedon (often reduced, wrongly, to the final epitome) as the culmination of patristic Christology, but it is better seen as a watershed, for much water was still to flow after the council."[41]

What must be understood is that the Council of Chalcedon did not really settle many theological debates, nor did it create the Church unity that it sought. Louth explains that,

> Although the aim of the Fathers of the Council of Chalcedon was to secure unity, the result was the opposite: the council opened divisions in the Church that have never been healed between those who accepted the council (who called themselves Orthodox or Catholic, but were called by their opponents dyophysites, if not "Nestorians") and those who rejected the council (who also called themselves Orthodox or Catholic, and were called by their opponents monophysites, if not "Eutychians," and by modern scholars "miaphysites"—a barbarous coinage).[42]

According to G. L. Prestige, the success of the Council "was only negative; they defined what was false but provided no positive and convincing rationalisation of the right faith."[43] This lack of a convincing rationalization can be felt in the aftermath of Chalcedon. After the Formula of Chalcedon was framed, many Christians in the East remained unconvinced that there was a clear difference between the defenders of Chalcedon and the Nestorians.[44] Christopher Beeley explains that the Chalcedonian definition, which was enforced under governmental pressure, "left the basic identity of Christ and the nature of the union disastrously ambiguous from the point of view of the more unitive traditions. It is no wonder that Nestorius reportedly felt vindicated by the result."[45] According to Prestige, Nestorius thought that Chalcedon and Pope Leo's *Tome* "expressed exactly what he himself had always believed."[46] Given this, the inability of Eastern Christians to see a difference between Chalcedon and Nestorianism is understandable.

The monophysite Christians in the East at that time were not happy with "the sickness of Chalcedon" that declared that Christ had two natures.[47] These Christians are called monophysites because they denied that Jesus Christ had two natures, claiming instead that Jesus had only one nature. Like Chalcedon, many monophysites held to a three-part composite Christology, but unlike Chalcedon they held that in the incarnation the two natures became one nature—without confusion—through the composition. These monophysites maintained that if there are two natures in the incarnation, then there are two people. That would be Nestorianism. Again, we see the Two Sons Worry at play here. Hence, why the monophysites called the Chalcedonian Christology a "sickness."

While many in the East were distraught over the result of Chalcedon, various Nestorian parties felt that they were able to interpret Chalcedon in such a way that they could agree to the formula. In fact, one of the main motivations for the Fifth Ecumenical Council, Constantinople II (553), was to give a proper interpretation of Chalcedon that fully excluded Nestorianism.[48] The Eastern Church made a serious push to get single-nature Christologies included in the scope of orthodoxy, and the emperor Justinian was keen to make peace with these groups from the East.[49] Though the monophysite effort did not succeed in getting single-nature language incorporated into the ecumenical tradition, Constantinople II does make some much needed strides toward ridding ecumenical Christology of its Nestorian tendencies. The Christology that eventually emerges from this is often called "neo-Chalcedonian." It is this Christology that is adopted by the Council of Constantinople II, and has left a huge mark on the way subsequent generations of Christians have thought about the incarnation. Most Christology today is not in fact Chalcedonian because it bears the marks of the much needed clarifications that neo-Chalcedonian Christology developed.

One of the most important developments during this time period for this discussion is the *anhypostasia* and *enhypostasia* distinction. This is a distinction that developed in the aftermath of Chalcedon leading up to the Fifth Ecumenical Council. Though the terms anhypostasia and enhypostasia are not used by the Council, the theology is adopted and affirmed by the Fifth Ecumenical Council as the proper interpretation of Chalcedonian Christology.[50]

The deep concern to avoid saying that there were two sons, or two persons, in the incarnation is one issue that led to the development of the an/enhypostasia distinction. The Fifth Ecumenical Council took place because of a controversy over Adoptionism, Nestorianism, and Origenism—views that many at the time believed entailed two persons. These views seemed to entail the possibility of the human nature of Christ being a complete, separate person apart from God the Son. Hence, these views naturally fall under the Two Sons Worry. In order to avoid the Two Sons Worry, the neo-Chalcedonian Christology of the Council claims that the human nature of Christ cannot have a *hypostasis* (person) of its own. Christ's human nature is *anhypostasis,* thus avoiding the Two Sons Worry.[51] The hypostasis of the Son is brought to the assumed human nature thus giving the human nature a hypostatic and personal reality.[52]

This en/anhypostasia distinction needs some unpacking because it gets a bit muddled in contemporary discussions due to the mutually confirming nature of each claim.[53] The enhypostasia claim is that the Son's human nature only exists because of the incarnation. The anhypostasia claim seems to contain two conditions. First, the Son's human nature would not have existed if it were not for the incarnation. This is incredibly similar to the enhypostasia claim, but anhypostasia adds a further condition. The second condition is that the human nature is only personal because it is assumed by a divine person—namely, God the Son. In other words, the Son brings His personhood to the assumed human nature. The human nature is not, nor could have been, a person independent of the Son's assumption.[54]

Fred Sanders explains that this is where the strength of the distinction comes into play in ridding ecumenical Christology of Nestorianism. It excludes the very possibility that the human nature of Christ could have formed some person from coming into existence if the Son had not assumed this nature.[55] The human nature of Christ cannot form a person apart from the incarnation. The human nature is only a person because it is assumed by the person of the Logos. The human nature only exists because of the incarnation. Wolfhart Pannenberg sums up the neo-Chalcedonian theology as follows: "By itself Jesus' humanity would not only be impersonal in the modern sense of lacking self-conscious personality, but taken by itself Jesus' human being would be non-existent."[56]

I must emphasize that the en/anhypostasia distinction is a constraint on Christological theorizing. Nothing about this constraint, by itself, gives us an actual Christological model. This constraint gives us a way to test Christological models for any underlying Nestorian tendencies. If a model cannot satisfy this constraint, it is not up to the task of satisfying the Christological desiderata listed above.

THE TWO SONS WORRY REVISITED

Now that we have the anhypostasia and enhypostasia distinction before us, I will return to the Two Sons Worry. In this section, I shall examine some further Christological problems for the dualist, turning then to articulate a way for the dualist to escape the Two Sons Worry. I will end this section by arguing that the physicalist cannot escape the Two Sons Worry.

Dualism and the Two Sons Worry

It might seem that the physicalist now has even more incarnational ammunition against the dualist. The physicalist could argue that the standard three-part dyothelite Christology cannot obviously satisfy the anhypostasia/enhypostasia constraints on Christological theorizing. The Son's human nature could quite obviously be a complete human person without the occurrence of the incarnation. The Son's human nature is a soul and a body with its own distinct will—that is, a complete human person. In fact, certain contemporary proponents of this Christology readily admit that the Son's human nature would be a complete human person without the incarnation.[57] That is a straightforward violation of the anhypostasia/enhypostasia constraint on

Christology.[58] In other words, this three-part Christology does not escape the Two Sons Worry.

The dualist might respond in several ways. One attempt comes from Oliver Crisp. Crisp will say that there never was a time when the human nature of Christ existed apart from God the Son. When the Holy Spirit conceived the human nature of Jesus in Mary's womb, the Son joined Himself to that human nature. So there never was a moment when the human nature existed without being joined to the Son. The "human nature is never in a position to form a supposit distinct from God the Son."[59] In other words, the human body and soul that the Son assumed never had a chance to form a person distinct from the Son.

Does this help the dualist escape the Two Sons Worry? The physicalist can complain that this does not obviously avoid the Two Sons Worry. The move that Crisp and others make avoids the charge of Adoptionism, but not the Two Sons Worry. As noted earlier, Adoptionism is one Christological heresy that often falls under the category of the Two Sons Worry. On Adoptionism, Jesus exists for a certain stretch of time and is later united to God the Son. In this scenario, we clearly have two persons. But what must be understood is that Adoptionism isn't the only way to fall victim to the Two Sons Worry. As the Council of Constantinople II understood, all one needs to do in order to fall subject to the Two Sons Worry is to offer a Christological model that *entails* two persons in Jesus Christ.

It is instructive to note that Theodore of Mopsuestia held that Jesus "had union with the Logos straightaway from the beginning when he was formed in his mother's womb."[60] This is the exact claim that Crisp wishes to make to avoid the Two Sons Worry. However, Theodore was explicitly condemned by the Council of Constantinople II for holding a view that entails two persons in Christ. So, more needs to be said in order to avoid the Two Sons Worry.

The physicalist can say that the dualist is in serious trouble at this point. She can complain that Crisp's response is missing something important in the en/anhypostasia constraint. The underlying intuition of the en/anhypostasia constraint seems to be that persons are necessarily identical to themselves. Necessarily, a person cannot exist apart from, or separate from, herself. If P and P* could possibly exist apart from one another, then P and P* are two different persons. On the three-part dyothelite Christology under consideration, it is possible that the human soul and body of the Son could have existed apart from the Son. It does not matter that the human nature of Christ never got a chance to become a separate person. All that is needed to violate the en/anhypostasia constraint is the metaphysical possibility of the human nature being a complete human person apart from the incarnation. The three-part dyothelite Christology makes this a very real metaphysical possibility. So the physicalist can maintain that it violates the en/anhypostasia constraint.[61]

There is, however, another move open to the dualist. Andrew Loke has recently articulated an alternative version of the three-part dyothelite Christology called the Divine Preconscious Model (DPM).[62] Loke's alternative account is, in part, motivated by the Two Sons Worry because he sees no way for the traditional version to avoid Nestorianism. While I don't agree with everything in Loke's alternative proposal, I think a strong case can be made that his Christology can help the dualist avoid the Two Sons Worry.

On Loke's DPM, substance dualism is assumed. Further, Loke postulates that a mind includes the conscious and the subconscious "parts" of the soul. (Recall that the "parts" language is being used here rather loosely.) According to Loke, "The conscious is that which, when it is active, exhibits a mental condition characterized by the experience of perceptions, thoughts, feelings, awareness of the external world and, often in humans, self-awareness."[63] He further explains that, "the subconscious is defined as mental contents which exist outside of consciousness."[64] The "outside" here is not to be taken literally. It means the mental content that one is not consciously aware of at a particular time.

The subconscious can be further subdivided into the preconscious and the unconscious. "The preconscious is defined as mental contents that are not currently in consciousness but are accessible to consciousness by directing attention to them."[65] This typically includes mental states of which one is neither currently aware, nor focusing on. These are sometimes referred to as nonoccurrent mental states. For example, I know a fair bit about heavy metal, but there are many times throughout the day where none of this knowledge is at the forefront of my mind. My subconscious contains all of the mental states that I am not currently focusing on.

The unconscious is somewhat different. According to Loke, the unconscious "is defined as an aspect of the mind containing repressed instincts and their representative wishes, ideas and images that are not accessible to direct examination." They are not accessible to direct examination because these are repressed within our minds. The "operation of repression prevents the contents of the unconscious from entering either the conscious or the preconscious."[66]

With these distinctions before us, one can start to unpack the DPM account of the incarnation. Prior to the incarnation, God the Son has/is a mind. At the incarnation, the Son's mind comes to include a consciousness that is divided into, or possesses, two preconsciouses: one divine preconscious and one human preconscious. Certain divine properties, like omniscience, are located in the divine preconscious, thus retaining the consciousness of the Son's divine properties. The Son takes on various human properties by becoming embodied in human flesh. These human properties are located in the human preconscious and human body of the Son. In particular, the Son's consciousness comes to have the disposition to functionally depend upon the brain of His newly acquired human body.[67] In accordance with the above discussion, Loke could appeal to the perichoretic binding relation between minds and bodies. According to Loke, the result of the Son's consciousness being embodied is that the Son's consciousness has the capacity to experience physical sensations and bodily desires.[68]

On the DPM, Loke maintains that the Son's mind does not contain an unconscious because he believes that would entail the Son giving up certain divine attributes such as omnipotence. He claims that this does not entail that the Son has an incomplete human nature since, although an unconscious is a common human property, it is not an essential human property.[69]

You might be wondering how exactly this is a three-part Christology since the DPM only contains one mind (i.e., the Son), and one body. Isn't that a two-part Christology? Loke will disagree. Loke explains that "at the Incarnation the Logos had a consciousness (which included access to the divine preconscious), a preconscious that had two parts (part A having the properties of divinity and part B having the properties of a human preconscious), and a human body."[70] If I understand DPM correctly, we have one mind/one person in the incarnation: the Son. The Son is a self-conscious being having a preconscious that is divided into two parts: a part that contains the divine properties, and a part that contains the human properties. This one center of consciousness (the Son) comes to have two immaterial "parts" at the incarnation by having His mind divided into a divine preconscious and a human preconscious. Further, at the incarnation, the Son acquires a physical part: the human body.[71] So, we have three parts: one divine part, one immaterial human part, and one material human part. (It should be recalled again that "parts" is being used loosely.)

The DPM avoids the Two Sons Worry because it only involves one person, or one center of consciousness. Further, it can easily account for the an/enhypostatic constraint. The human nature of Christ simply would not count as a full human person apart from the incarnation. The Son only divides His mind into a divine and human preconscious at the incarnation for the purposes of becoming human. This human preconscious of the Son cannot possibly exist separated from the Son. Loke can concede that the human body of Jesus could be created without an incarnation taking place, but he can easily say that this body would not constitute a complete human person given his prior commitments to dualism.

Before moving on, I must concede that Loke's DPM has a great deal more nuance than I can capture here. Further, the DPM might face other theological and philosophical problems that some Christians might raise. All I can say at this point is that the DPM provides a seemingly clear way for the dualist to avoid the Two Sons Worry. The physicalist is thus prevented from saying that dualism falls victim to some version of Nestorianism. As I will argue, the physicalist has no clear way to avoid the Two Sons Worry, thus leaving the dualist in a better position on Christological grounds.

Physicalism and the Two Sons Worry

As I have noted before, the suggestion that a physicalist Christology will fall victim to the Two Sons Worry will be surprising to most readers. This is why I have chosen to focus my critique of physicalism on the Two Sons Worry instead of the standard critiques of physicalist Christology.[72] The world of theology can always use more surprising discussions.

How does the physicalist fall prey to the Two Sons Worry? It does so because a physicalist Christology violates the en/anhypostasia constraint. On physicalism, a human person is identical to a human body. A complete human person is a living human body. On the physicalist incarnation endorsed by Merricks, God the Son becomes completely identical to a human body. It seems like it is metaphysically possible for the Son's body to exist without the incarnation. If this body were to exist without the incarnation, it would be a complete human person.[73] The human body of Jesus would be a complete human person without the Son. So this physicalist Christology has violated the en/anhypostasia constraint.

How might the physicalist respond? Perhaps she will say that, given the virgin birth, the particular body of Jesus would never have come into existence without the incarnation of the Son. She will go on to say that the very coming into existence of this particular body is a miracle performed by the Holy Spirit. However, I don't find this move persuasive because it is metaphysically possible for the Holy Spirit to perform the miracle of a virgin birth without an incarnation. If this is a metaphysical possibility, then it is possible that the particular body of Jesus could come into existence without being incarnated by the Son. As such, it is a metaphysical possibility that the particular body of Jesus be a person without the incarnation. So we still have violated the an/enhypostasia constraint. The physicalist needs something stronger than the mere fact that the particular body of Jesus only came into existence with the incarnation of the Son. The physicalist will have to say that it is metaphysically impossible for the particular body of Jesus to come into existence without the incarnation. I'm not certain what the physicalist can offer to explain this metaphysical impossibility, but perhaps she can come up with some sort of story. That story remains to be told.

Without any sort of story here, the dualist now has a new reason to reject physicalism on incarnational grounds. As noted before, the dualist already believes that it is metaphysically impossible for an immaterial person to become wholly identical to a physical object. Thus, the dualist can say that physicalism makes the incarnation impossible. What I have offered is an additional reason for the dualist to reject physicalism—that is, physicalism cannot escape the Two Sons Worry.

CONCLUSION

In this chapter, I have examined the prospects for a physicalist Christology and found it wanting. I have focused my discussion on an unexplored area for physicalist Christology in hopes that the debate can be advanced beyond the standard complaints against the physicalist. I have also offered a way for the dualist to avoid complaints that a physicalist might throw her way. However, I must confess that I have not addressed all of the complaints that a physicalist has against a dualist Christology. Yet I do hope that my work here has at least helped advance the case for seeing dualism as playing an important role in Christian theology, and perhaps placed a bit more disparagement on the potential of physicalism for future Christian thought.

NOTES

1. Trenton Merricks, "The Word Made Flesh: Dualism, Physicalism, and the Incarnation," in *Persons: Human and Divine*, eds. Peter van Inwagen and Dean Zimmerman (Oxford: Oxford University Press, 2007), 281–300.
2. Merricks, "The Word Made Flesh," 282.
3. Eric T. Olson, *What Are We: A Study in Personal Ontology* (New York: Oxford University Press, 2007), chapters 2 and 4.
4. Nancey Murphy, *Bodies and Souls, or Spirited Bodies?* (New York: Cambridge University Press, 2006).
5. Richard Swinburne, *Mind, Brain, and Free Will* (New York: Oxford University Press, 2013), 141.
6. Swinburne, *Mind, Brain, and Free Will*, 141–142.
7. Merricks, "The Word Made Flesh," 294.
8. E. J. Lowe, "Substance Dualism: A Non-Cartesian Approach," in *The Waning of Materialism*, eds. Robert C. Koons and George Bealer (Oxford: Oxford University Press, 2010), 441.
9. Charles Taliaferro and Alison J. Teply, eds., *Cambridge Platonist Spirituality* (New York: Paulist Press, 2004), 17.
10. Charles Taliaferro and Stewart Goetz, *A Brief History of the Soul* (Oxford: Blackwell Publishing, 2011). Swinburne, *Mind, Brain, and Free Will*, 173.
11. Nicole Oresme, *Le Livre du ciel et du monde*, ed. and trans. by Albert D. Menut and Alexander J Denomy (London: The University of Wisconsin Press, 1968), 285–289.
12. Merricks, "The Word Made Flesh," 293.

13. Robin Le Poidevin, "The Incarnation: Divine Embodiment and the Divided Mind," *Royal Institute of Philosophy Supplement* 68 (2011): 269–285. Richard Swinburne, *The Coherence of Theism* (Oxford: Oxford University Press, 1977), 102–104. T.J. Mawson, "God's Body," *The Heythrop Journal* 47 (2006): 171–181.

14. Le Poidevin, "The Incarnation: Divine Embodiment and the Divided Mind," 273.

15. For more on supervenience see Jaegwon Kim, *Supervenience and Mind: Selected Philosophical Essays* (New York: Cambridge University Press, 1993).

16. Leftow, "The Humanity of God," in *The Metaphysics of the Incarnation*, eds. Anna Marmodoro and Jonathan Hill (Oxford: Oxford University Press, 2011), 21–22. Robin Le Poidevin, "The Incarnation: Divine Embodiment and the Divided Mind," 276.

17. Merricks, "The Word Made Flesh," 296. Merricks develops an account based on a rejection of kind essentialism. However, for a reply, see Kevin W. Sharpe, "The Incarnation, Soul-Free: Physicalism, Kind Membership, and the Incarnation," *Religious Studies* 53 (2017): 117–131.

18. I follow Swinburne's account here. Poidevin makes some minor revisions to this account, and Mawson's account lacks several of the conditions.

19. Merricks, "The Word Made Flesh," 284–287.

20. Luke Van Horn, "Merricks' Soulless Savior," *Faith and Philosophy* 27 (2010): 334.

21. Linda Zagzebski, *Omnisubjectivity: A Defense of a Divine Attribute* (Milwaukee, WI: Marquette University Press, 2013).

22. Robin Collins, "A Scientific Case for the Soul," in *Soul Hypothesis: Investigations into the Existence of the Soul*, eds. Mark C. Baker and Stewart Goetz (New York: The Continuum International Publishing Group Inc., 2011).

23. Van Horn, "Merricks' Soulless Savior," 335.

24. For discussion on the differences in the Trinity and the incarnation, see my *The End of the Timeless God* (Oxford: Oxford University Press, 2016), 176–179.

25. John Anthony McGuckin, ed., *The Westminster Handbook to Patristic Theology* (Louisville, KY: Westminster John Knox Press, 2004), 260.

26. McGuckin, *Westminster Handbook to Patristic Theology*, 260.

27. Mullins, *The End of the Timeless God*, 156–194.

28. Crisp, "Incarnation," in *The Oxford Handbook of Systematic Theology*, eds. John Webster, Kathryn Tanner, and Iain Torrance (New York: Oxford University Press, 2007), 161.

29. I will be following the taxonomy of incarnation models that Oliver Crisp uses in *Divinity and Humanity: The Incarnation Reconsidered* (New York: Cambridge University Press, 2007).

30. Anna Marmodoro and Jonathan Hill, "Composition Models of the Incarnation: Unity and Unifying Relations," *Religious Studies* 46 (2010): 469–488.

31. William Hasker, "A Compositional Incarnation," *Religious Studies* (forthcoming).

32. Thomas V. Morris, *The Logic of God Incarnate* (London: Cornell University Press, 1986).

33. William Lane Craig and J. P. Moreland, *Philosophical Foundations for a Christian Worldview* (Downers Grove, IL: InterVarsity Press, 2003), 597–613.

Garrett J. DeWeese, "One Person, Two Natures: Two Metaphysical Models of the Incarnation," in *Jesus in Trinitarian Perspective: An Introduction to Christology*, eds. Fred Sanders and Klaus Issler (Nashville, TN: B&H Publishing Group, 2007), 114–152.

34. Richard Norris, ed., *The Christological Controversy* (Philadelphia: Fortress Press, 1980), 110.

35. Oliver Crisp, "Incarnation," in *The Oxford Handbook of Systematic Theology*, 162–163.

36. John of Damascus, *An Exposition of the Orthodox Faith*, book III.14.

37. Crisp, *Divinity and Humanity*, 49ff. Also, Crisp, "Incarnation," *The Oxford Handbook of Systematic Theology*, 163.

38. Ian A. McFarland, "Willing Is Not Choosing: Some Anthropological Implications of Dyothelite Christology," *International Journal of Systematic Theology* 9 (2007): 3–23.

39. Merricks, "The Word Made Flesh," 282.

40. Richard Paul Vaggione, *Eunomius: The Extent Works* (New York: Oxford University Press, 1987), 155–157. Christopher A. Beeley, *The Unity of Christ: Continuity and Conflict in Patristic Tradition* (New Haven, CT: Yale University Press, 2012), 176–182.

41. Andrew Louth, "Christology in the East from the Council of Chalcedon to John Damascene," in *The Oxford Handbook of Christology*, ed. Francesca Aran Murphy (Oxford: Oxford University Press, 2015), 139.

42. Louth, "Christology in the East from the Council of Chalcedon to John Damascene," 139–140.

43. G. L. Prestige, *Fathers and Heretics* (New York: Macmillan, 1940), 239.

44. See J. N. D. Kelly, *Early Christian Doctrines* (London: Adam and Charles Black Limited, 1958), 340–342.

45. Beeley, *The Unity of Christ*, 284.

46. Prestige, *Fathers and Heretics*, 269.

47. Sergius the Grammarian, for example, in *Christology After Chalcedon: Severus of Antioch and Sergius the Monophysite*, ed. Iain Torrance (Norwich, UK: The Canterbury Press Norwich, 1988), 144.

48. Fred Sanders, "Introduction to Christology: Chalcedonian Categories for the Gospel Narrative," in *Jesus in Trinitarian Perspective*, eds. Fred Sanders and Klaus Issler (Nashville, TN: B&H Publishing Group, 2007), 27–35.

49. Beeley, *The Unity of Christ*, 294.

50. Richard Price, *The Acts of the Council of Constantinople of 553 with Related Texts on the Three Chapters Controversy*, vol. 1 (Liverpool, UK: Liverpool University Press, 2009), 73. Fred Sanders, "Introduction to Christology: Chalcedonian Categories for the Gospel Narrative," 30.

51. Demetrios Barthrellos, *Byzantine Christ: Person, Nature, and Will in the Christology of Saint Maximus the Confessor* (Oxford: Oxford University Press, 2005), 34–35.

52. Jaroslav Pelikan, *The Christian Tradition*, vol. 2 (Chicago: University of Chicago Press, 1974), 84–85, 88–89.

53. Thomas F. Torrance, *Incarnation: The Person and Life of Christ* (Downers Grove, IL: InterVarsity Press, 2008), 84. Crisp, *Divinity and Humanity*, 74.

54. David Brown, *Divine Humanity: Kenosis and the Construction of a Christian Theology* (Waco, TX: Baylor University Press, 2011), 24.

55. Fred Sanders, "Introduction to Christology: Chalcedonian Categories for the Gospel Narrative," 30–35.

56. Pannenberg, *Jesus—God and Man* (Philadelphia: The Westminster Press, 1964), 338.

57. Thomas P. Flint, "The Possibilities of Incarnation: Some Radical Molinist Suggestions," *Religious Studies* 37 (2001): 307–320.

58. R. T. Mullins, "Flint's 'Molinism and the Incarnation' is Too Radical," *Journal of Analytic Theology* 3 (2015): 109–123. R.T. Mullins, "Flint's 'Molinism and the Incarnation' is Still Too Radical—A Rejoinder to Flint," *Journal of Analytic Theology* (forthcoming).

59. Crisp, "Compositional Christology without Nestorianism," *The Metaphysics of the Incarnation*, 59. Cf. Peter Lombard, *The Sentences*, Book 3, Dist. 2, 3.

60. Theodore of Mopsuestia, in Norris, *Christological Controversies*, 117.

61. For a deeper examination of this problem for a three-part dyothelite Christology, see my *The End of the Timeless God*, chapter 7.

62. Andrew Ter Ern Loke, *A Kryptic Model of the Incarnation* (Farnham, UK: Ashgate, 2014).

63. Loke, *A Kryptic Model*, 65.

64. Loke, *A Kryptic Model*, 65.

65. Loke, *A Kryptic Model*, 66.

66. Loke, *A Kryptic Model*, 66.

67. Loke, *A Kryptic Model*, 69.

68. Loke, *A Kryptic Model*, 69.

69. Loke, *A Kryptic Model*, 69.

70. Loke, *A Kryptic Model*, 69.

71. Loke, *A Kryptic Model*, 70–71.

72. Cf. Oliver Crisp, *God Incarnate: Explorations in Christology* (New York: T&T Clark International, 2009), chapter 7.

73. The physicalist Glenn Andrew Peoples concedes this point. "The Mortal God: Materialism and Christology," in *The Ashgate Research Companion to Theological Anthropology*, eds. Joshua R. Farris and Charles Taliaferro (Farnham, UK: Ashgate, 2015), 338. Though, it should be noted that Peoples's physicalist account of the incarnation has some differences from Merricks's, but on this issue, they will agree: the human nature of Christ would be a fully functioning human person apart from the incarnation.

SUGGESTED FURTHER READING

Beeley, Christopher A. *The Unity of Christ: Continuity and Conflict in Patristic Tradition*. New Haven, CT: Yale University Press, 2012.

Crisp, Oliver. *Divinity and Humanity: The Incarnation Reconsidered*. New York: Cambridge University Press, 2007.

Le Poidevin, Robin. "The Incarnation: Divine Embodiment and the Divided Mind." *Royal Institute of Philosophy Supplement* 68 (2011): 269–285.

Loke, Andrew Ter Ern. *A Kryptic Model of the Incarnation.* Farham, UK: Ashgate, 2014.

Marmodoro, Anna, and Jonathan Hill, eds. *The Metaphysics of the Incarnation.* Oxford, UK: Oxford University Press, 2011.

Merricks, Trenton. "The Word Made Flesh: Dualism, Physicalism, and the Incarnation." In *Persons: Human and Divine*, edited by Peter van Inwagen and Dean Zimmerman. Oxford, UK: Oxford University Press, 2007.

Mullins, R. T. "Flint's 'Molinism and the Incarnation' is Too Radical." *Journal of Analytic Theology* 3 (2015): 1–15.

Murphy, Francesca Aran, ed. *The Oxford Handbook of Christology.* Oxford, UK: Oxford University Press, 2015.

Norris, Richard. *The Christological Controversy.* Philadelphia: Fortress Press, 1980.

Price, Richard. *The Acts of the Council of Constantinople of 553 with Related Texts on the Three Chapters Controversy*, vols. 1 and 2. Liverpool, UK: Liverpool University Press, 2009.

Sanders, Fred, and Klaus Issler. *Jesus in Trinitarian Perspective.* Nashville, TN: B&H Publishing Group, 2007.

Chapter 9

Physicalism and the Death of Christ

Charles Taliaferro

"God is dead." Such a claim has been made by various philosophers to broadcast shockingly what they believed to be the erosion of theism in European culture, but *what is even more shocking* is the classical, enduring Christian belief that Jesus of Nazareth, fully God and fully human, actually died—*literally*. It is not surprising that there are dynamic changes in the history of religion and culture; religious traditions flourish, sometimes atrophy, and then die out or undergo revitalization, reformation, or revival. But how could it be that Jesus as God (the second Person in the imperishable, everlasting, uncreated Triune Godhead) die? The phrase "God is dead," as used by Hegel and Nietzsche (among others), describes a comparatively prosaic, nonmysterious shift in Europe compared to the utterly mysterious, bold, confident Christian claim that Jesus qua God and man died.

Other chapters in this volume address physicalism and the life of Jesus Christ. It is impossible, however, to address the death of Christ in this chapter without *some* reflection on His life. So, with an apology for any overlap, this chapter begins with contrasting physicalist and dualist accounts of the incarnate life of Jesus of Nazareth. Section two compares physicalist and dualist accounts of the death of human persons. Section three examines the advantage of dualism in addressing a traditional belief about the death of Jesus. I propose that dualists are in a better position in accounting for belief in the harrowing of hell (the descent of Christ into hell or the *Descensus Christi ad inferos*), as affirmed in the Apostles' Creed with some scriptural support (1 Peter 3:19–20 and Eph. 4:9). Reasons are advanced as to why this traditional belief may be both credible and convey theologically a vital Christian truth: God's loving, redeeming power is stronger than death.

THE METAPHYSICS OF GOD INCARNATE

There are many forms of physicalism today. For present purposes, I assume that the most radical version of physicalism—one that denies the existence of mental states (conscious awareness, subjective experience, beliefs, and desires)—is not only in tension with the evident fact of consciousness, it is so profoundly dismissive of the Christian worldview that it can offer little constructive material for Christian philosophers and theologians.[1] That said, one might radically reinterpret Christianity as advancing a way of life or a set of powerful metaphors to live by, without any metaphysical commitments (that is, commitments to the reality of, say, God). But if we keep with traditional Christian teaching, more moderate forms of physicalism are preferable, such as those that acknowledge the reality of consciousness, beliefs, desires, and so on, but identify these with bodily states, processes, or properties that we have as physical, living human animals.

One of the challenges facing nonreductive or moderate physicalism is how to square the belief that Jesus of Nazareth (the physical, embodied person or animal) can be in any way identical with the incorporeal second Person of the Trinity. If you accept the idea that the second Person of the Trinity exists prior to the existence of the body of Jesus of Nazareth, it seems that you are committed to there being two distinct realities (the second Person of the Trinity and the body of Jesus). According to what philosophers refer to as Leibniz's law or the law of the indiscernibility of identicals, identity relations entail that if A is B, then whatever is true of A is true of B. If it is true that Jesus of Nazareth (the physical animal or his animal body) came into being in the first century, and Jesus of Nazareth is the second Person of the Trinity, then the second Person of the Trinity came into being in the first century. Arguably, this would be preposterous. Such an outcome may well motivate physicalist Christians to adopt a "lower Christology." Some, for example, have contended that the divinity of Jesus of Nazareth should be understood as the affective, conscious identity of Jesus and God. In this chapter, though, I will be primarily addressing the high Christology found in the Chalcedonian Creed.

Does substance dualism do any better? I think the answer is "yes," but before making that point, allow me to offer a brief account of what I believe to be the most plausible form of substance dualism.

Critics long have proposed that substance dualism offers us an excessively bifurcated understanding of the soul (or mind or person) and the body. Dualists are caricatured as supposing that the soul is a ghost inhabiting its body. If dualism is true, you never observe persons, only their containers. The soul is actually some immaterial, ethereal mind attached to the body. On the contrary, I defend a form of *integrative dualism* according to which in normal, healthy conditions the person and body function as a singular reality.[2]

To see me and to heal my wound and for us to share a meal is for us to interact without making any metaphysical distinction between person and body. Integrative dualism rejects the caricatures of dualism one finds in, for example, the following portrayal of dualism by Peter van Inwagen:

> If dualism is true, our relation to our bodies is analogous to the relation of the operator of a remotely controlled device . . . to that device. Now consider Alfred, who is operating a model airplane by remote control. Suppose that something . . . strikes a heavy blow to the model in midair . . . the blow will have no effect on Alfred, or no effect beyond his becoming aware of its effects on the performance of the model and his ability to control it.[3]

The above scenario is perhaps intended to be comic, but as a serious engagement with contemporary dualism it is not successful. Dualists do not posit anything like a remote relation between person and body. To shoot my body is to shoot me or, to use a more pleasant example, to kiss my face, is to kiss me. The very idea of the soul as Alfred, standing some distance away from his controlled device might accurately be an illuminating way to describe the way some persons are able to emotionally "distance" themselves from their bodily states (for example, engaging in "mind control" to ignore or to mentally block wounds that would normally create great pain), but this is not the normal state of affairs. When I am wounded, the feelings involved are not "remote." My body is not even vaguely like a remote model airplane; it is more like a pilot in an actual airplane, struggling to control her thinking and reactions to having her airplane being struck by antiaircraft fire.

I have argued for dualism elsewhere involving two lines of reasoning, both arguments involving cases wherein the unity of person and body comes apart. In one argument, often called the "knowledge argument," if a person is identical with their body, then to know all about their body would be to know all about them. But if we restrict ourselves to what we can know about their body to only physical properties (anatomy, the nervous system, the brain, and so on), we would only know about a person's conscious states if we could correlate what we know physically with what the subject reports about their experiences.[4] The argument employs Leibniz's law and I refer you elsewhere for its development, the formulation of objections, and replies. The second is a modal argument, as it refers to different possibilities or modes that dualism, but not physicalism, can account for. I have argued for the reasonability of believing that persons can exist without their physical bodies, and vice versa. One's physical body cannot exist without itself, so reasonably believing in the bare possibility of a person existing without their body (surviving death, for example) would be a reason to believe in the nonidentity of person and body. Like the knowledge argument, this one faces lots of objections, which I have addressed elsewhere.[5] Let us return to the topic of the incarnation.

Because dualists uphold the strict nonidentity of person and body, they are not committed to believing that the second Person of the Trinity is numerically identical with the body of Jesus of Nazareth. Nor do they believe that, strictly speaking, the person Jesus is numerically identical with his body. What is open to dualists is to contend that in the incarnation, the second Person of the Trinity became embodied as a finite human being. Embodiment involves coming to be so causally sustained and shaped by a human body, that the person and boy function as a single being. In the incarnation, the person Jesus breathes with his lung, feels by means of his nervous system and brain, is nourished by food, sees with his eyes, hears with ears, smells with his nose, retains his balance with proprioception, and undergoes events and acts in and as the functional unity of Jesus qua embodied person. Contra van Inwagen, this embodiment involves taking on fully the many ways in which our mental states are intertwined with our physical sates (my intentionally shaking your hand is a single, unified event) and vice versa (a blow to my head is an injury to me as a conscious, embodied person, and my abusing alcohol is my abusing the integrity of my embodiment). Contrast an integrative dualist portrait of the incarnation with what van Inwagen proposes dualists offer us.

> What effects should dualism lead us to expect from a blow to the body? . . . The blow at the base of Alfred's skull that in fact produces unconsciousness should, according to dualism, produce the following effect on Alfred: he experiences a sharp pain at the base of his skull; he then notes that his body is falling to the floor and that it no longer responds to his will; his visual sensations and the pain at the base of his skull and all of the other sensations he has been experiencing fade away; and he is left, as it were, floating in darkness, isolated, but fully conscious and able to contemplate his isolated situation. . . . Here is another wrong prediction: if dualism were correct, we should expect that the ingestion of large quantities of alcohol would result in a partial or complete loss of motor control but leave the mind clear.[6]

This portrait of dualism seems wide of the mark with respect to ourselves, and would leave us with a fragmentary view of the incarnation. Why wouldn't a dualist allow that Alfred falls to the floor, rather than Alfred simply noting "that his body is falling to the floor"? Why wouldn't the dualist refer to Alfred as being motionless, rather than observing that his body no longer is responsive to his will? Why expect that, if dualism is true, Alfred would have serene (floating?) consciousness after a head trauma? And why on earth would dualists expect alcohol not to cloud or even destroy a person's conscious life? To put the integrative dualist position in succinct terms: integrative dualism understands you and I, and Jesus of Nazareth, as profoundly different from Alfred.

Regarding the incarnation, I accept what may be called the *life within a life model* of the incarnation. Our lives can be highly complex and conflicted; we can play different roles separately or simultaneously; we can sometimes dedicate ourselves to living one life within a greater overall life. For example, as a professor I may have a niece as a student. As Uncle Charles I have known her and loved her from birth, but as her professor, I need to see, act, and be with her as one among, say, a hundred students whom I care for passionately but equally and not showing my niece any preferential regard. My life as a professor would be taking place within my broader life as Charles. This may seem like "role playing," but to see that this kind of life within a life can be not a kind of game, consider severe cases in which there are very real, perhaps even life-threatening conditions. Imagine, as a dedicated physician you are the only one who can aid a patient whom you personally detest; he betrayed your friendship and was cruel to those you love. You are, however, the only one in a hospital who can save him from certain death from wounds inflicted by criminals. We can imagine how in those circumstances, you might well feel (and be) obliged to concentrate solely on rescuing him. While not denying your broader understanding of his character and history, you may well (admirably) set it to one side to do what you are bound to do as a doctor. Cases like that can be multiplied in which the life within one's life actually costs you your life. Imagine a similar case in which you, as a firefighter, are ethically obliged to rescue a person in a fire and die in the process, even though you (as a person) believe the person you rescued is deplorable from a moral point of view. According to the life within a life model, God the Son as not-incarnate retains the divine attributes of omniscience, omnipotence, omnipresence, necessary existence, essential goodness, and yet dedicates Its life to become *the One who lives the life of Jesus of Nazareth*, living under conditions of ignorance yet with the ability to grow in knowledge and awareness; limited power yet with the ability to grow in efficacy; and living in functional unity with a contingent body with all the dignity and frailty that involves.[7] The life within a life model does not require dualism; it is open to some Christian physicalists, but it comports well with a dualist anthropology.

Does this dualist account of the incarnation fall short of what traditional Christians believe about the authenticity and fullness of the humanity of Jesus Christ? After all, when I finish teaching as a professor I can step away from that role and resume my personal life, whereas you and I cannot step away from our role as being the human beings that we are. I suggest that one of the ways to flesh out the dualist understanding of the fullness of the humanity of Jesus Christ is to take note that the incarnation is so thorough that not only did Christ experience the agonizing suffering involved in the tortuous flogging and crucifixion, but he actually died. Arguably Christ could have stepped away from this passion and death (as we see in the narrative of the

Garden of Gethsemane), but He did not. One way to stress the robust nature of the incarnation, from a dualist perspective, is to once again contrast it with van Inwagen's Alfred. The New Testament gives no hint that the second Person of the Trinity would be like Alfred controlling his body as though it were some remote model airplane. The blows to Jesus's body are naturally understood to be blows to him. On the cross, when Jesus is given sour wine to drink, we are to imagine this to involve as real a bodily exchange as any we undergo.

Before turning to the death of Christ, let us reflect on the philosophy of the death of human persons and the prospects of physicalism and dualism. While this first section raised a prima facie problem for physicalism from the standpoint of traditional Christology, I shall not assume it is decisive at this stage of our inquiry. Apart from adopting a lower Christology, another option for the physicalist would be to adopt Lynn Baker's constitutional account of persons.[8] On her view, you are not identical with your physical body, but you are constituted by it. Identity relations are symmetrical (if A is B, B is A), but the constitution relationship is not (if A is constituted by B, it does not follow that B is constituted by A). If the statue of David is constituted by (that is, the statue is made up of) marble, it does not follow that the marble is made up of the statue of David. So, going back to the problem facing physicalism, on the constitution account, you may hold that the second Person of the Trinity came to be constituted by the living animal body which comes to form the unified, embodied person, Jesus of Nazareth. I only mention this as a possible position available to physicalists. I personally think that Baker's account, when fully worked out, entails (or, as it were, teeters on the brink of) what I am referring to as *integrative dualism*, but that is a matter for another occasion.

THE DEATH OF HUMAN PERSONS

In approaching this topic we confront a major difference between some (but not all) physicalists and some (but not all) dualists. Some physicalists contend that we have a problem-free understanding of what is physical. We know that animals, plants, and planets are physical. The physical world is what is revealed in the physical sciences. I am not so sure these claims are straightforward or survive close inspection. What is it to be an animal? When it comes to human animals, should we count ideas, thoughts, feelings, tastes, and sounds as just as recognizable as cells? When you examine a human's brain, do you thereby (ipso facto) examine a person's thinking or is the thinking some additional activity? As for the sciences, I propose that you cannot have any physical science, or any science at all, without scientists. And scientists are individual persons, subjects of experience who reason, think, devise theories, carry out experiments, argue with each other in laboratories and conferences, and so on. These involve highly complex, purposive, mental or psychological intentions.[9] Rather than seek to develop a lucid or strict demarcation of what should count as physical, I suggest a different move, one that does not beg the question against physicalism but one that does give some initial advantage to dualism. I propose that whether or not we are thoroughly physical, we should recognize that what we can be most sure about is the reality of ourselves as individual persons who think, reason, have ideas, feel pleasure and pain, act, speak, and so on. This is sometimes referred to as *the first-person point of view*. It is perhaps most vividly brought to light when you contemplate that one day, you will die. This is not just a realization that someone or other will die or an awareness that some animal somewhere will die, or in my case, a realization that Charles Taliaferro will die, but a realization that *I myself will die*.[10]

Many physicalists do not recognize the primacy of the first-person point of view. For example, some animalists (philosophers who believe that human persons are identical with animals) propose framing the key move in philosophy of mind with two equally credible assertions: the animal occupying this chair is thinking. I am thinking. They then go on to raise this question: am I identical with that animal? The very idea that there might be two persons (myself as well as this animal) occupying the same space seems preposterous, thus paving the way for the more plausible belief that all persons (myself included) are identical with their animal bodies. An important reply to this strategy is to question whether we can have as clear a conception of *the animal occupying this chair is thinking* as the conception that *I am thinking*. For reasons suggested in section one on the knowledge argument, it is quite open to question whether it is reasonable to believe that thinking itself is a physical property. After all, we cannot observe a subject's thinking in any way akin to the way in which we can observe any of an animal's brain, anatomical processes, activities. I would also go further in claiming that we would have no idea how to conceive of "this animal thinking" without a prior concept of who we are as self-aware subjects who endure over time, have perception, sensations, and so on. When we use the indexical "this" or "that," we essentially mean the thing I, as a subject, am drawing (your) attention to. I suggest that a more philosophically sound first step in forming a philosophy of mind is to take note of the primacy of the first-person and the mental: we are self-aware thinking subjects who endure over time, have experiences, and so on. We can be certain, too, that we have ideas such as the idea of being an animal, the concept of brains, cells, bodies, plants, sensations, and so on. This, then, is a further question: Are you and I and are thinking, experiencing, and so on, the very same thing as our brains, their activities, our body as a whole, or not?

In any case, let us consider what is involved with a person dying or being dead. For both dualists and physicalists, I assume that there would be little dispute that a person has died when the person has undergone what may be referred to as D: a complete cessation of all brain functions and the complete cessation of respiratory and circulatory functions, none of which can be reversed or restored given the (current) laws of nature. I believe D adequately describes a case of what may be thought of as the biological death of a human person, without having to address borderline cases (what about persons who experience cessation of brain activity but have other biological functions such as breathing?). So, D does not advance as an analysis of the concept of death, but as only one, perhaps paradigmatic case of when it is (relatively) uncontroversial to claim that a human person is dead. So, because D is not an analysis of what counts as a person's death, one may readily allow that a person may be dead even if his conditions could be reversed but simply are not. Note that D does not explicitly refer to consciousness. Presumably, the reason why many of us give primacy to cessation of brain activity is that this is the sign of the loss of consciousness, but D does not explicitly require that the person who has died has ceased to be conscious; indeed, in the next section, I suggest that the traditional teaching about the harrowing of hell suggests that Christ survived as a conscious subject following his biological death.

We come now to what I suggest is a problem for at least some forms of physicalism when it comes to a philosophy of death, which has been called the Corpse Objection. Consider the following challenge. Hypothesis: you are the very same thing as your animal body. When you die, you cease to be. If you are identical with your animal body, your body would cease to be when you ceased to be. But it does not. Your body (except under conditions when your death involves your body being annihilated) persists in being. Yes the body is no longer living, and we may not even refer to it as *you* or as *a person*. That is, we might prefer to say, "we buried Skippy's corpse" or "the remains of Skippy" as opposed to saying, "we buried Skippy." And yet, if animalism (the view that you are a human animal) is true, the animal still exists. Dualism does not face this problem, for it does not hold that persons are either identical to or a mode of an animal body. Because dualists hold that there is a metaphysical distinction between person and body, they hold that (strictly speaking) a person is neither identical with her living or dead body, though at death dualists claim that there is a radical ceasing of the functional, healthy unity of person and body.

Consider two ways physicalists may reply to the Corpse Objection.

Persons are modes of their animal bodies: one might concede that persons are actually modes of their animal bodies. On this view, your animal body (at conception, say) was not you, but it became you (when, for example, the organism reached a unified, functioning state), and at death will cease to be you.

The difficulty with this approach is that *modes or phases of things do not think, feel, experience, act, intend things, and so on.* Thinking and the like are done by substantial individuals, not the phases or modes of life in which they pass through or obtain. Right now, I am living a phase of my life or mode of being in which I am a professor. But my being a professor does not teach, write, lead discussion groups, and so on. That is what I do as a professor.

Consider a second, more promising alternative. There is a version of hylemorphic animalism, according to which you as the living, physical animal are a different substance from your corpse. Patrick Toner has constructed this inventive, promising alternative. On his view, when you are alive, you are a unified substance that has no parts as substances.

> The hylemorphic account of substance tells us that any substance will have one and only one substantial form. This means that substances like you and me are not made out of littler substances like cells or atoms. It does not mean that you and I have no cells or atoms as parts. Sure we do. They're just not substances. We have no substances as parts: that doesn't mean we have no parts. It means, rather, that if we have cells or atoms as parts, those things are not substances while they're our parts. They may very well be substances at times at which they are not our parts, however.[11]

The reasoning is, then, that when you are alive, you are a substance that has no substances as parts; your corpse does have substances as parts; therefore your corpse is a different substance than your living body. Toner sees this as not at all counter-intuitive or embarrassing but an implication of his view.

> The hylemorphic animalist's response is simple. When we die, a new body (or new bodies) does come into existence. That's part of what it means for something to die. For death is a substantial change. In a substantial change, one substance goes out of existence, and one or more new substances come into existence. Neither the body itself, nor any of its atomic parts, existed while the animal was alive. This just follows from the account of substance I've given, according to which substances have no substances as parts—there is only one substance here in my boundaries, and it's an animal. When the animal dies, whatever is left over is not the same thing that was there before. This answer to the corpse problem simply falls out of hylemorphic animalism. It's not a bullet we have to bite.[12]

Toner rightly notes that what many of us see as death is a substantial change. The difference between a substantial and accidental change is that in a substantial change, you lose a substance or substantive individual, whereas individuals persist through accidental change, as when I cease to be a philosopher and become a circus clown. I think there is an admirable plausibility in Toner's depiction of the integration of embodiment. When healthy, we identify our bodily parts as forming a functioning, whole organism, whereas after we have died, our hearts, lungs, and so on, no longer function as hearts and lungs. Changing the example to farm animals, when a pig is slaughtered we often do regard its detachable, bodily parts as something other than the animal from which they came (we have pork, bratwurst, and so on).

While I believe Toner's position is ingenious and, if the overall case for hylemorphism is plausible, the preferred solution to the Corpse Objection, I still think it does not outweigh the positive case for dualism and its more common-sense approach to the corpse, which is to claim that it is the very same thing as one's body, only it is dead. Consider "Romeo and Juliet," act 5, scene 3. The scene is complicated, for Romeo thinks Juliet is dead, whereas she is not. But it would be very odd for Romeo to lament and kiss Juliet's dead body if he believed it to be a substance other than the body of his beloved when she was alive.

How oft when men are at the point of death
Have they been merry, which their keepers call
A lightning before death! Oh, how may I
Call this a lightning?—O my love, my wife!
Death, that hath sucked the honey of thy breath,
Hath had no power yet upon thy beauty.
Thou art not conquered. Beauty's ensign yet
Is crimson in thy lips and in thy cheeks,
And death's pale flag is not advancèd there.—
Tybalt, liest thou there in thy bloody sheet?
O, what more favor can I do to thee,
Than with that hand that cut thy youth in twain
To sunder his that was thine enemy?
Forgive me, cousin.—Ah, dear Juliet,
Why art thou yet so fair? Shall I believe
That unsubstantial death is amorous,
And that the lean abhorrèd monster keeps
Thee here in dark to be his paramour?
For fear of that, I still will stay with thee,
And never from this palace of dim night
Depart again. Here, here will I remain
With worms that are thy chamber maids. Oh, here
Will I set up my everlasting rest,
And shake the yoke of inauspicious stars
From this world-wearied flesh. Eyes, look your last.
Arms, take your last embrace. And, lips, O you
The doors of breath, seal with a righteous kiss
A dateless bargain to engrossing death.
(kisses JULIET, *takes out the poison)*[13]

The tragedy has its poignancy, I suggest, from Romeo's mourning over what he takes to be the body of his beloved, not a new substance. I do not advance Romeo as a philosopher; in fact he seems rather exasperated at the limitations of philosophy (he exclaims, "Hang up philosophy!" in act 3, scene 3). But I suggest he does display what I think is a common-sense approach to the body as enduring as the self-same substance after the death of a person.

I will not press the point further here. These first two sections may be read as preliminary, raising prima facie difficulties for physicalism, but not decisive, and laying the ground work for the alternative dualist account of ourselves and the incarnation. I now turn to the main topic of this chapter.

THE DEATH OF JESUS CHRIST

At this stage, let us assume (if only for the sake of argument) that dualism and physicalism are equally well placed in terms of philosophical and theological integrity and credibility. Both can account for why the passion, tormenting, scourging, and crucifixion led to the death of Jesus Christ. Both can account for the interning of the body in a tomb and its remaining there from the Sabbath to the first Easter morning when Christ rises from the dead (Luke 23:50–56). But can they both account for belief in the harrowing of hell?

Traditionally, many (but not all) Christians believe that between the crucifixion and the Resurrection, Christ descends into hell (or into Sheol or Hades) to preach to and free "imprisoned spirits" (I Peter 3:19–20). The significance of this belief is supported by its being referenced in our records of the first preaching by the apostles (Acts 2:24, 27, 31). According to the *Catechism of the Catholic Church*, part one, section two:

> The frequent New Testament affirmations that Jesus was "raised from the dead" presuppose that the crucified one sojourned in the realm of the dead prior to his resurrection. This was the first meaning given in the apostolic preaching to Christ's descent into hell: that Jesus, like all men, experienced death and in his soul joined the others in the realm of the dead. But he descended there as Savior, proclaiming the Good News to the spirits imprisoned there. Scripture calls the abode of the dead, to which the dead Christ went down, "hell"—*Sheol* in Hebrew or *Hades* in Greek—because those who are there are deprived of the vision of God. Such is the case for all the dead, whether evil or righteous, while they await the Redeemer: which does not mean that their lot is identical, as Jesus shows through the parable of the poor man Lazarus who was received into "Abraham's bosom": "It is precisely these holy souls, who awaited their Savior in Abraham's bosom, whom Christ the Lord delivered when he descended into hell." Jesus did not descend into hell to deliver the damned, nor to destroy the hell of damnation, but to free the just who had gone before him. "The gospel

> was preached even to the dead." The descent into hell brings the Gospel message of salvation to complete fulfillment. This is the last phase of Jesus' messianic mission, a phase which is condensed in time but vast in its real significance: the spread of Christ's redemptive work to all men of all times and all places, for all who are saved have been made sharers in the redemption.[14]

The belief in this descent among the dead has been powerfully represented in Christian iconography (in Resurrection icons used in the Orthodox Church, when Christ rises from the dead, He brings with Him Adam, Eve, Moses, and others) and in the Christian imagination (Charles Williams's novel, *The Descent Into Hell*).

Belief in the harrowing of hell is a belief in the awesome power of God to overcome the enthralling, tyrannical power of death. It essentially professes that after the death of Christ, Christ himself, prior to the Resurrection, shared with deceased human persons the good news of His victory. It implies that the saving Redeemer of the world was conscious of his immanent victory over death and actively sought the lost. This seems to me a powerful testimony of the limitless power of God's redemption through the life, teaching, and even during the death of Christ. It comports well with various scriptural passages of how God's love is more powerful than death (Rom. 8:38) and that even in hell or Sheol, we cannot escape God's love. "You will not abandon my soul to Sheol" (Ps. 16:10).

Physicalists might well challenge whether contemporary Christians can or should accept the descent into hell narrative historically. Why not treat the belief figuratively or as a metaphor or parable (like the parable of the Good Samaritan) or as a dramatic expression of the awesome love of God? After all, we do not feel constrained to treat the dragon, beasts, and horsemen of the Apocalypse realistically or as purported future stages of history involving a real dragon. Besides, the language of descent into hell, like the language of ascent into heaven, suggests a three-storied universe that we know to be false. The idea that hell is literally below us, heaven above (among or beyond the clouds?), while we are in the middle is not an idea that is acceptable to modern, educated Christians.

Granted that the notion of "descent" (as in "descent into hell") is a metaphor, just as it would be a metaphor when someone claims that after the divorce, *Smith descended into reckless relationships and drug use*. This need not involve a literal, spatial lowering of himself; we might even picture his moral and spiritual descent while he (literally) is a passenger on a soaring jet. But I do not think that modern, educated Christians need to treat as figurative belief in life after this life of persons. There are ample defenses of the coherence of life after life of persons, especially from a dualist perspective.[15] And for those of us who deny the unity of space (there can be spatial objects that are not spatially related to our world), it is not absurd to suppose there can be indefinitely spatial worlds where there may be persons in a myriad of sites.[16] A Christian who believes God is omnipotent has a reason to think that such worlds are not beyond the limits of God's power to create, sustain, and to providentially act within them. So, I do not think that the positing of such worlds involves a metaphysical impossibility.

If we know of no reason why Christ could not act redemptively between the crucifixion and Resurrection, then it seems that scriptural (and perhaps creedal) testimony that He did should carry the day. It may also count as additional reason for accepting it on the grounds that such a redemptive act is something we would expect of the person Jesus Christ as depicted in the New Testament accounts of His life. Scriptural references to Christ's descent seem very different from the dragons and beasts in Christian apocalyptic literature, which scholars have interpreted as representing different empires such as the Babylonian, Persian, and Roman Empires. If, on the other hand, we take the narrative of Jesus Christ harrowing hell as a real event, we may see it as an actual (not merely metaphorical) display of divine love. It would teach us that not even death can conquer Jesus Christ, the Incarnate Lord of life.

Thomas Aquinas may be interpreted as depicting the death of Jesus Christ as a severing of his soul and body that then come to be reintegrated at the Resurrection; he thereby sees the descent into hell as itself an integral part of Christ's redeeming work (taking on the penalty for sin, death, freeing captives).

> As Christ, in order to take our penalties upon Himself, willed His body to be laid in the tomb, so likewise He willed His soul to descend into hell. But the body lay in the tomb for a day and two nights, so as to demonstrate the truth of His death. Consequently, it is to be believed that His soul was in hell, in order that it might be brought back out of hell simultaneously with His body from the tomb.[17]

I propose that between dualism and physicalism, dualism is better placed to accept and give praise for this powerful, redeeming act. Note, too, how the dualist account would avoid the difficulties of thinking that the dead body of Christ is a new substance, different from his living body. On the hylemorphic account, discussed earlier, the living body would have ceased to be at death, a new substance come into being (the dead body), and a yet new (or recovered and transformed) body come into being at the Resurrection. The hylemorphic physicalist account would also face the challenge of supposing that anyone would be in hell (or another life) requiring deliverance (given that the human bodies of the dead are not resurrected by being dispersed throughout the earth). In all, I suggest that integrative dualism offers a less complex, refined philosophical framework within which to address the death of Jesus Christ.

NOTES

1. I address the problems facing eliminativism in *Consciousness and the Mind of God* (Cambridge, UK: Cambridge University Press, 1994). Also see *A Brief History of the Soul* by Stewart Goetz and Charles Taliaferro (Oxford: Blackwell, 2011).

2. See my *Consciousness and the Mind of God* and "The Virtues of Embodiment," *Philosophy* 76 (2001): 111–125.

3. Peter van Inwagen, *Metaphysics* (Boulder, CO: Westview Press, 1993), 179.

4. See "Substance Dualism" by Charles Taliaferro, *The Blackwell Companion to Substance Dualism,* ed. J. J. Loose, et al. (Oxford: Wiley-Blackwell, forthcoming).

5. See "Substance Dualism" and *Consciousness and the Mind of God.*

6. Van Inwagen, *Metaphysics*, 180.

7. The life within a life model is akin to the two minds model of T. V. Morris and Richard Swinburne.

8. See Lynn Baker, *Persons and Bodies* (Cambridge, UK: Cambridge University Press, 2000).

9. See *Contemporary Philosophical Theology* by Charles Taliaferro and Chad Meister (London: Routledge, 2016), chapter 1.

10. See Lynn Baker, *Naturalism and the First-Person Perspective* (Oxford: Oxford University Press, 2013).

11. Patrick Toner, "Hylomorphic Animalism," *Philosophical Studies* 155 (2011): 67.

12. Toner, "Hylomorphic Animalism," 71.

13. Shakespeare, "Romeo and Juliet," in *The New Cambridge Shakespeare*, ed. G. Blakemore Evans (New York: Cambridge University Press, 2003).

14. *Catechism of the Catholic Church* (Cincinnati, OH: Benziger Publishing Company, 1994), 164–165.

15. See William Hasker and Charles Taliaferro, "Afterlife," *The Stanford Encyclopedia of Philosophy* (Winter 2014 Edition), ed. Edward N. Zalta, https://plato.stanford.edu/archives/win2014/entries/afterlife/

16. See, for example, Richard Swinburne on space not being unified, *Space and Time* (New York: Palgrave Macmillan, 2014).

17. Thomas Aquinas, *Summa Theologica*, vol. IV, pt. III (New York: Cosimo Classics, 2007), 2299.

SUGGESTED FURTHER READING

Baker, Lynne Rudder. *Persons and Bodies*. New York: Cambridge University Press, 2000.

———. *Naturalism and the First-Person Perspective.* New York: Oxford University Press, 2013.

Goetz, Stewart, and Charles Taliaferro. *A Brief History of the Soul*. Hoboken, NJ: Wiley-Blackwell, 2011.

Taliaferro, Charles. *Consciousness and the Mind of God.* New York: Cambridge University Press, 1994.

Van Inwagen, Peter. *Metaphysics*. Boulder, CO: Westview Press, 1993.

Chapter 10

Christian Materialism Entails Pelagianism

Matthew J. Hart

Christian philosophy has not escaped the influence of the current materialist paradigm, and consequently many Christian philosophers are materialists. Peter van Inwagen,[1] Hud Hudson,[2] Lynne Rudder Baker,[3] Nancey Murphy,[4] and Trenton Merricks[5] are notable examples. In this chapter, I argue that Christian philosophers should not be materialists, because such a commitment is in tension with, among other things, the church's condemnation of Pelagianism and Semi-Pelagianism.

But what is materialism? Materialism might mean many things, but it nearly always at least means this: that the mental supervenes on the physical with a force that is at least as strong as that of nomological necessity. Now, if a Christian is a materialist, then he will of course be a *local* materialist: a materialist about human beings. The Christian cannot be a *global* materialist because, even if he is sufficiently liberal to reject the existence of angels and demons, he will nevertheless believe that there is at least one immaterial mind that operates quite independently of the physical world, namely God's.[6] So, we can formulate the Christian materialist's supervenience claim as follows:

(CM) Human mental properties supervene on human physical properties.

How is the supervenience relation to be understood? I think it must be at least as strong as this:

(N) For any human mental property M, it is nomologically necessary that if any human individual x has M at time t, then there exists a physical (subvenient) property P such that x has P at t, and it is nomologically impossible for any human individual to have P at a time and lack M at that time.[7]

Some materialists might claim that the connection between the physical and the mental is closer than that described in (N). They may make the stronger claim that the impossibility in question is metaphysical, not merely nomological. Call this claim (M).[8] It will not be necessary for me to deal with (M) explicitly, because any reason to reject (N) will also be a reason to reject (M), for (M) entails (N) (though the converse is false).

So Christian materialists will subscribe (at least) to (N) and also, I shall assume, adhere to Christian doctrine in a manner that is sufficiently conservative such that they will agree with the church's condemnation of Pelagianism and Semi-Pelagianism. It is therefore a commitment to (N) that I intend to show is in tension with the church's condemnation of those views. Along the way, I will also make two additional arguments: (i) that Christian materialism makes possible an unhealthy approach to evangelism, and (ii) that Christian materialism has untoward consequences for the security of the believer.

PELAGIANISM AND SEMI-PELAGIANISM

But what are Pelagianism and Semi-Pelagianism? Pelagianism was at its peak around the beginning of the fifth century AD. Warfield describes the "central and formative principle of Pelagianism" as "the assumption of the plenary ability of man; his ability to do all that righteousness can demand—to work out not only his own salvation, but also his own perfection."[9] This central claim that it is possible for a merely human being, of his own free will, by sheer dint of effort, to secure his own salvation, or to live a sinless life, led to two other claims from the Pelagians.

The first concerned *original sin.* It was supposed by many at that time that Adam's sin had so affected his progeny that they now labored under a great corruption of nature and weakness of will that amounted to an inability, relative to man's natural powers, to choose the good. Such a view was in obvious contradiction to Pelagian doctrine. Accordingly, they supposed that the effect Adam had on his posterity was not transmission of a corrupt nature, but merely that of a bad exemplar: "they denied that Adam's sin had any further effect on his posterity than might arise from his bad example."[10]

The second concerned the *nature of grace.* They denied that any supernatural grace from God was necessary for salvation or for holiness. Given their supposition of the full and complete ability of man to live righteously, it was a natural consequence that they supposed that whatever grace came from God to help man live a holy life came in the form of external and natural aids, such as Scripture, the encouragement of the brethren, and the supreme example of the holy life of the Lord Jesus Christ. As Warfield puts it, "they meant by 'grace' the primal endowment of man with free will, and subsequent aid given him in order to its proper use by the revelation of the law and the teaching of the gospel, and, above all, by the forgiveness of past sins in Christ and by Christ's holy example. Anything beyond this external help they utterly denied."[11] There was no need for any internal renovation, no need for any supernatural rolling back of the native darkness and depravity of the human mind—once the relevant information was understood, the natural capacities of the person were by themselves quite up to the task of holy living. According to Warfield, "Pelagius consistently denied both the need and the reality of divine grace in the sense of an inward help (and especially of a prevenient help) to man's weakness," instead holding that "man has no need of supernatural assistance in his striving to obey righteousness."[12]

Pelagianism was condemned at the Council of Carthage in 418 AD and at the Council of Ephesus in 431 AD.

Semi-Pelagianism then sprang up. Semi-Pelagianism is the view that although it is not possible for a man, of his own natural powers and without divine aid, either to secure his salvation or to live a sinless life, yet it is possible for a man to *initiate* the process of faith by virtue of his natural powers alone, and in this manner to appropriate the divine aid necessary for living a holy life. In other words, although the sin of Adam did damage his progeny, this damage was not so great that it prevented them from calling out to God of their own power.

Semi-Pelagianism was condemned by the Second Council of Orange in 529 AD.

I wish to focus on the condemnations of this council, taking them as normative in what follows. It is meet to do so, since the council carries an ecumenical appeal: Roman Catholics appeal to it in their rejection of Semi-Pelagianism (Pope Boniface II gave his papal commendation of the council);[13] and the Reformed approve of it also, perceiving in the council's declarations a strong doctrine of original sin, human inability, and the necessity of grace that antedated their own later emphasis on these matters. As we shall see, the canons of the Second Council of Orange are extensive in their claims about mankind's native inability to do good and be righteous.

Now, it is true that I intend to show that Christian materialism does not merely imply Semi-Pelagianism, but full-blown Pelagianism, and therefore more needs to be done than to show that Christian materialism is at odds with the council. But it should become plain when my arguments are presented that they commit the Christian materialist to Pelagianism and not merely Semi-Pelagianism.

Here follows a representative sample of the declarations of this council. Canons 3–5 insist that grace from God must precede praying, willing, and the initial stages of having faith. Canon 6 tries to leave no room for escape, broadening matters by listing yet more items that cannot occur without God's help.

> CANON 6. If anyone says that God has mercy upon us when, apart from his grace, we believe, will, desire, strive, labor, pray, watch, study, seek, ask, or knock, but does not confess that it is by the infusion and inspiration of the Holy Spirit within us that we have the faith, the will, or the strength to do all these things as we ought; or if anyone makes the assistance of grace depend on the humility or obedience of man and does not agree that it is a gift of grace itself that we are obedient and humble, he contradicts the Apostle.

In case that list was not broad enough, canon 7 generalizes the point to concern anything that relates in a positive way to salvation:

> CANON 7. If anyone affirms that we can form any right opinion or make any right choice which relates to the salvation of eternal life, as is expedient for us, or that we can be saved, that is, assent to the preaching of the gospel through our natural powers without the illumination and inspiration of the Holy Spirit, who makes all men gladly assent to and believe in the truth, he is led astray by a heretical spirit, and does not understand the voice of God.

Finally, here is a brief excerpt from the conclusion of the council:

> The sin of the first man has so impaired and weakened free will that no one thereafter can either love God as he ought or believe in God or do good for God's sake, unless the grace of divine mercy has preceded him.[14]

So, according to the Second Council of Orange, whatever else may lie within man's natural powers, the power to seek after salvation is not one of them. For that, supernatural intervention is required, as canons 6 and 7 show with their emphasis on the necessity of the Holy Spirit's prior working. The council also claims this inability is a consequence of "the sin of the first man," and in this way, affirms a strong view of original sin.

But how can we summarize the central contention of this council? A good stab at that might be what I shall call INABILITY:

> INABILITY: All good acts that relate to salvation, and all good mental states relating to salvation, are beyond man's natural powers (post-Eden) to perform or bring about—supernatural aid is required for such things.[15]

This rules out both Semi-Pelagianism, which we might characterize as being the denial of INABILITY, namely ABILITY. It also rules out Pelagianism, which we might characterize as implying FULL ABILITY.

> ABILITY: It is not the case that all good acts that relate to salvation, and all good mental states relating to salvation, are beyond man's natural powers

(post-Eden) to perform or bring about—supernatural aid is not required for at least one such thing.[16]
FULL ABILITY: To meet conditions (whether states or acts) sufficient for salvation, is within man's natural powers (post-Eden) to perform or bring about—supernatural aid is not required for these things.

To agree with FULL ABILITY, I shall suppose, is to be Pelagian, and to disagree with FULL ABILITY but agree with ABILITY is to be a semi-Pelagian of some description.

I want to begin the central argument of my chapter by discussing faith. I wish to discuss other aspects of Christian soteriology as well, but I begin with faith because it offers the fuller discussion and also enables me to introduce the argument from the security of the believer. We can see how the place of faith relates to the foregoing discussion by noting that INABILITY implies FAITH, defined as follows.

FAITH: Coming to faith in Christ is beyond man's natural powers (post-Eden) to perform or bring about—supernatural aid is required for that.

I shall argue that Christian materialism implies the falsity of FAITH, and therefore of INABILITY, and therefore that the Christian materialist, whether he realizes it or not, is adrift in Pelagian waters.

FAITH

Faith is of course of prime importance in Christian soteriology. Salvation, on the Christian scheme, requires it. Indeed, Protestants tend to say that the possession of faith is not only necessary but sufficient for salvation, while Catholics demur, thinking that baptism and good works (or the disposition to such) are also required. Therefore, if we assume that the Christian materialist is a Protestant, then he will hold to the sufficiency of faith at death for salvation:

(F1) For any human individual, if that individual instantiates faith at the moment of his death, then that individual dies saved.

With regard to the necessity of faith for salvation, the following proposition invites acceptance from both Catholics and Protestants.

(F2) For any human individual, only if that individual instantiates faith at the moment of his death does that individual die saved.[17]

It might be thought that a Calvinist, and therefore an adherent of the doctrine of the perseverance of the saints ("once saved, always saved"), might be inclined to deny (F2) because he may construe that portion of his theology as the following claim:

> (F3) For any human individual, if that individual instantiates faith at any moment in his life, then that individual dies saved.

This claim is consistent with people lacking faith at death and nevertheless entering heaven. But I think (F3) is too weak a claim for the Calvinist to accept as a full statement of the doctrine. I believe most Calvinists also would hold that an *enduring faith* is the means by which God secures the perseverance of his saints such that

> (F4) For any human individual, if that individual instantiates faith at any moment in his life, then that individual will instantiate faith at all times from that moment until his death.[18]

So, I take it that the Calvinist is also motivated to hold (F2).

Now, faith is a mental state (it is a trust or a belief or an inner perception or some such thing), and so Christian materialists—those who subscribe to (N)—will claim that faith states supervene on physical states, more precisely, surely, on brain states. So, from (N) and (F1) we can therefore derive (B1), where B_1 is a reasonably determinate type of brain state.

> (B1) For any human individual, if that individual instantiates B_1 at the moment of his death, then that individual dies saved.

This is derived as follows. Take any individual with faith. Given (N) there will be a subvenient physical property—this we can presume to be a neural property of some sort—that is such that whenever that property is instantiated, then faith is also instantiated. Call this property B_1 and the truth of (B1) is secured.

But note that no parallel reasoning secures the truth of (B2).

> (B2) For any human individual, only if that individual instantiates B_1 at the moment of his death does that individual die saved.

This is so because (N) is compatible with the multiple realizability of mental states—(N) says it is nomologically impossible to have *P* without *M*, but it doesn't say it is impossible to have *M* without *P*. So, there might be many ways to bring about the instantiation of some mental state M_1: perhaps each of the physical states P_1, P_2, or P_3 could individually do the job. Because the subvenient property doesn't have to be the same every time, there may be many types of brain state that bring about faith. But we can say this: take all of the brain states the individual presence of which is nomologically sufficient for faith, gather them into a set $\{B_1, B_2, B_3 \ldots B_x\}$, and call this set S_b. Then our Christian materialist appears committed to the truth of (B3).

> (B3) For any human individual, only if that individual instantiates one of the members of S_b at the moment of his death does that individual die saved.

Crucially, the Christian dualist is not committed to (B1) and (B3). Although he is committed to (F1) and (F2), he is not committed to (N): the dualist need not think there are any brain states which only occur when faith occurs—he can believe that faith states and brain states vary independently of one another. It is this ability to evade (B1) and (B3) which, I believe, gives him significant theological advantage.

Accepting (N) is necessary to be a materialist, and rejecting (N) is sufficient to be a dualist. However, rejecting (N) is not necessary for dualism. David Chalmers is an example of a property dualist who accepts (N).[19] Although he rejects (M), he thinks there are contingent psychophysical laws which guarantee the nomological supervenience of conscious states on certain states of functional organization.[20] But the arguments of this chapter tell against any sort of view, dualist or materialist, that implies (N). I also believe that, if these arguments are sound, then they show that the Christian should be a substance dualist. But I shall postpone discussion of what sort of dualism is required until all the relevant evidence has been laid out.

NEUROSCIENCE

What is the problem with the Christian materialist believing (B1) and (B3)? How does that commitment issue in a Pelagian conclusion? I will illustrate how with two scenarios: *Chance* and *Design*.

Chance: I slap a fellow in the face. The molecules that I strike are arranged in such a way that the perturbations I cause to this man's face continue to have an effect upon his brain. Moreover, the neural structure of this man's brain is arranged in such a way that when it encounters these perturbations it is altered by them such that one of the members of S_b comes to be instantiated. In this way, I cause another man to have faith without divine aid. To be sure, this is doubtless exceedingly improbable, but that is not the point: there is no reason to think that such an occurrence is impossible on materialist assumptions, and therefore no reason to think supernatural aid is necessary.

But then FAITH is false, and the Christian materialist has fallen afoul of the church's condemnation of Pelagian thought.

Design: Neuroscience is a field that has proceeded in leaps and bounds, and the tools employed are increasingly sophisticated. If materialism is true, then with greater knowledge and understanding of the brain comes the prospect of discovering some or all of the subvenient bases of faith states—some or all of the neural structures contained in S_b. Indeed, neuroscientists already have set their eyes on such a prize. The discipline of neurotheology is a discipline devoted precisely to discovering and understanding the neurological underpinnings of religious belief and experience.[21] Furthermore, surely not far from this discovery will be the ability artificially to bring about some of these states. So, suppose a neuroscientist discovers some of these states in the future. You are an unbeliever and he offers to alter your brain for you. He will perform an operation that changes your brain such that you come to instantiate one of the members of S_b. Or perhaps the technology has advanced so far that nothing so invasive as an operation is necessary; perhaps you merely have to sit under a scanner, or just swallow a pill containing tiny robots that will do all the required brain alteration, if mere chemicals are insufficient. Such a neuroscientist would, if materialism is true, have the power to bring about faith, and therefore something would lie within his natural powers that is supposed to be the exclusive purview of God. Again, FAITH is false in such a case.

It does not follow straightaway from the Christian materialist's being committed to the falsity of FAITH, and therefore of INABILITY, that he is committed to Pelagianism, though it does follow that he is committed to Semi-Pelagianism (as I have defined it). However, if he is a Protestant, and therefore committed to the sufficiency of faith for salvation (F1), then it *will* follow that his commitment to the possibility of faith without divine aid commits him to FULL ABILITY, the full Pelagian belief that it is possible for man to meet conditions sufficient for salvation without divine aid. What if the Christian materialist is a Catholic? But even then we can suppose, using parallel reasoning, that the neuroscientist has developed the capacity to bring about whatever other psychological elements Catholics believe are necessary for salvation (a desire to be baptized, a disposition to good works, etc.), and again FULL ABILITY will follow.

It is the great advantage of the dualist that by believing in the soul, or at least denying (N), he can suppose that there is no brain state that guarantees the presence of faith. The dualist has the luxury of supposing that it is just as impossible for man to bring about faith as it is (currently believed to be) for him to travel faster than light—the psychophysical laws that govern the relationship between the physical and the mental will not permit it.

Before I move to consider objections, I would like to note that the discussion has set the scene for another argument against Christian materialism. Recall the possibility of a neuroscientist installing faith in someone. Notice the unwelcome effects that such a possibility would have on the evangelistic enterprise. For as soon as it were obvious that the discovery of one of the members of S_b were not far away, it would become prudent for great swathes of funding ordinarily the preserve of missionary outreach to go instead to neuroscience, for as soon as a medical procedure is available to bring about B_1, say, then thousands could be converted. And if, as suggested above, the treatment could be orally administered, and some sort of "Faith Pill" were available, then there would be no need any more to persuade people of a conviction of the truth of Christianity just so long as they are sufficiently motivated to swallow a tablet. Furthermore, consent might be thought unimportant. The issue is of such great importance that if a faith drug could be mixed into the drinks or meals of unbelievers then, it might be argued, it should be done, whether they would wish it or not. Christian restaurants might become the vehicles for great religious revivals: the hungry unbeliever enters and orders a sandwich and coffee, only to emerge singing psalms and making hymnody.

The bad effects *Design* would have for evangelistic practice are a separate consideration, though they are a consequence of the fundamental problem here: that Christian materialism appears to imply the possibility of salvation taking place outside of the usual means. This issues in the possibility not only of salvation taking place in the absence of divine aid (Pelagianism) but of it being prudent to ignore the biblically prescribed methods of evangelism.

OBJECTION FROM LIBERTARIANISM

But there are responses that the Christian materialist might make, because some Christian materialists are libertarians (van Inwagen is a prominent example),[22] and the libertarian may mount an objection to the possibility of the above scenarios. He will say it is not possible to bring about another person's conversion through a neuroscientist's surgery, making them swallow a pill, or slapping them, because of the more general fact that it is not possible to cause anyone's conversion at all. Conversion requires the free response of the prospective convert, and if the response is caused then it isn't free.[23]

The libertarian might put it this way: proper conceptual analysis of faith reveals "being freely chosen" as an essential component, in which case (F1) and (F2) are still true although the move from them to (B1) and (B3) would fail because (N) does not hold for mental properties that are characterized at least in part historically. (N) holds good only when it ranges over mental properties that are characterized purely intrinsically.

But I do not think this response helps much. We should realize that while there is an occurrent aspect to faith, namely the decision at a time to place one's faith in Christ, there is also a categorical or dispositional element too: when one places one's faith in Christ, one's mental configuration changes from one state to another. What makes it the case that one is a believer in Christ isn't merely a decision made in the past. Rather, there are present facts intrinsic to one that make it true that one is a believer. These intrinsic facts will still bedevil the Christian materialist.

We might conceive of the categorical aspect of faith as a switch in the mind. The believer's switch was set to the "off" position before he was converted, but when he was converted the switch was moved to the "on" position, remaining at that position thereafter. The libertarian will insist that it must be moved by a free decision, to be sure, but nevertheless the switch must remain at the "on" position if one is to remain a believer—this state must persist long after the decision to acquire it has been and gone. Then we respond to the libertarian as follows: while it may be granted that the neuroscientist cannot duplicate the first, occurrent, decisional aspect properly (for the decision will not be free) there is no reason to suppose that he cannot bring about the categorical element,[24] the state of faith which persists long after the decision has gone away, and this remains problematic.

It is problematic because INABILITY covers "all good mental states relating to salvation" and having the categorical aspect of faith is surely a most excellent state to be in as far as salvation goes. Even though the libertarian Christian materialist might avoid Pelagianism through libertarianism (salvation cannot be secured without supernatural aid), he would not avoid Semi-Pelagianism (a good step in the direction of salvation can be secured without supernatural aid).

Lest anyone should think INABILITY is a stronger claim than the council would wish to put its stamp on, I invite them to reconsider canon 6. It says: "If anyone says that God has mercy upon us when, apart from his grace, we believe, will, desire, strive, labor, pray, watch, study, seek, ask, or knock, but does not confess that it is by the infusion and inspiration of the Holy Spirit within us that we have the faith, the will, or the strength to do all these things." Both acts and states of mind are listed here, and it is surely the intent of the author(s) to include any mental state or action of which he can think that draws a man closer to God. Were we to suggest it, doubtless the categorical aspect of faith would be included in the list. Semi-Pelagianism is not, therefore, avoided.

Furthermore, I think it plausible to believe that Pelagianism is not avoided either. Consider those individuals who have the categorical element of faith artificially created in their minds. Are they to count as saved? Presumably not, if the libertarian is right (and "faith" in (F2) is such that it requires free choice). But to suppose they would be damned is surely an unpalatable consequence given the trust in God they would exhibit when they come before him for judgment. Their faith in Christ need be no less sincerely felt or fervently clung to than the greatest of Christians we have read about in history, despite its curious origin. It seems to me a very strange thing for God to damn those who come before him full of faith and joy at meeting him, only because their faith was not freely brought about.[25] If my intuitions are a reliable guide here, then it will follow that Pelagianism has not been avoided at all, because to bring about the categorical ground of faith in a man is to effect that man's salvation.

Nor will the appeal to libertarianism avoid the suggested absurd consequences materialism might have for the evangelistic enterprise. Although the libertarian Christian materialist might insist that those who undergo an operation (or swallow a "Faith Pill" or whatever) that brings about the categorical aspect of faith are not really saved, presumably these people could still be saved through a free recommitment to Christ. But it looks like this would be a trivial matter to obtain once the neuroscientist has secured the initial unfree commitment. It is to be expected that they would be happy to recommit given that they already have a commitment, because it is easy to reaffirm something you already believe. So, the possibility of such an operation would still result in the trivialization of the evangelistic enterprise.

OBJECTION FROM EXTERNALISM

There is another line of response the Christian materialist may take: endorse an externalism about faith. What if what constitutes faith is not entirely a matter of things intrinsic to a person? What if it involves external relations, causal relations perhaps, to things outside the person? In this case, we might say that faith is only truly faith if it involves causal relations to God. The power of the neuroscientist, it may be thought, is therefore stayed. It is not possible for a neuroscientist to bring about faith because such causal relations lie beyond his capacity to secure—he cannot force the hand of God. Therefore, in this way the Christian materialist avoids the commitment to both Pelagianism and the problematic effects that neuroscience might have for the evangelistic enterprise.

Note first that such an externalist account would violate (N): suppose we had two individuals, both of whom were in the same brain state, namely one of the brain states necessary for faith. Assume, however, God causally relates in the required way with one but not with the other. In such a case we would have physically identical individuals with a mental difference: one has faith and the other does not. Nevertheless, I think we can grant the materialist this externalism—I think such a position would remain true to the spirit of Christian materialism if not the letter.

For an instance of an externalist account of faith, we might turn to Alvin Plantinga.[26] His account appears to run close to, if not satisfy, this description. Plantinga writes,

> The believer encounters the great truths of the gospel; by virtue of the activity of the Holy Spirit, she comes to see that these things are indeed true. . . . [Faith] resembles memory, perception, reason, sympathy, induction, and other more standard belief-producing processes. It differs from them in that it also involves the direct action of the Holy Spirit, so that the immediate cause of belief is not to be found just in her natural epistemic equipment. There is the special and supernatural activity of the Holy Spirit.[27]

Plantinga here appears to endorse the operation of the Holy Spirit as essential to the faith process. If the direct activity of the Holy Spirit is not present, then we know ipso facto that faith is not present either. The neuroscientist cannot force the Holy Spirit to work, ergo he cannot bring about faith.

But then I should like to know *what* the neuroscientist is bringing about, if not faith. Suppose the neuroscientist brings about the belief in the great things of the gospel in a way that does not involve the operation of the Holy Spirit, by surgery, say. Has faith been brought about? Plantinga's account appears to commit one to the view it has not. On the one hand, this results in the happy conclusion that the neuroscientist cannot bring about faith; on the other hand, we are once again confronted with the disconcerting spectacle of sincere believers in Christ that are not saved, because what appears to be faith in them does not count as such because of its nondivine initiation. Again, I suggest that such a picture is so disconcerting that it is more sensible to say that they do in fact have faith. But then it will follow that Pelagianism has not been avoided.

Even if one bites the bullet in that regard, it remains plain that Semi-Pelagianism has not been avoided. Merely by bringing about belief in the great things of the gospel, even if faith is nowhere to be seen, is sufficient to prove ABILITY, for such a state is certainly a big step toward salvation.

Perhaps the problem is that the externalism suggested was insufficiently extreme. Perhaps we need a causal-content externalism about faith. On this view, what makes faith in God faith *in God* are the causal relations that state bears to God. If the causal relations were borne to something else, it would be faith in that something else. Again, this would mean it is not possible for a neuroscientist to bring about faith because the requisite causal relations to

God lie beyond his capacity to secure. But it would nevertheless issue in some strange consequences. It would remain possible for the neuroscientist to bring about a state that is, although not faith, phenomenologically and dispositionally identical to it. If he tries to bring about faith in the lab, then, despite his best efforts, and although he may produce what it is to all appearance a cheerful and devout believer, that supposed believer's faith will actually be, unbeknownst to the "believer" themselves, in the neuroscientist and not in God. Once more, I find it very implausible to suppose that these supposed pseudobelievers are not in fact faithful believers. They would be just as sincere or devout as their genuine counterparts, and they may be introspectively and intrinsically just as we genuine believers are. For God to give them radically different eternal destinies is counterintuitive. But to follow our intuitions here is to realize that Pelagianism has not been avoided.

Let us also note that, externalism or no, faith in God will never be an entirely extrinsic matter. It will not be the case that an individual comes to faith and the "change" has been entirely extrinsic and nothing has altered about that person intrinsically. Focusing our attention on these intrinsic aspects helps us see that causal-content externalism does not help the Christian materialist avoid Semi-Pelagianism either. Call these extrinsic aspects receptor states. Although bringing these about will not be sufficient to bring about faith, they remain necessary for faith and are an important step in the process of coming to Christ. But if it remains possible for the neuroscientist to bring these about, then INABILITY is false and ABILITY is true.

I wish to introduce a different argument at this juncture. Note that the removal of the externalist receptor-states is sufficient for the removal of faith. But is not the ability to remove faith just as problematic as the ability to install it? It may well be, and I would therefore like to, in the next section, press the argument that Christian materialism, if true, would threaten the security of the believer.

REMOVING FAITH

Christian materialists of every stripe would hold to (F2), that is, the necessity of faith for salvation. But then, even if it is conceded that materialism does not permit the possibility of faith *obtaining* via illicit means, there remains the question of whether it permits the possibility of faith being *removed* via illicit means.

I gave two scenarios when it came to bringing faith about: *Chance* and *Design*. We can rework them both to deal with the removal of faith rather than its initiation. With *Chance*, rather than the slapped fellow's brain being rearranged so that religious faith comes to be, we suppose that the slap so rearranged his brain that his religious faith ceased to be, and maybe also replaced it with a rejection of religious belief.

With *Design*, we might suppose that the neuroscientist, at the prompting of militant atheists perhaps, decides to prey on elderly Christians. As these Christians near the end of their lives, the neuroscientist sneaks into their houses or care homes and performs his operation while they sleep, removing their faith just before their death. Not merely that, we might also suppose he brings about a state sufficient for a positive rejection of God. In this way, he brings it about that these elderly folk enter into eternity as determined unbelievers. Rather than deathbed conversions, we might have deathbed deconversions! I discussed before the possibility of "Faith Pills"—pills that, through chemicals or tiny robots, put your brain in a state such that faith obtains. We can instead consider the possibility of "Deconversion Pills," pills that will rearrange a believer's brain such that no subvenient base for religious belief remains. Suppose an enthusiastic secularist manages to mix these pills into a church's postservice brew, turning the usually helpful tea ladies into ministers of apostasy. Is this something we want to acknowledge as a possibility? Yet Christian materialism says we must.

Chance and *Design* cases that deal with the removing of faith, rather than its inception, we can append a minus sign to, like so: *Chance–* and *Design–*. This should keep matters clearer.

The reasoning behind *Chance–* and *Design–* is parallel to that which we used when it came to *Chance* and *Design*, but with the acknowledgment that removing faith is a harder task than installing it, because, by (B3), all members of S_b must be removed from the target, as opposed to bringing about just one of them. But that is not a principled obstacle, and therefore by removing these brain states, possession of at least one of which is necessary for faith, the slap or the neuroscientist removes faith.

The possibility of such a thing is not condemned by the Second Council of Orange, but it is at odds with what we might call the security of the believer. Consider verses such as John 10:28–29: "I give them eternal life, and they will never perish, and no one will snatch them out of my hand. My Father, who has given them to me, is greater than all, and no one is able to snatch them out of the Father's hand."[28] Calvinists will see here a declaration of the perseverance of saints (F4). Arminians may see here only the weaker claim that believers cannot lose their salvation under any circumstances save by personal repudiation of it. Either way, *Chance–* and *Design–*, by describing the removal of faith in a way that does not involve personal repudiation, show how Christian materialism threatens the assurance Jesus wants to give the believer. How could it be impossible for the believer ever to fall away from the faith (except, perhaps, by voluntary choice), if it is possible for all the relevant brain states to be removed? And even replaced by ones which guarantee a rejection of God?

One objection to *Chance–* and *Design–* (and *Chance* and *Design*) is that the scenarios described are rather far-fetched. One might think that God will so order matters as to prevent them from happening. I deal with that objection in more detail at the end of the chapter. When it comes to the removal of faith, though, there is no need to appeal to future technological advances or unlikely happenings, for it looks as if the problematic scenarios are already here. Call these cases *Accident* cases. I invite you to consider, as one example, the case of Terri Schiavo. This was a famous case in the United States for the issues it raised over the right to life.[29] The details are as follows. Terri Schiavo collapsed of a heart attack on February 25, 1990 and remained in a coma for two and a half months. When she emerged from the coma, she regained a sleep-wake cycle, but did not exhibit repeatable and consistent awareness of herself or her environment. She had gone a long period without oxygen before being taken to the hospital, and this led to profound brain injury, severely damaging those parts of the brain concerned with cognition, perception, and awareness. Indeed, the matter from these regions of the brain had more or less gone: her brain weighed only 615 grams (21.7 ounces)—half the weight expected for a female of her age, height, and weight, an effect caused by the loss of a massive number of neurons. She eventually died on March 31, 2005.

A case where someone is put in a coma but later recovers with no ill effects is no problem for the materialist: the brain matter which the materialist supposes grounds the categorical aspect of faith remained throughout. But in *Accident* cases, such as the Schiavo case, the brain matter which would ground faith is gone—it has died. Here consciousness appears to be retained, but anything that would ground higher-order conceptual function is gone. But if it is possible for someone to lose such higher-order capacities and remain alive then surely, if this were to happen to a believer, that believer would have lost their faith while remaining alive.[30] But then, by (F2), they are no longer saved. But it is a severe doctrine indeed that implies that such a person now faces damnation because they no longer have faith—after all, they may have been in no way culpable for its loss.

Moreover, even if they were culpable to some extent for their injury, nevertheless, as noted above, it is part of Christian doctrine that salvation is not supposed to be capable of being lost by injury. Paul gives us in Romans 8:38–39 the stirring and powerful declaration that "neither death nor life, neither angels nor demons, neither the present nor the future, nor any powers, neither height nor depth, nor anything else in all creation, will be able to separate us from the love of God in Christ Jesus our Lord." It would be unfortunate indeed if materialism implied that a mere blow to the head could do the trick!

I mentioned that I think substance dualism the best solution to these problems. Why is that? Why not just be property dualists? One important reason is this: in cases such as the Schiavo case, it seems the relevant physical substrate has gone. So, if faith is to remain, then it must have an immaterial substratum, namely the soul. The substance dualist may hold that faith is always locked away in a compartment of the soul, invulnerable to harm. It then becomes easy for him to suppose that believers in *Accident* cases lose all the relevant portions of their brain without losing faith, because faith is instantiated in the soul, not in any portion of corporeal matter.

ORIGINAL SIN AND REGENERATION

After that detour concerning the removal of faith, I would like to turn attention back to the Second Council of Orange, in particular, to its affirmation of a doctrine of original sin. Recall the conclusion of the council: "The sin of the first man has so impaired and weakened free will that no one thereafter can either love God as he ought or believe in God or do good for God's sake, unless the grace of divine mercy has preceded him." Ott summarizes Catholic thought on original sin as follows: "it is the absence of . . . sanctifying grace" and this has given mankind over to mortality, ignorance, malice, infirmity of will, and concupiscence.[31] The Reformed view is stronger: original sin has left man with a fundamental deformity of nature. His will is bent toward evil in a way that is unnatural, and there is a darkness over his entire psychology. Berkhof speaks of a "contagion of . . . sin . . . spread through the entire man, leaving no part of his nature untouched, but vitiating every power and faculty of body and soul."[32]

This darkness under which we labor is not supposed to be capable of being humanly removed. This weakness which is part of original sin is surely something which only the supernatural power of the Holy Spirit can overcome. This is why canon 7 says we cannot "form any right opinion or make any right choice which relates to the salvation of eternal life through our natural powers without the illumination and inspiration of the Holy Spirit"—our minds are too darkened for that now. The crookedness of our hearts is now so great that only God can fix it. We might say this:

> ORIGINAL SIN: The wounding of man's nature, and the wicked desires which assail him, are beyond his natural powers (post-Eden) to cure or remove—supernatural intervention is required for that.

But this is false on Christian materialism. If Christian materialism is true, then this great problem is humanly soluble. By pouring all our energy and resources into neuroscience and the study of the body we can discover the material subvenient bases of these various malaises which afflict man's nature and then remove or fix them, restoring man in large measure to his prelapsarian state, and rolling back the effects of the fall.[33] But it is surely a grave mistake to think we have such a thing within our power. The curse can only be undone through the grace and mercy of God. Materialism encourages what the Christian will think is an unwholesome attitude to the spiritual predicament of humanity; the answer is not necessarily bound up with God, but within man's grasp, if he is sufficiently enterprising and ingenious. To block this idea we must deny (N), and therefore be Christian dualists.

INABILITY and ORIGINAL SIN deal with matters before salvation. What about things after? There too we encounter another problem for Christian materialism. Consider the doctrine of regeneration. This is the renewal of mind and spiritual empowerment which a man receives upon conversion. The Westminster Confession describes it thus: an "enlightening [of] their [the elect's] minds, spiritually and savingly, to understand the things of God, taking away their heart of stone, and giving unto them an heart of flesh; renewing their wills, and by his almighty power determining them to that which is good; and effectually drawing them to Jesus Christ; yet so as they come most freely, being made willing by his grace."[34] Catholics prefer to call it "sanctifying grace."[35] Now, it is not merely the coming to faith that is beyond man's power to secure, but also, surely, the great psychological blessings given and experienced by the believer when he comes to Christ. Again, we want to say that such blessings from God are things only the Holy Spirit could bestow.

> REGENERATION: All the spiritual virtue and delight in God and renewal of mind that are the typical result of conversion are beyond man's natural powers (post-Eden) to secure for himself—supernatural intervention is required for those.

But this is also false on Christian materialism. All those great states of the Christian—the delight in God, the hatred of sin, the strength to resist temptation, the love of the brethren, and "the peace of God, which surpasses all understanding" (Phil. 4:7)—are within man's power to acquire for himself, if he can locate their subvenient brain states. Then he can bring about an artificial regeneration, bypassing the power of the Holy Spirit. But this is not something we want to be able to say, and if Christian materialism permits us to say it, then we should reject Christian materialism.

SOME OBJECTIONS

Now for some objections.

Objection 1: Why do we not take INABILITY, FAITH, ORIGINAL SIN and REGENERATION to be concerned with practical possibility? In other words, FAITH expresses the *practical* impossibility of a man effecting his own or another's salvation without divine aid. REGENERATION expresses the *practical* impossibility of a man giving himself the "peace of God, which surpasses all understanding," and so on. We might concede that these things are possible in a nomological sense, but deny their possibility in a practical sense. Just as it is nomologically possible that I hit a hole-in-one, but it is not practically possible for me—I am a hopeless golfer, say—so it may be nomologically possible that I bring about faith without the Holy Spirit, but not practically possible. *Chance*, for instance, would then be no counterexample to FAITH so construed, for it involves a great fluke, and such unlikely flukes do not make you practically able to bring about whatever it was that the great fluke brought about.

Response: The problem here is precisely that something being practically impossible is compatible with it happening as a fluke. But does one really think that the theologians at the Second Council of Orange would have been happy to say that "salvation is beyond the practical ability of man to secure for himself, but nevertheless it might occur through a great fluke"? Surely they intend to be stricter than that.

Objection 2: Perhaps the restriction in INABILITY, FAITH, ORIGINAL SIN and REGENERATION is better thought of in terms of *providential* impossibility. In other words, we can read these claims as expressing that, relative to the *providential purpose of God*, it is not possible for an unbeliever to acquire faith by chance or by the ingenuity of the neuroscientist, nor for man to roll back the curse, and so on. Again, we may well grant that it is nomologically possible for human beings to do these things (give themselves faith, cure original sin, grant themselves the regeneration that is thought of as the exclusive purview of the Holy Spirit, etc.), but the point is that God will so order history that none will ever achieve it. In any case, won't it be a long time before the technology has advanced enough as these scenarios describe? Won't the Lord have returned by then? Either way, we can suppose the scenarios you describe just aren't going to happen—God will make sure of it.[36]

Response: First, this does not sound like something we should say either. Would those who take seriously the sort of theology that lies behind the Second Council of Orange be prepared to say the following? "Of course, it is quite within our natural powers to bring about salvation by ourselves, you know! Only God won't let us. He'll always make sure it won't come to pass." I do not think so, and I do not find it ideal.

Second, dualism remains the more probable explanation of the data. We can suppose the dualist and the materialist agree on the following proposition:

> (P) God has made it the case that none of the objectionable scenarios I described above will come to pass.

But the dualist and the materialist will disagree on how God has secured the truth of (P). The dualist hypothesis is that he has done this by locking the relevant mental states away in the soul. The materialist hypothesis is that he has done this by spatiotemporal means: structuring the physical world or history to foreclose the objectionable scenarios, perhaps by ensuring that the technology will not or cannot develop that far.

The problem with this latter suggestion is that we know that many ways God could have used to prevent this sort of thing spatiotemporally have not been used, and this lowers the probability that he has foreclosed the objectionable scenarios in that manner. For example, God could have made the brain entirely inscrutable to scientific investigation, yet neuroscience proceeds apace; he could have surrounded the relevant part of the brain with an impenetrable wall of force that departs on death, but there is no such thing; or he could have deprived us of the concepts needed to study the brain, and so forth. Yet none of these spatiotemporal means of blocking the objectionable scenarios has happened; they are ways God could have used, but we find he has not used them; and the more we learn about the brain, and the closer history gets to one of the problematic scenarios I described, the less likely it is that God is using spatiotemporal means.

Indeed, every advance we make in neuroscience is evidence against the idea that God is ensuring by physical or historical means that (P) is true, because with every advance we discover another way in which the brain becomes accessible to us and another way which God has not used to prevent (P) from occurring, namely, by preventing neuroscience from getting as far as *that*. Because every advance in neuroscientific technology becomes evidence against (P)'s being secured by contingencies of world and history, they therefore become evidence against Christian materialism.

CONCLUSION

Let me draw this chapter to a close with a brief summary. I have argued that Christian beliefs about original sin, the nature of faith, and regeneration as represented by the Second Council of Orange are inconsistent with Christian materialism. The Christian materialist, because of his commitment to the nomological supervenience of mental states on brain states, is forced to think that many of the mental states of central importance in Christian soteriology lie, in principle, within man's power to bring about. These include faith itself, an undoing of the psychological effects of the fall, and the renewal of the mind that is part of regeneration. I noted the untoward consequences for evangelism, if such things lie within man's natural powers.

I also made the argument that Christian materialism threatens the security of the believer. Not only does Christian materialism imply the capacity to bring about faith artificially, but also to remove it artificially, it appears. This conclusion is in tension with the believer's assurance of safety in Christ (and especially with the doctrine of the perseverance of saints). Christian materialism raises the worrying question of whether it is possible to cause someone to lose their salvation irregardless of their wishes. Because Christian dualism can straightforwardly guarantee the nomological impossibility of this, we have further reason to prefer dualism over materialism.

Do I think the arguments developed here are decisive? I do not. Were I a materialist, I would take the suggestion that I developed in *Objection 2*. I would grant that it is in principle possible for man to give himself faith, et cetera, but that God would, as sovereign Lord, simply prevent such things from ever coming to pass. That said, I would also concede that it would be better overall for the problematic scenarios described in this chapter to be nomologically impossible, and I would therefore grant that the unavailability of such a response would be a cost to Christian materialism. I am not a materialist, though, and so I will strike a triumphalist note: the unavailability of such a response is one more nail in the coffin of Christian materialism.

NOTES

1. Peter van Inwagen, "A Materialist Ontology of the Human Person," in *Persons: Human and Divine*, ed. Peter van Inwagen and Dean Zimmerman (Oxford: Oxford University Press, 2007); "Dualism and Materialism: Athens and Jerusalem?" *Faith and Philosophy* 12, no. 4 (1995): 475–488; *Material Beings* (Ithaca, NY: Cornell University Press, 1990).
2. Hud Hudson, *A Materialist Metaphysics of the Human Person* (Ithaca, NY: Cornell University Press, 2001).
3. Lynne Rudder Baker, "Material Persons and the Doctrine of Resurrection," *Faith and Philosophy* 18, no. 2 (2001): 151–167; "Need a Christian Be a Mind/Body Dualist?" *Faith and Philosophy* 12, no. 4 (1995): 489–504.
4. Nancey Murphy, *Bodies and Souls, or Spirited Bodies?* (Cambridge, UK: Cambridge University Press, 2006).
5. Trenton Merricks, "How to Live Forever without Saving Your Soul: Physicalism and Immortality," in *Soul, Body, and Survival*, ed. Kevin J. Corcoran (Ithaca: Cornell University Press, 2001).
6. See William Vallicella, "Could a Classical Theist Be a Physicalist?" *Faith and Philosophy* 15, no. 2 (1998): 160–180 for issues this raises.
7. Cf. Jaegwon Kim, *Mind in a Physical World* (Cambridge, MA: MIT Press, 1998), 9.
8. Just replace "nomologically" in (N) with "metaphysically" to get (M).

9. Benjamin B. Warfield, *Two Studies in the History of Doctrine* (New York: The Christian Literature Company, 1897), 6.

10. Warfield, *Two Studies*, 8.

11. Warfield, *Two Studies*, 8.

12. Warfield, *Two Studies*, 7–8.

13. Aidan Nichols, *The Conversation of Faith and Reason: Modern Catholic Thought from Hermes to Benedict XVI* (Chicago: Liturgy Training Publications, 2011), 9.

14. For this text and the text of Canons, see "The Canons of the Second Council of Orange (529)," accessed July 15, 2017, http://www.crivoice.org/creedorange.html

15. "Good acts relating to salvation" and "good states relating to salvation" might seem too broad. If an unbelieving mayor agrees to let a church host a gospel meeting in the town hall, then wouldn't the mayor be performing a good act relating to salvation? Do we want to insist that he could only have done so if a supernatural power lay behind his act? I'm inclined to think not. But, although it may well be a tricky affair to demarcate precisely the two classes of act and state for which the intervention of the Holy Spirit is always required, I hope it is clear from the council the sort of thing that is meant.

16. I realize that Semi-Pelagianism is really a narrower claim than this, namely that it is within man's natural powers to perform an act, or be in a state, that initiated, or made available the initiation of, the process of salvation without supernatural aid. But I find the characterization I give more useful for my purposes, logically more perspicuous, nevertheless tending toward Pelagianism, and coming under the condemnation of the council in any case.

17. Bracketing, if you like, the case of infants.

18. Wayne Grudem, a prominent Calvinist theologian, defends this view in his *Systematic Theology* (Grand Rapids, MI: Zondervan, 2000), 792–793, on the basis of his exegesis of 1 Peter 1:5 and other passages.

19. David Chalmers, *The Conscious Mind* (Oxford: Oxford University Press, 1996), 125–126.

20. Chalmers, *Conscious Mind*, 248–249.

21. See Andrew Newberg, *Principles of Neurotheology* (Farnham, UK: Ashgate, 2010) for a survey of the field.

22. Peter van Inwagen, *An Essay on Free Will* (Oxford: Oxford University Press, 1983).

23. Some compatibilists can make a similar complaint. Fischer and Ravizza's view, for instance, requires for moral responsibility that the relevant decision be caused by a reasons-responsive mechanism (John Martin Fischer and Mark Ravizza, *Responsibility and Control: A Theory of Moral Responsibility* [Cambridge, UK: Cambridge University Press, 1998]). Perhaps the neuroscientist's procedure or "Faith Pills" don't satisfy this criterion. However, the response I make to the libertarian will hold good for this sort of compatibilist also.

24. He can do this by bringing about the categorical element directly or by causing an unfree decision to have faith—it should not matter which.

25. This possibility also risks conflicting with texts such as Acts 16:31, Rom. 10:9–11, and Psalm 25:3, which promise salvation to all who believe. To read all such

passages as involving only freely inaugurated belief places an unwelcome pressure on biblical interpretation. Of course, if the only nomologically possible way to come to the categorical ground of faith was by a free choice, then there would be no such interpretative pressure (because if you believed, then you would have freely believed, and the whole idea of "unfree believers" would be nomologically impossible), but to secure that the Christian libertarian will have to reject (N) and therefore be a Christian dualist.

26. I do not wish to suggest that Plantinga is a materialist. He is not. See Alvin Plantinga, "Materialism and Christian Belief," in *Persons: Human and Divine*.

27. Alvin Plantinga, *Warranted Christian Belief* (Oxford: Oxford University Press, 2000), 256.

28. All Scripture quotations are from the ESV.

29. See Hessel Bouma, "Challenges & Lessons from the Terri Schiavo Case," *Perspectives on Science and Christian Faith* 57, no. 3 (2005): 212–220, and T. Koch, "The Challenge of Terri Schiavo: Lessons for Bioethics," *Journal of Medical Ethics* 31 (2005): 376–378 for summaries of the affair.

30. Was Terri herself a believer? Her parents said she was a devout Roman Catholic. "Terri was a devout Catholic all her life" is a quote from "Terri Schiavo Would Not Want To Go Against The Catholic Church, Her Parents Argue (Sept. 15, 2004)" accessed July 19, 2017, http://mn.gov/mnddc/news/inclusion-daily/2004/09/091504fladvschiavo.htm. See also Bouma, "Challenges & Lessons," 212.

31. Ludwig Ott, *Fundamentals of Catholic Dogma* (Cork: Mercier Press, 1955), 110, 113.

32. Louis Berkhof, *Systematic Theology* (Edinburgh: Banner of Truth, 1958), 223–224.

33. As far as the moral aspect goes, at any rate. We might not be able to restore ourselves to bodily immortality.

34. Philip Schaff, *The Creeds of Christendom,* vol. 3, 3rd ed. (New York: Harper & Brothers, 1882), 624.

35. Ott, *Fundamentals*, 254.

36. Note that this isn't an adequate response to *Accident* cases, for they are already happening.

SUGGESTED FURTHER READING

Baker, Lynne Rudder. "Need a Christian Be a Mind/Body Dualist?" *Faith and Philosophy* 12, no. 4 (1995): 489–504.

Bouma, Hessel. "Challenges & Lessons from the Terri Schiavo Case." *Perspectives on Science and Christian Faith* 57, no. 3 (2005): 212–220.

Grudem, Wayne. *Systematic Theology*. Grand Rapids, MI: Zondervan, 1994.

Hudson, Hud. *A Materialist Metaphysics of the Human Person.* Ithaca, NY: Cornell University Press, 2001.

Kim, Jaegwon. *Mind in a Physical World.* Cambridge, MA: MIT Press, 1998.

Merricks, Trenton. "How to live forever without saving your soul: Physicalism and Immortality." In *Soul, Body, and Survival*, edited by Kevin J. Corcoran. Ithaca, NY: Cornell University Press, 2001.

Murphy, Nancey. *Bodies and Souls, or Spirited Bodies?* New York: Cambridge University Press, 2006.

Newberg, Andrew. *Principles of Neurotheology*. Farnham, UK: Ashgate, 2010.

Vallicella, William. "Could a Classical Theist be a Physicalist?" *Faith and Philosophy* 15, no. 2 (1998): 160–180.

van Inwagen, Peter. "A materialist ontology of the human person." In *Persons: Human and Divine*, edited by Peter van Inwagen and Dean Zimmerman. New York: Oxford University Press, 2007.

Warfield, B. B. *Two Studies in the History of Doctrine.* New York: The Christian Literature Company, 1897.

Chapter 11

Sanctification and Physicalism

R. Scott Smith

For the Christian, sanctification is the process by which we grow into the likeness of Jesus Christ. As image bearers, Christians can become like Christ through being his disciple, and this process includes (at least) moral, intellectual, and volitional transformation. Since the Christian is to become like Christ and live as he did, this growth is developed through deep, intimate union with him, living just as he did in a deep heart and mind unity with the Father in the power of the Spirit.

Now, on the received Christian tradition, these abilities have been understood as involving a body-soul unity in humans, as well as various mental states (such as thoughts and beliefs) and immaterial properties (such as the moral and intellectual virtues). However, are these abilities to become like Christ truly workable on a physicalist view of humans?

To explore this question, first, I will develop a biblical portrait of what is involved in becoming like Christ morally, intellectually, and volitionally, and how that requires a deep unity with the Lord's heart and mind, all in the power of the Spirit. I will sketch how this process and its relationship to our nature as image bearers has been understood traditionally to involve a dualism of body and soul, as well as a dualism of properties, in human persons.

Second, I will sketch some main points of Christian physicalism. Third, I will explore to what extent we can become like Christ in these ways on that ontology. I will develop three main lines of argument that physicalism undermines sanctification. For one, I will argue that relationships with God and other humans are impossible on physicalism. However, that result undermines our being able to fulfill the many "one another" obligations in Scripture, as well as our being able to live in relationship with God. For another, physicalism disrupts our being able to have knowledge, which undermines our ability to grow into having the mind of Christ. Furthermore, physicalism makes it impossible to grow in virtue, or even make sense of what it means to be virtuous. A crucial reason for these results is that physicalism cannot accommodate intentionality, the ofness or aboutness of our thoughts, beliefs, experiences, etc. In conclusion, I will draw out some further implications from this study.

BIBLICAL PORTRAIT OF THE LIFE UNITED WITH CHRIST

While Scripture portrays the unfolding of God's great plan of redemption, it also presents that within a larger, overarching theme, which spans the entirety of Scripture, even before the fall, and into eternity. This can be helpfully stated as follows, from God's perspective: "I will be your God, you will be my people, and I will dwell in your midst" (cf. Rev. 21:3). God wants to be intimate and personal with his people who are set apart (sanctified) for him.

Several passages address aspects of this theme, such as Jeremiah 31:33c: "I will be their God, and they shall be My people."[1] In God's plan to accomplish his salvation, Jesus "became flesh, and dwelt [tabernacled] among us" (John 1:14). Moreover, in fulfillment of the new covenant, God has given believers his Spirit to be with them forever (John 14:16). Jesus also explains the connection of love for him and his word with God's making his home with us: "if anyone loves Me, he will keep My word; and My Father will love him, and We will come to him and make Our abode [home] with him" (John 14:23). The Spirit of Christ *indwells* Christians forever (John 14:17; see also Rom. 8:9 and Eph. 1:13).

There are many beautiful pictures in Scripture of God's intimate presence with his people. For instance, consider the tender care and affection of the Father in Revelation 21:3–5a:

> And I heard a loud voice from the throne, "Behold, the tabernacle of God is among men, and He will dwell among them, and they will be His people, and God Himself will be among them, and He will wipe away every tear from their eyes; and there will no longer be any death; there will no longer be any mourning, or crying, or pain; the first things have passed away." And He who sits on the throne said, "Behold, I am making all things new."

Another depiction occurs in Exodus 33, after Moses has interceded with God not to utterly destroy the Israelites for their idolatry with the golden calf. In verse 11, Moses's relationship with God is described as enjoying intimate friendship: God "used to speak to Moses face to face, just as a man speaks to his friend." After their sin, God states that he does not want to go before the people. Nevertheless, Moses persists and pleads with God: "Now therefore,

I pray You, if I have found favor in Your sight, let me know Your ways, that I may know You, so that I may find favor in Your sight" (v. 13). God responds in grace, that Moses has found favor in his sight, and God knows him personally, by name. Then, Moses asks for something amazing, which for many years has expressed my own heart's cry: "I pray You, show me Your glory!" (v. 18), a request that delights God, and he honors it (34:6–8).

David is called a man after God's own heart, and he too had intimate acquaintance with the Lord. In Psalm 27:4, David's one desire of the Lord is to dwell in his house all his days, to behold God's beauty, and seek his face (v. 8). In Psalm 25:14 he knows God's secret, intimate counsel is for those who fear him. The narrative of his life illustrates that he knew this experientially, such as when he asked God directly for knowledge of what Saul, who was seeking David's life, would do (1 Sam. 23:9–13).

So, God's desire is to make a people for himself, a people in the midst of whom he may dwell and be intimate. God wants us to be with him, glorify him, and enjoy him forever. Although for now we see him in a mirror dimly, one day we will see him face to face (1 Cor. 13:12; Rev. 22:4), even better than what Moses experienced in Exodus 33:11.

Now, that theme is one to be lived in deep, intimate unity with Christ in light of the benefits from the new birth through the new covenant. Consider Ezekiel 36:25, in which God promises to give us a new heart after being cleansed of our sin, by being sprinkled with water (a symbol of forgiveness). Moreover, when Jesus tells Nicodemus of the need to be born by the Spirit (John 3:3–8), he is thinking of Ezekiel 36:26–27 and God's promise to give us a new heart, that he would put his Spirit in us so that we would obey him.

In addition, through the new birth, he has given us the mind of Christ (1 Cor. 2:16). This does not mean that our minds have been replaced by his; rather, we now have access to Christ's very mind. Consider how Christians each have a first-person access to their own thoughts: I know mine by reflecting upon them immediately. My wife, however, does not know my thoughts in that same kind of way; she has access to my thoughts in a third-person way, by my communicating them to her verbally and nonverbally. Paul puts this concept as follows: "For who among men knows the thoughts of a man except the spirit of the man which is in him?" Then, he applies the principle to God: "Even so the thoughts of God no one knows except the Spirit of God" (1 Cor. 2:11).

Then Paul makes a stunning claim: "Now we have received, not the spirit of the world, but the Spirit who is from God, so that we might know the things freely given to us by God" (v. 12). For Old Testament believers, God's laws were written on scrolls and tablets, and they had access to his mind in a third-person way. Now, in our being united with Christ, we have access through the Spirit in us to Jesus's mind in a first-person kind of way, as the Spirit discloses Jesus's thoughts, plans, etc. This too is another amazing facet of the kind of intimacy God wants with his people.

We may summarize the kind of relational intimacy God wants with his set-apart ones as involving our hearts living in a deep, intimate unity with his heart; and similarly for our minds with his mind; and all this by abiding in Christ, that is, in the life and power of the Spirit. It should be no wonder, then, that the greatest commandment focuses on exactly these "elements": "you shall love the Lord your God with all your heart, and with all your soul, and with all your mind, and with all your strength" (Mark 12:30). Here, in light of Jesus's other specifications, I understand "soul" to refer to the whole person, as an emphasis that we are to love him with all we are and all we have. "Strength" brings to mind the Spirit's life and power, for he gives us power (e.g., Acts 1:8), and apart from abiding in him we can do nothing (John 15:5). Paul reiterates these emphases in his charge to Timothy, that God has not given us a spirit of timidity, but his Spirit, and through him a "threefold cord" that binds us close to the Lord: one of "power [from the Spirit] and love [from the heart] and discipline [or, a sound judgment, from the mind]" (2 Tim 1:7, inserts mine).

In Jesus, the perfect God-man, we see this kind of unity with the Father lived out through the Spirit. The deep love relationship between the Father and the Son shows us how Jesus loved the Father with all his heart and mind. He knew experientially the beauty and fulfillment of the Father's love, knowing him as Abba. And, he always lived in the fullness of the Spirit.

These emphases are of crucial importance because of our core need. Before the fall, Adam and Eve lived in a deep heart, mind, and spirit unity with the Lord, as well as with one another and even within themselves. But, at the fall, that unity was severed; as Paul explains, apart from Christ, we are spiritually dead to God (Eph. 2:1). This can be seen in Genesis 3:1–13, where the serpent tempts Eve (and Adam) to eat of the fruit from the tree of the knowledge of good and evil. He claims that they will not die, as God had warned. Rather, God was withholding something good from them, that they would become like God, knowing (choosing, defining) good and evil. In effect, they could usurp God's place, deciding for themselves what is real, even morally (Gen. 3:5).

It seems Eve was mesmerized (cf. 1 Tim. 2:14), but Adam simply disobeyed. It seems he chose to suspend what he knew God had said in order to have a moment of perspective from his own thoughts. Yet, both were united in their will to become powerful over God Almighty, by escaping the death sentence (2:17) and diminishing his absolute love.

Here, sin is portrayed as the attitude and willful choice to usurp God and worship a god of our own making, even ourselves. Moreover, this same attitude is displayed throughout Scripture (e.g., Mark 12:1–12, 29; 15:10).

So, the default human condition after the fall is that our hearts are deeply idolatrous and intent on usurping his throne. Like our hearts, our minds have become united with Satan's, so that we do the deeds of our father, the devil (cf. John 8:37). Moreover, without the Spirit living in us, we are dead spiritually.

So, our deep need is to be born from above, so that in fact we are united with the Lord's heart and mind, and we have his Spirit living in us. Still, we also must abide in him (John 15), so that we actually live in that deep unity of heart, mind, and strength he has made possible. These are necessary conditions for sanctification.

Now, typically, these qualities that Scripture attributes to the heart and mind have been thought to be immaterial kinds of properties of the soul. That is, the traditional, received theological view of human ontology is that we are a unity of body and soul (substance dualism). For instance, we may see this in Augustine. For him, the soul is not a bodily kind of thing; nor can it be transformed into a body. Moreover, the entire soul is present at the same time throughout all of the body.[2] Aquinas also affirmed substance dualism of human beings.[3] Moreover, arguably on Thomism, the soul is the essence of the person.[4] John Calvin also accepted a substance dualist view. He clearly distinguishes the soul from the body as a created, incorporeal essence, which is the seat of the image of God.[5]

Now, God is nonphysical, and he has beliefs, thoughts, willings, emotions, and more, all of which are important qualities to have interpersonal relationships. At least on the received substance dualist view of humans, since we bear God's image, it would seem that these properties would be ones of our immaterial essence: our soul.[6] These qualities, along with experiences, often have been called *mental* states, which terminology reflects a dualist ontology. Now, almost every mental state has a common quality: they are *of* or *about* things.[7] This quality is known as *intentionality*. Also, moral character qualities (virtues) often have been thought to be qualities of the soul.

However, with the rise of forms of physicalism advocated by various modern Christians, such qualities could not exist as such. While as Christians they would want to hold onto the need to grow in sanctification, they would need to be able to explain the biblical qualities involved therein in terms of the ontology of physicalism. Moreover, they would need to be able to preserve our abilities to have interpersonal relationships, to know what is on the mind of Christ, and to be able to become like him morally. In what follows, I will sketch briefly the main points of physicalism advocated by a Christian, and I will use the Christian philosopher and theologian, Nancey Murphy, to lay out a model. Others' versions (e.g., from Kevin Corcoran, Lynne Rudder Baker, and Joel Green) will need to fit into a basic physicalist ontology, too. Having made that sketch, I will explore to what extent physicalism can explain and preserve these aspects of sanctification, including to grow in intimacy of relationship with Christ, and to live in deep unity of heart, mind, and spirit with him.

MURPHY'S CHRISTIAN PHYSICALISM

Murphy observes that many versions of physicalism try to reduce causation to just the lowest level of parts that determine behavior at higher levels.[8] In contrast, she argues for a *nonreductive* physicalism, in which there is both "bottom-up" and "top-down," or "whole-part," causation. For example, she argues properties or processes emerge from the physical, and these features can be described only by concepts from a higher level of analysis than physics.[9] For another, she believes that "emergent laws (laws relating variables at the higher level) are coming to be seen as significant in their own right and not merely as special cases of lower-level laws."[10]

Additionally, laws at higher levels restrain lower-level processes, and higher-level states are multiply realizable. This means an act that can be described at the biological level also can be described at still other, higher levels (such as psychological, moral, or sociological) by speaking from the standpoint of their respective languages. For instance, biologically, a person may kill some fish, the Santa Ana River sucker. Psychologically, we consider intentions, which involve the circumstances of the event, making the event an action (ordering a diversion of the river's water for a city's use). Socially, a different description could arise, such as how this was a prudent move to offset drought conditions for citizens. This social-level description involves a different set of circumstances. There also could emerge moral descriptions and claims; for example, "The act was immoral because it threatened an endangered species." Or, a legal claim could arise: "the move violated Environmental Protection Agency regulations." Following Wittgenstein, these linguistic usages involve different languages, which are governed by the rules of their respective language games.[11]

According to Murphy, at each level, different circumstances, and thus different languages, come into play. For her, certain lower-level properties (biological or physical, for instance) can constitute a kind of higher-level property (social, moral, etc.) under proper circumstances.[12] Thus, these higher-order properties are not identical to the lower-order ones.

However, while she approves of higher-level properties, and she rejects causal reductionism, nonetheless she clearly embraces *ontological* reductionism. For her, ontologically, creation is physical.[13] Accordingly, ontologically any so-called "mental" properties are physical properties. Yet, we may speak (or conceive) of them *as* mental from a different language-game, such as psychology.

Likewise, for Murphy, there is no need philosophically or scientifically for a substantial, immaterial soul in her ontology.[14] For her, the soul simply is a "functional capacity of a complex physical organism"; it is not a separate, immaterial essence.[15] If there is no neocortex, then there is no capacity for thought, and there would not be persons.[16] Or, as she puts it elsewhere, "The nonreductive physicalist view . . . attributes mental and spiritual properties to the entire person, understood as a complex physical and social organism."[17] Moreover, scriptural uses of *psuche* do not require the existence of the (immaterial) soul. Rather, these uses could be translated as "life," which does not seem to require substance dualism.[18]

With this sketch of her Christian physicalism in hand, let us explore to what extent it will fit well with the biblical model that we have seen is needed for sanctification.

PHYSICALISM AND SANCTIFICATION

As I have argued, sanctification necessarily involves believers growing in relationship with the Lord, and also with one another. We need to know what is on the Lord's heart and mind. He also wants us to have a deep, rich, experiential relationship with him. The same, of course, would apply to others, too, for we are to grow into Christlikeness in how we relate to others. Thus, interpersonal relationships are crucial for sanctification, and, as I discussed above, they involve the kinds of states that I called "mental" ones.

Let me illustrate some of the qualities needed for interpersonal relationships by considering my relationship with my wife. To be in a good relationship with her involves my having experiences that are of her, as well as many thoughts, beliefs, desires, and more. When beginning to get to know her and what she is like, I could observe her words as well as her nonverbal behaviors. Then I could form some tentative thoughts of what would please her, what she would enjoy doing, etc. Over time, I could form beliefs about her based on what I observed of her responses to questions, situations, and my actions. She also needs to be able to do the same in regards to me. Now, years later, we have desires for our family, as well as intentions—that is, our goals for which we act.

Notice that these qualities all involve what I mentioned before: intentionality, the ofness or aboutness of our mental states. Now, if physicalism is correct, it seems that intentionality would have to be able to be accounted for in a way in keeping with physicalism. But, can it?

There are two main ways that philosophers have tried to explain intentionality on a physicalist ontology. Despite the fact that these ways have been suggested by naturalists, I think they could apply just as well to Christian physicalists, for these naturalists also embrace a reductive physicalist ontology. The first option is from Michael Tye, who argues that the ofness or aboutness of a mental state is just a matter of causal covariation under optimal conditions.[19] On his kind of view, my belief is of the tree because the tree is causing in me that belief through a long, causal process of light waves bouncing off the tree, impinging on my retina, and ultimately causing that belief in my brain.

Now, surely Tye is correct in that there is a causal story to be told in sensory perception, and beliefs formed on that basis. Even so, this view of intentionality has some significant problems.[20] The causal chain of physical states is potentially infinitely long, and I would seem to be able to have access only to the last state (the brain state that was caused by the preceding physical state in the causal chain). Moreover, the immediately prior physical state that causes that belief *modifies* the brain. It is not a simple reproduction of the same physical set of originating conditions (the tree) that is passed down through the chain; rather, *each state modifies the subsequent one*. If so, by the time the brain state occurs (which we are *calling* the belief about the tree), it does not seem we should conclude that we are aware of the tree itself, as it really is.

Not only that, but the *entire chain* of causes stands between me and the tree. In light of these issues, how then could I know that my belief actually is of the tree? It seems that I, as a bundle of physical states, cannot transcend the last state in the chain and arrive at the original cause (the tree). I can interact only with the last one, it seems.

However, it seems we need to be able to make epistemic contact with that originating object itself, in order to be able to tell the difference between veridical and false beliefs. Moreover, we need to be able to do this in a way that does not somehow modify the object's properties, lest we never be able to access the object itself, but only as it is modified. Yet, on this causal view, this seems dubious, for a physical, causal chain inevitably modifies its object.[21]

Moreover, on this view of intentionality as causal covariation, it assumes there is a relation that obtains between the originating object and the person (or, perhaps better, the brain state in the person). Thus, to have mental states requires that *both* relata obtain in reality. But, this seems false. Consider cases of *intentional inexistence*. In these, what we are thinking of does not obtain in reality. It seems we can think of such cases; for example, we can think of Pegasus, or the present-day king of France. Clearly, there are no such things in reality. Thus, these thoughts cannot be caused by their intentional objects. Moreover, even in the case of hallucinations, on this view they must be caused by something that exists physically. However, since we can think of such things, it seems intentionality is not a matter of causal covariation under optimal conditions. Though we can have such thoughts, which have intentionality, yet those thoughts get no further.

Yet, an objector could claim that while I think nothing physical caused such thoughts, something else physical actually did cause them. If that were the case (and even considering this very scenario seems to undermine the claim), then consider what would happen in ordinary life situations. Suppose I cannot find my eyeglasses at home. I can have a thought of what would be the case if I had left my eyeglasses on the coffee table. Then I can investigate by going and looking at the coffee table, but notice that they are not there (for in fact I left them on my dresser). So, then I can look elsewhere and confirm (or disconfirm) my new thought of where I left them by matching it with what I notice in experience. Moreover, it seems that in science, we also explore such hypotheticals frequently, to test if things are as we think or not.

Yet, if something physical always must cause our thoughts, experiences, and other mental states, then whether my experience of my glasses was caused by my glasses or something else radically different, I will not know. That is, it seems we would not be able to tell the difference between veridical and nonveridical cases. However, such inabilities would undermine not only significant aspects of daily lives, but science as well.

Also as a naturalist, Daniel Dennett offers a second way to treat intentionality in a physicalist ontology. For him, if we take naturalistic evolution seriously, there are no real mental entities, and there is no real intentionality. Dennett considers his theory of mental content (the intentional content of beliefs, desires, fears, hopes, etc.) to be functionalist, by which he means "all attributions of content are founded on an appreciation of the *functional roles* of the items in question in the biological economy of the organism (or the engineering of the robot)."[22]

Dennett uses a tactic he calls the *intentional stance*. This is a strategy "of interpreting an entity by adopting the presupposition that it is an approximation of the ideal of an optimally designed (i.e., rational) self-regarding agent."[23] That is, we treat humans, chess-playing computers, robots, and more as if they have real beliefs, thoughts, purposes, etc, with intentionality. This is a tactic we adopt in order to predict behavior efficiently. For instance, suppose Mr. Spock is playing chess against the computer onboard the starship *Enterprise*. According to Dennett, neither one has real intentionality, for there is none according to naturalistic evolution and physicalism. Nevertheless, if we treat them as though they have it, we can predict their behaviors by ascribing to them desires and beliefs as to how they likely will move their pieces to checkmate each other. When we adopt the intentional stance, we have made a "decision to conduct one's science in terms of beliefs, desires, and other 'mentalistic' notions," and Dennett thinks that is not unusual to do in science.[24]

Now, on his naturalism, there is no room for any essential natures, whether to humans themselves or even their beliefs, thoughts, experiences used to make observations, and other states we label as mental. Nor would there be any essence to intentionality. But, Dennett realizes an important implication of there not being any essences: without them, there are no "deeper facts" (i.e., beyond mere behavior) of the matter of what our thoughts (or beliefs, experiences, etc.) are *really* about.

This claim, however, seems demonstrably false. Suppose we are enjoying some frozen yogurt at a shop. When I take a spoonful, I can experience the taste of my chosen yogurt (e.g., birthday cake flavor). Indeed, it seems I can notice, if I pay attention, that my experience is *of* the taste of that yogurt. Moreover, it does not seem that that experience could be the one that it is and has turned out instead to be about the sound of the music in the store, some other flavor of yogurt, or anything else. That would be a different experience altogether. Thus, it seems that my experience of the taste of that yogurt has an essence to it.

Yet, following Dennett, if there are no essences, then there would not be any "deeper fact" (i.e., the intentional quality) of that experience that defines what it is about and makes it the experience that it is, and not something else. However, this has a significant and detrimental result for Dennett's own view. Dallas Willard rightly observes that Dennett seems to be left only with events of "taking as," in which we *take* some input *as* something else.[25] Indeed, Dennett comes close to stating this point himself, when he discusses how brains process their raw input: "there is no place where 'it all comes together,' no line the crossing of which is definitive of the end of pre-conscious processing and the beginning of conscious appreciation."[26] Nevertheless, if we are left with only takings, and nothing is just given to us in conscious awareness, then it seems everything is interpretation, including not only intentionality but even Dennett's naturalism itself.

From what we have seen, it seems intentionality cannot be preserved on these physicalist views. Further, it seems hard to conceive of another option for how adequately to treat intentionality on physicalism. The problem in both Tye's and Dennett's reductive ontological views is that there does not seem to be any room for intentionality to exist. Moreover, at least from Murphy's example, Christian physicalists, even of a (causally) nonreductive kind, will not have ontological room for real intentionality either. After all, they are physicalists. Therefore, it seems three disastrous kinds of implications follow for Christian physicalists in regards to sanctification.

First, without real intentionality, it seems that interpersonal relationships will not be able to happen. We would not be able to be in a deep heart and mind unity with the Lord, or experience his voice in intimate communion with us. We could not experience his love poured out in our hearts, nor could we experience his presence and power. In contrast to the Lord, who is an immaterial being with intentional states, we could not come to form justified true beliefs about what the Lord is like. Yet, that seems necessary for good, intimate relationships. For instance, I can trust my wife since I know her very well; I have formed over time deeply justified beliefs about her character from much experience and evidence. However, without the ability even to have beliefs, it seems then intimacy with any person, including God, would be undermined.

Since ideas, once acted upon, do tend to have consequences, these implications easily could lead to a mindset that though God exists, he is distant from us now, since he cannot be intimate with us. Moreover, that tragic result would be due to how he has made us, which conflicts with God's explicit desire to dwell intimately with his people. In short, since we would lack any real mental states with their intentionality, we would not be able to truly love God with all our hearts, souls, minds, and strength.

However, of course, the relational problems do not stop with God. For, if we lack the requisite ontological qualities to love and be intimate with God, we also will not be able to love one another. We will be divided and (at best) relationally distant from one another, unable to communicate. While we could stand in physical relations with one another (e.g., in terms of proximity, or origins), it seems that interpersonal relations will not happen.

Not only that, it seems that we ourselves will not be whole, well-integrated beings. How so? If we are to become like Jesus, then it seems that our hearts and minds should be deeply united with each other. In Jesus's case, it seems he defined the mind of Christ by his love (from his heart) for the Father. But, without any real mental states with intentionality, it seems this deep unity within us will be impossible, for there would not be any real willings, beliefs, or thoughts.

Second, there will not be any propositional knowledge (knowledge that something is indeed the case), for the standard definition of propositional knowledge involves true beliefs as well as justification.[27] Clearly, then, if there are no beliefs, there will not be any propositional knowledge. Unfortunately, that means that we cannot have knowledge of what is in the mind of Christ. We could not know what Scripture teaches, and we could not know anything else that is on Jesus's mind, such as his plans for us as individuals.

Moreover, to see that there is justification for a belief requires, at least, that the knowing subject can direct one's attention to, access, and even notice (empirically or rationally) the evidence for a belief; and consider how well that belief is supported thereby. These abilities also seem to require intentionality, such as with directing one's attention to the evidence. For all intents and purposes, then, it seems that physicalism undermines our having the mind of Christ (1 Cor. 2:16), and even our minds' being able to be transformed (Rom. 12:1–2).

Third, physicalism undermines our abilities to be like Christ morally. In a key way, moral transformation involves knowledge of what pleases the Lord, what he expects, and what he is like. Nevertheless, as we have seen, physicalism undermines the knowledge necessary for the moral aspects of sanctification.

Moreover, moral virtues, and even moral principles, do not seem susceptible to being reduced to what is just physical.[28] Crucially, being virtuous involves more than just bodily behaviors that we label as virtuous. Suppose a young person (A) extends his or her arm to an elderly person (E) who does in fact have difficulty walking without assistance, and the elderly person holds A's arm with a hand. The two proceed to cross a street together. Now, we probably would think that that was a kind act by A. Yet, someone could go through all the motions that we might normally associate with being kind, and yet not have the requisite attitude of intending to be kind. Instead, any number of possible scenarios could in fact be the case. A might have done these actions in order to give the appearance of being kind to F, A's watching friend, so as to gain F's trust more, so that in turn F would do what A wanted (say, give A the answers to an exam that A still has to take). Or, perhaps A thinks that the best way to get an ice cream cone is to "help" E so that E will be willing to buy A a cone.

In sharp contrast, Aristotle argues that virtues are concerned with passions and actions.[29] So, for instance, we can feel "fear and confidence and appetite and anger and pity and in general pleasure and pain" too much or too little.[30] He argues that we are to feel our passions toward the right person, to the right extent, and at the right time. Moreover, intentions are essential for moral acts, and so acting on the right motive, and in the right way, also are crucial.[31] Additionally, an agent needs knowledge, must choose the acts for their own sake, and perform them from a "firm and unchangeable character."[32] But, these passions, intentions, and knowledge all require intentionality, which undermines physicalism for becoming morally virtuous.

However, now let me consider two counterexamples to my claims that physicalism undermines sanctification. First, some could suggest that while people like Murphy endorse an ontologically reductive physicalism, still it is conceivable that on a "pluralistic" physicalism, mental states and intentionality could be emergent properties of the brain in a suitably complex structure.

That is to say, on such a view, there could be real, irreducible intentionality and mental states.

As reasonable as it might seem to suggest this alternative, I do not think it will suffice to overcome the problems I have surfaced. For even if we have such states and intentionality, there will be nothing real that can *have* or *use* them. There is not a duality of substances on physicalism; there is just the brain, not an irreducible mind (or soul). In that case, how would a physical brain interact with, much less use, these nonphysical states? The qualities involved in interpersonal relationships (and propositional knowledge) involve more than just the mere existence of mental states in us that genuinely are about their intended objects. They also require our being able to use them in various ways, such as in my example of getting to know my wife. On the basis of my experiences of her, I can form concepts and thoughts of what she is like, and then, with more evidential support, I can form and accept a belief that she is indeed (for instance) a very compassionate and accepting person.

The same abilities to use these states with their intentionality are involved in growing in relationship with the Lord. In a crucially important way, I get to know the Lord through reading Scripture, which involves my experiencing the words in the text and then thinking (meditating) about them. I address my thoughts and desires to him through prayer, and I get to experience his pleasure as I obey and please him (e.g., John 14:21). I have experienced events that I recognize as answers to prayer, which demonstrate his presence and care. Furthermore, in keeping with the examples and themes we observed earlier from Scripture, I have experienced God's speaking intimately with me, through his guidance in my studies and in his personalizing his love for me. Yet, all these examples involve my being able not merely to have, but also to *use*, these mental states and their intentionality in order to get to know the Lord better and better.

Moreover, there is a related problem. Interpersonal relationships require time to develop. Yet, without an essential nature (i.e., a soul), how can we remain the same through time and change, including growth in Christlikeness? Christian physicalists, such as Joel Green, realize this problem, and so they must have another basis for our strict sameness (or numerical identity) of person through the process of sanctification. For on an ontologically reductive physicalism (or even a pluralistic version in which there are emergent mental properties and intentionality), it seems we are constantly changing, simply because the body's parts are constantly changing. Even the brain itself is not the same through time, for it can undergo (for instance) changes in levels of serotonin, or new "grooves" can be developed as new psychological habits are cultivated.

It seems to me that Green's solution is the most obvious option for Christian physicalists, who lack any essential natures to which to appeal to ground personal identity. Green argues that instead of sameness of soul, what grounds our personal identity through change is the sameness of our individual *narratives*. Our narratives tell the story of our respective lives, and the unity we have is that of an embodied character. Our story unfolds and develops, which tells the story of our growth in Christlikeness. Moreover, for Green, while a "mystery," our personhood is preserved *in* Christ or *with* Christ. He thinks "the relationality and narrativity that constitute who I am are able to exist apart from neural correlates and embodiment only insofar as they are preserved in God's own being, in anticipation of new creation."[33] So, upon death, God remembers and then unites our story with a resurrected body. Thus, we can inherit eternal life.

It surely is true that our narrative tells the story of our lives. Yet, on a physicalist ontology, what kind of thing is a narrative? It seems it would be a bundle of sentences (which are sense perceptible) that I, as the main author of my story, and others ascribe to me. Still, can a narrative *itself* remain the same through time and change, given that it too is physical? It seems it cannot; the sentences are changing. More and more ascriptions are added as I continue to live, and thus the whole narrative would have different properties than it did before. Thus, my narrative is not the same through change, and therefore it cannot be the basis for *my* remaining the same through change, even if that is into Christlikeness. Change (including into Christlikeness) presupposes that there is a fundamental, even essential, sameness to the person, lest there not be growth in sanctification, but only succession and replacement of many persons as the narrative changes.[34]

As the second counterexample to my claim that physicalism undermines sanctification, consider Murphy's suggestion that God communicates with us through the quantum level of our brains.[35] If so, then it might seem that communication and relationships could take place. Yet, whatever else quantum-level phenomena are, for her they must fit within ontologically reductive physicalism. Accordingly, there will not be any real intentionality. Thus, based upon what I have argued already, quantum phenomena simply will not be able to supply the needed ontological resources for relationships, knowledge, and moral transformation, all of which are vital for our sanctification.

CONCLUSION

I do not see Christian physicalists suggesting that sanctification is somehow unnecessary, or a relic of bygone, Greek, dualistic thinking. Rather, they want to uphold that crucial doctrine and see us engage in the practices needed to grow in Christlikeness. This good goal is well-intended, but I have argued that physicalism undermines sanctification. Indeed, it seems that on physicalism, we would be like *philosophical zombies*. That is, though we would be alive, yet we would lack any conscious experiences, as well as thoughts, beliefs, and other such states.[36] Such creatures, however, clearly cannot have and enjoy interpersonal relationships or knowledge, or grow in moral transformation. Thus, at least in these crucial respects, physicalism seems incompatible with Christianity.

NOTES

1. All Scripture references are from *New American Standard Bible* (Anaheim, CA: Foundation Press Publications, 1995).

2. St. Augustine, De Animae Quantitate 13.22. See also De Civitate Dei V.10 (e.g., ch. 6). On the soul being present throughout the body, see *On the Immortality of the Soul* 16.25.

3. See, for example, Thomas Aquinas, *Summa Theologica* 1.Q75.

4. See the discussion in J. P. Moreland and Scott Rae, *Body & Soul* (Downers Grove, IL: InterVarsity Press, 2000), 206.

5. John Calvin, *Institutes of the Christian Religion*, book I, chapter 15.2, 3.

6. See also J. P. Moreland, *The Recalcitrant Image Dei: Human Persons and the Failure of Naturalism* (London: SCM Press, 2009).

7. It seems that *some* experiences, like a painful feeling, might not be of or about something.

8. See Nancey Murphy and Warren Brown, *Did My Neurons Make Me Do It? Philosophical and Neurobiological Perspectives on Moral Responsibility and Free Will* (Oxford: Oxford University Press, 2007), 47–48, where she provides five distinctions (methodological, epistemological, causal, ontological, and atomist) concerning the "many faces of reductionism."

9. Murphy and Brown, 78–84.

10. Nancey Murphy, *Anglo-American Postmodernity* (Boulder, CO: Westview Press, 1997), 21.

11. Murphy, *Anglo-American Postmodernity*, 24. See also *Did My Neurons Make Me Do It?* generally, 28–29, 151, 181–90; and specifically on "forms of life," 165, 187; and "language games" 165, 181–85, 188, 190.

12. Murphy, *Anglo-American Postmodernity*, 199.

13. See Nancey Murphy, "Human Nature: Historical, Scientific, and Religious Issues," in *Whatever Happened to the Soul?* ed. Warren S. Brown, et al. (Minneapolis, MN: Fortress Press, 1998), 18, where she argues against a need for humans to have a soul.

14. Murphy, "Human Nature," 18.

15. Ibid., xiii. See also her *Did My Neurons Make Me Do It?* chapter 1.

16. Nancey Murphy, *Beyond Liberalism & Fundamentalism: How Modern and Postmodern Philosophy Set the Theological Agenda,* ed. Werner H. Kelber (Harrisburg, PA: Trinity Press International, 1996), 93.

17. Murphy, *Beyond Liberalism & Fundamentalism*, 150.

18. See Nancey Murphy, *Bodies and Souls, or Spirited Bodies?* (New York: Cambridge University Press, 2006), 16–22. See also Joel B. Green, *Body, Soul, and*

Human Life: The Nature of Humanity in the Bible (Grand Rapids, MI: Baker Academic, 2008).

19. Michael Tye, *Ten Problems of Consciousness: A Representational Theory of the Phenomenal Mind* (Cambridge, MA: Bradford Books, 1995), 42–43.

20. For a more detailed discussion, see my *Naturalism and Our Knowledge of Reality* (Farnham, UK: Ashgate/Routledge, 2012), chapters 1 and 2.

21. Fred Dretske, another naturalist, has responded to my objection that if "knowledge that x (some external object) is f is a reliably caused belief of x that it is f (a belief that is caused by the information that x is f), then you don't have to 'traverse' the causal chain resulting in the belief in order to have knowledge of the external cause. All that is required is that the belief, in fact, be the result of some reliable process" (e-mail message to author, Febraury 10, 2007). Dretske elaborates that one is not directly aware of what is going on in one's head. However, he claims we are directly aware of the external object, for "information [about the tree] . . . is being transferred in the perceptual process to the representation (experience) of the [tree]" (*op. cit.*). In response, there is some truth to his reliabilism; our cognitive faculties generally do function reliably in appropriate circumstances. Nevertheless, Dretske's reply does not seem to alleviate the problems I have raised against causal chain accounts of intentionality.

22. Daniel C. Dennett, "Dennett, Daniel C.," in *A Companion to the Philosophy of Mind: Blackwell Companions to Philosophy*, ed. Samuel Guttenplan (Oxford: Basil Blackwell, 1994), 239 (emphasis in original).

23. Dennett, "Dennett, Daniel C.," 239.

24. Daniel Dennett, *The Intentional Stance* (Cambridge, MA: MIT Press, 1990), 239.

25. Dallas Willard, "Knowledge and Naturalism," in *Naturalism: A Critical Analysis*, ed. J.P. Moreland and William Lane Craig (New York: Routledge, 1999), 40.

26. Dennett, "Dennett, Daniel C.," 242.

27. Knowledge by acquaintance also seems to dissolve, for there will not be any experiences that are *of* something. Likewise, know-how (practical knowledge of how to do something) will not work, for it too seems to require making observations, forming beliefs about how to accomplish a task (e.g., how to purchase a train ticket from a vending machine), and so on, all of which require intentionality.

28. For a more complete treatment of this topic, see my *In Search of Moral Knowledge: Overcoming the Fact-Value Dichotomy* (Downers Grove, IL: InterVarsity Press, 2014), chapters 5–6, and 7 (on Christine Korsgaard).

29. *In Search of Moral Knowledge*, Kindle locations 770–74.

30. Aristotle, *The Nicomachean Ethics*, trans. David Ross, rev. by J. L. Ackrill, et al. (Oxford: Oxford University Press, 1980), book II.6, 38.

31. *Nicomachean Ethics*, book II.6, 9.

32. *Nicomachean Ethics*, book II.4, 34.

33. Green, 180.

34. Of course, this also has the disastrous result that we will not be able to inherit life everlasting.

35. Murphy, *Bodies and Souls, or Spirited Bodies?* 131–32.

36. See Robert Kirk, "Zombies," *Stanford Encyclopedia of Philosophy* (2015), accessed April 13, 2017, https://plato.stanford.edu/entries/zombies/

SUGGESTED FURTHER READING

Dennett, Daniel C. "Dennett, Daniel C." In *A Companion to the Philosophy of Mind: Blackwell Companions to Philosophy*, edited by Samuel Guttenplan, 236–43. Oxford, UK: Basil Blackwell, 1994.

———. *The Intentional Stance*. Cambridge, MA: MIT Press, 1990.

Green, Joel B. *Body, Soul, and Human Life: The Nature of Humanity in the Bible*. Grand Rapids, MI: Baker Academic, 2008.

Moreland, J. P. *The Recalcitrant Image Dei: Human Persons and the Failure of Naturalism*. London: SCM Press, 2009.

Moreland, J. P., and Scott Rae. *Body & Soul*. Downers Grove, IL: InterVarsity Press, 2000.

Murphy, Nancey. *Anglo-American Postmodernity*. Boudler, CO: Westview Press, 1997.

———. *Beyond Liberalism & Fundamentalism: How Modern and Postmodern Philosophy Set the Theological Agenda*, edited by Werner H. Kelber. Harrisburg, PA: Trinity Press International, 1996.

Murphy, Nancey, and Warren Brown. *Did My Neurons Make Me Do It? Philosophical and Neurobiological Perspectives on Moral Responsibility and Free Will*. Oxford, UK: Oxford University Press, 2007.

Smith, R. Scott. *In Search of Moral Knowledge: Overcoming the Fact-Value Dichotomy*. Downers Grove, IL: InterVarsity Press, 2014.

———. *Naturalism and Our Knowledge of Reality*. Farnham, UK: Ashgate, 2012.

Tye, Michael. *Ten Problems of Consciousness: A Representational Theory of the Phenomenal Mind*. Cambridge, MA: Bradford Books, 1995.

Willard, Dallas. "Knowledge and Naturalism." In *Naturalism: A Critical Analysis*, edited by J. P. Moreland and William Lane Craig. New York: Routledge, 1999.

Chapter 12

Neuroscience, Spiritual Formation, and Bodily Souls

A Critique of Christian Physicalism

Brandon Rickabaugh and C. Stephen Evans

The link between human nature and human flourishing is undeniable. "A healthy tree cannot bear bad fruit, nor can a diseased tree bear good fruit" (Matt. 7:18). The ontology of the human person will, therefore, ground the nature of human flourishing and thereby sanctification. Spiritual formation is the area of Christian theology that studies sanctification, the spirit-guided process whereby disciples of Jesus are formed into the image of Jesus (Rom. 8:28–29; 2 Cor. 3:18; 2 Peter 3:18).[1] Talk of such transformation permeates the New Testament.[2] As a natural part of salvation, spiritual formation includes every aspect of our being, "spirit, soul, and body" (1 Thess. 5:23–24), as it is the whole person that is sanctified.[3] Until the nineteenth century, there was an overwhelming consensus among Christian thinkers that some form of mind-body (or soul-body) dualism is true of human beings. Recently, that consensus has eroded, and with it the availability of a shared body of knowledge about spiritual formation.[4]

Two views dominate this discussion. Substance dualism (hereafter referred to as dualism) is the thesis that we consist of soul and body. The following distinction is helpful:

> *Minimal Dualism*: The self and its body are distinct entities.[5]
> *Significant Minimal Dualism*: The self and its body are distinct entities, while the self is an agent with causal powers such that it can affect the physical world and be affected by that world.[6]

Christian dualists are at least committed to Significant Minimal Dualism. According to the alternative view, Christian physicalism, we are physical bodies or at least parts of physical bodies, like the brain and central nervous system.[7] Some Christian physicalists hold that these physical bodies have nonphysical emergent properties. Some Christian physicalists argue that dualism is incompatible with central elements of spiritual formation. Neuroscientist Warren Brown and psychologist Brad Strawn offer the only substantive account of spiritual formation from the view of Christian physicalism and its accompanying objections to dualism.[8] Hence, it is on their arguments that this chapter focuses.

Although some argue that dualism is the biblical backdrop which informs and makes intelligible Christian spiritual formation, Brown and Strawn argue that contemporary neuroscience proves this false while supporting Christian physicalism.[9] On their view, spiritual formation is illuminated by a set of neuroscientific data. This data, they claim, supports a view of spiritual formation that requires special attention to the physical nature of our spiritual life. As such, it emphasizes our embodiment and neurological and social development, which they claim is incompatible with dualism.[10] Call this the incompatibility thesis.

We argue that Brown and Strawn fail to support their incompatibility thesis. Additionally, we argue that Christian physicalism stands in tension with important philosophical and theological foundations of Christian spiritual formation. In doing so, we offer a specific form of dualism, the bodily soul view, and explain how this view illuminates the importance of embodiment, our neurological and social development, and hence the important physical aspects of Christian spiritual formation.

WHY DUALISM IS COMPATIBLE WITH EMBODIED SPIRITUAL FORMATION

Just how is dualism supposed to be incompatible with a neuroscientifically informed account of spiritual formation? The strongest statement of the incompatibility thesis is one of logical impossibility. This view is impossible to defend for one simple reason: there is no logical contradiction between dualism and the neuroscientific data. The truth of dualism and the importance of the physical nature of our spiritual life do not entail a contradiction. God could have created natural laws uniting soul and body, such that neuroscience studies the bodily aspects of this unity. Likewise, the incompatibility thesis cannot be stated in terms of metaphysical impossibility. This thesis entails that God could not create a world where dualism is true and the neuroscientific data of this world obtain. That is a considerable constraint on God's creative capacity. This strikes us as highly implausible, and Brown and Strawn do not give us reason to think otherwise. Hence, the incompatibility thesis can make only the much weaker claim, that the conjunction of dualism and the neuroscientific data is improbable or less probable than the conjunction of Christian physicalism and the neuroscientific data. So, how do they defend this thesis?

We must recognize that Christian physicalists cannot make use of popular objections to dualism, especially the conservation of energy[11] and causal closure arguments, which rely on an in-principle rejection of causation between the physical and the nonphysical and the causal closure of the physical. Christianity is necessarily committed to causation between the nonphysical and the physical, at least with respect to God and his creation.[12] Sanctification, for example, requires the causal interaction of the Holy Spirit with human persons (e.g., Gal. 5:16–24). Hence, Brown and Strawn must object to dualism in other ways, to which we now turn.

WHAT HAS DUALISM TO DO WITH GNOSTICISM AND INDIVIDUALISM?

The main objection from Brown and Strawn is that dualism leads to Gnosticism, which is incompatible with biblical and neuroscientific data. Gnosticism, they explain, is the view that the material world is evil, while nonmaterial reality is good. Human souls are saved from this material world only by embracing the fact that we belong in a heavenly realm of light.[13] "The inward focus on the soul, fostered by dualism," they say, "creates a strong magnet drawing modern religious perspectives almost inevitably toward Gnosticism."[14] From this they conclude that dualism is false.

We find this objection unconvincing. Brown and Strawn offer no empirical support for this hypothesis, much less an explanation as to how dualism leads "almost inevitably" to Gnosticism. It isn't clear what is their argument, as they can be read in several ways. If taken in the anthropological or psychological sense, their conclusion does not follow. That many people believe or are caused to believe *x* does not tell us if *x* is true or false, or if *x* is unreasonable to hold. Furthermore, conflicting empirical evidence is easy to furnish. After surveying the main Christian proponents of dualism, one is hard-pressed to find a single Gnostic among them. Instead, we find outright rejections of Gnosticism. For example, Dallas Willard, a dualist and spiritual formation scholar, explicitly rejects the Gnostic view that what is immaterial and spiritual is inherently good, while the body and other material things are inherently bad.[15] On Willard's view, the soul and body are both in a ruined condition in need of redemption.[16] The body is central to Willard's detailed account of how the entire person is sanctified in Christ.[17]

If taken in the philosophical sense, the Gnostic thesis faces other problems. First, dualism is not and does not entail a thesis about what is or is not valuable, the nature of sanctification or salvation. Brown and Strawn admit that we cannot equate dualism with Gnosticism.[18] Contra Gnosticism, Christian dualists hold a very high view of the body. Charles Taliaferro, for example, argues that given dualism, embodiment allows for the exercise of six types of virtue: sensory, agency, constitutional, epistemic, structural, and affective.[19] Richard Swinburne argues that embodiment makes possible great goods that souls otherwise couldn't have, such as the ability of free choice between good and evil and the ability to influence others and the inanimate world.[20] Howard Robinson defends a robust view of the soul's dependence on the brain and body for the great good of psychological development.[21] Contrary to what Christian physicalists claim, Christian dualism maintains that the *telos* of the human soul, as created by God, is embodiment.[22] This alone entails the rejection of Gnosticism.

Of course, it is true that dualists have a history of valuing the soul more than the body. Augustine considers the soul as a much higher degree of reality and value than the body, with the soul surpassed only by God.[23] However, this does not mean that Augustine holds a low view of the body. That one takes *x* to be more valuable than *y* does not entail that *y* is not of great value. One could value their spouse more than their parents, and yet maintain a very high value of their parents. The Christian physicalist surely admits this when valuing God above creation although highly valuing creation.

Additionally, Brown and Strawn argue that belief in dualism leads to individualism, as dualists look inwardly at the soul rather than outwardly toward God and others. Like their previous Gnostic objection, Brown and Strawn fail to show a necessary connection between dualism and individualism. Even if there were a correlation, it would not follow that holding dualism leads to indifference toward others, the natural world, or historical events. Consider Kierkegaard, who, although a dualist, does not think of the self as merely a mental substance. For Kierkegaard, the self is a kind of synthesis of contrasting elements—finitude with infinitude, necessity with possibility. Human selves are a work in progress, involved in making themselves the persons they become, and doing so always in relationship to others. Far from being an individualist, Kierkegaard understands that we all are who we are by virtue of the relationships with others. He is interested in helping individuals develop a relation to God which relativizes those human relationships.[24] It is just false that dualism qua dualism leads to individualism.

Lastly, there are more plausible accounts of the turn away from embodiment and toward individualism that do not place the blame on dualism. Some have argued that the turn toward individualism is the result of theologians and then pastors abandoning the soul, which paved the way for the contemporary mental health movement.[25] This may be overstated, but it is relevant. Additionally, it seems far more plausible that the problem of contemporary individualism is with the conception of salvation as mere forgiveness of sins. Willard points out that such a view makes Paul's statement that we are "saved by his [Jesus's] life" (Rom. 5:10) unintelligible. Willard observes, "How can we be saved by his life when we believe salvation comes from his death alone? So if we concentrate on such theories exclusively, *the body and therefore the concrete life we find ourselves in are lost to the redemption process*."[26]

In fact, we find dualists, such as Willard, holding the exact opposite of Gnosticism and individualism.

> Spirituality in human beings is not an extra or "superior" mode of existence. It's not a hidden stream of separate reality, a separate life running parallel to our bodily existence. It does not consist of special "inward" acts even though it has an inner aspect. It is, rather, a relationship of our embodied selves to God that has the natural and irrepressible effect of making us alive to the Kingdom of God—here and now in the material world.[27]

The problem of individualism, escapism and rejection of embodiment is solved, not by rejecting dualism, but by embracing the Gospel of the Kingdom of God, that God's project of redemption has come and we are invited into that life of Kingdom community here and now. The Gnostic and individualistic objections to dualism are both unsupported, and fail to diagnose the real problem. Rejecting dualism isn't the solution, as dualism isn't the problem.

WHERE IS THIS SOUL OF THE GAPS?

The main argument from neuroscience proffered by Brown and Strawn against dualism is a soul-of-the-gaps objection. They write,

> However, three centuries ago, Descartes did not have access to what is known in modern neurology. Thus, he could not imagine how it could be that matter—that is, physical bodies and brains—could do anything rational or intelligent. So he concluded that these human capacities must be due to a nonmaterial thing.[28]

"Descartes," say Brown and Strawn, "was forced to the conclusion that we must have a nonmaterial soul due to the lack of knowledge during his time of the functioning of the human brain."[29] Brown and Strawn seem unaware of Descartes's extensive anatomy and physiology research.[30] Descartes knew quite well that mental states often depend on brain states.

Regardless, this soul-of-the-gaps objection fails to understand why Descartes and many others are dualists. Plato, Augustine, Descartes, Butler, and Reid held dualism in virtue of being aware of themselves from the first-person perspective as not reducible to or identical to their body.[31] Others, such as Aristotle and Aquinas, arrived at different kinds of dualism by analyzing positive arguments for the soul. After a detailed look at the literature, Stewart Goetz and Charles Taliaferro conclude, "There is not the least bit of evidence for the idea that they arrived at their belief in the soul's existence after failing to explain various experiences in terms of what goes on in the physical world."[32]

It is the awareness of self and metaphysical arguments that motivate dualism. This accords well with the prevailing view among cognitive scientists that dualism is a widespread, pretheoretical belief, shared across cultures, and developed in infancy.[33] This shouldn't be the case if dualism is simply maintained out of ignorance. We are, as Paul Bloom says, "natural Cartesians."[34]

Moreover, this soul-of-the-gaps objection presumes that dualists are ignorant of the neurosciences. This is certainly not the case today. Nobel prize winning neuroscientist John C. Eccles defended dualism,[35] as did Nobel Prize winning physicist, Eugene Wigner.[36] Likewise, neuroscientists Wilder Penfield[37] and Matthew Stanford,[38] research psychiatrist Jeffery Schwartz,[39] and psychologists Nancy Duvall,[40] Todd W. Hall,[41] Jeffrey H. Boyd,[42] Eric L. Johnson,[43] Sherwood Cole,[44] and Stephen Greggo are all dualists who take embodiment seriously.[45] Many dualists conversant with the relevant neuroscience make their case from neuroscientific data.[46] Even nondualists begrudgingly recognize that the neurosciences are often based on the conceptual framework of dualism.[47] Moreover, several dualists, after analyzing the data, argue that neuroscience fails to support physicalism over dualism.[48] While one might disagree with their arguments, one cannot claim these dualists are neurologically ignorant. It is simply false that dualism is the result of neuroscientific ignorance or soul-of-the-gaps reasoning.

Finally, this objection presumes, quite prematurely, that neuroscience has somehow undermined dualism. There is a growing skepticism, even among neuroscientists, about inflated claims from neuroscience.[49] This is certainly true regarding philosophical issues like free will and the mind-body problem.[50] Skepticism aside, several nonreductive physicalists, in accord with dualists, are convinced that even a complete understanding of all the physical facts about the universe could not explain consciousness. One reason for this is what Joseph Levine calls the explanatory gap, our inability to provide or even comprehend a plausible explanation of how consciousness could fully depend upon a nonconscious, physical substrate.[51] David Chalmers explains, "If this is right, the fact that consciousness accompanies a given physical process is a further fact not explainable simply by telling the story about the physical facts. In a sense, the accompaniment must be taken as brute."[52]

Secondly, nonreductive physicalists are in widespread disagreement over how to explain consciousness. This can be seen in the "new mysterians," who hold that although materialism must be true, we will never understand how it could be true.[53] Moreover, the recent turn toward panpsychism and panprotopsychism, which view consciousness as an irreducible, fundamental feature of reality, belies the devoted confidence of nonreductive physicalists who take neuroscience to fully explain consciousness. Neuroscience is far from proving physicalism or disproving dualism.

Curiously, Brown and Strawn fail to address the work of any dualist we've mentioned, much less interact with their arguments. How can one make the claims Brown and Strawn have about a view they don't seem to have seriously researched? Ironically, it is out of their own ignorance that Brown and Strawn presume dualism to be held out of ignorance or a soul-of-the-gaps thesis.

NO, NEUROSCIENCE HASN'T EXORCIZED THE SOUL

Brown and Strawn offer two types of argument from neuroscience against dualism. The first is that dualism is committed to a disembodied view of spiritual formation, which is incompatible with neuroscientific findings. According to Brown and Strawn, "We are formed into mature, virtuous, and wise persons, not by some disembodied mystical process, but by life together in a body of persons."[54] The assumption here is that dualism is somehow committed to a disembodied mystical process of spiritual formation. Of course, historically some dualists have embraced a mystical process of spiritual formation that denigrates the body. Ascetic Christians, such as the Desert Fathers, are often cited as examples. However, such a sweeping claim is naïve. In his seminal work on views of the body in early Christianity, Peter Brown observes,

> Yet to describe ascetic thought as "dualist" and motivated by hatred of the body, is to miss its most novel and its most poignant aspect. Seldom, in ancient thought, had the body been seen as more deeply implicated in the transformation of the soul; and never was it made to bear so heavy a burden. For the Desert Fathers, the body was not an irrelevant part of the human person, that should, as it were, be "put in brackets" . . . It was, rather, grippingly present to the monk: he was to speak of it as "this body, that God has afforded me, as a field to cultivate, where I might work and become rich . . . In the desert tradition, the body was allowed to become the discreet mentor of the proud soul.[55]

Dualism didn't always or even commonly lead to a mystical disembodied process. Most dualists embraced the body and its positive role in spiritual formation. However, even if some Christian dualists did neglect the body, a historical connection is not a logical connection. Brown and Strawn fail to demonstrate a logical link between dualism and a mystical disembodied approach to spiritual formation. Therefore, we have no reason to think that dualism must embrace a disembodied notion of spiritual formation.

While we can think of no contemporary Christian dualists who adopt such a view, there are many Christian dualists who reject it. For example, spiritual disciplines, a historic cornerstone of Christian spiritual formation, are not disembodied mystical processes.[56] J. P. Moreland, who is as dualist as anyone, explains this well:

> A Christian spiritual discipline is a repeated bodily practice, done over and over again, in dependence on the Holy Spirit and under the direction of Jesus and other wise teachers in his way, to enable one to get good at certain things in life that one cannot learn by direct effort.[57]

Willard arranges spiritual disciplines into two categories: abstinence/detachment (solitude, silence, fasting, frugality, chastity, secrecy, sacrifice) and engagement (study, worship, celebration, service, prayer, fellowship, confession, submission).[58] These disciplines, says Willard, "essentially involve bodily behaviors" as "whatever is purely mental cannot transform the self."[59] This is an outright rejection of a disembodied mystical process.

The second argument from Brown and Strawn is that neuroscience has made certain discoveries that present an understanding of spiritual growth that is incompatible with dualism. These discoveries support three theses.

> *Developmental Thesis*: Spiritual formation is a process that continues through adulthood.[60]
>
> *Interpersonal Thesis*: Spiritual formation takes place in virtue of interpersonal interactions such as imitation, shared attention, attachment, empathy, language, and story.
>
> *Bodily Process Thesis*: Human characteristics, such as rationality, relationality, morality, and religiousness are the outcome of the functioning of our bodies and brains, not a nonmaterial soul or mind.[61]

Let's begin with the developmental thesis. That sanctification is a gradual process has been known for quite some time (2 Cor. 4:16; Gal. 4:19), and is recognized by dualists.[62] It isn't clear how these are incompatible, and Brown and Strawn present no argument for us to analyze. Perhaps this objection is motivated by presuming that a soul is fully formed once it comes into existence and therefore cannot develop. But why should the dualist embrace this? There is nothing contradictory in holding that the soul psychologically develops in conjunction with the body over its lifetime. While most dualists hold that the soul is mereologically simple, as it has no parts, the soul is complex with respect to its modes or properties and causal powers. These features are what change as the person, body and soul, matures.[63] Hence, there is nothing about dualism that is incompatible with the developmental thesis.

Likewise, it is not at all clear how dualism is incompatible with the interpersonal thesis. Again, Brown and Strawn merely assert this incompatibility without explanation or argument. Rather, they presume that dualism leads to individualism, which undermines the importance of interpersonal relations for spiritual formation. However, we have shown that dualism does not necessarily lead to individualism. Moreover, there is nothing inconsistent about the dualist holding that the ontology of the soul is such that we require interpersonal relationships in order to grow spiritually.[64] Consequently, dualism is not incompatible with the interpersonal thesis.

The bodily process thesis, or something like it, is popular among Christian physicalists when objecting to dualism. The strength of this objection comes from what "the outcome of the functioning of our bodies and brains" means exactly. Presumably, that will be determined by the neurological evidence. Here are some examples they discuss.

> *Rationality*: fMRI studies show that brain activity increases in specific areas of the brain in conjunction with certain mental acts: the left side of the cerebral cortex when asked to perform language tasks, different but overlapping areas of the left cerebral cortex when listening to someone talk, and a different pattern of cerebral cortex areas when solving mathematical problems.
> *Relationality*: fMRI studies show that a participant's subjective experience of being shocked triggered a very similar pattern of brain activity that is triggered when they expect their friend to be shocked.
> *Morality*: fMRI studies show that the more complexity the moral reasoning the more intense is the brain activity in a particular region of the brain. Individuals with damage to the lower middle portions of their frontal lobes exhibit an inability to use moral guidelines.
> *Religiousness*: According to fMRI studies, when Buddhist monks and Catholic nuns reported reaching a state of "oneness" during meditation there was increased frontal lobe activity, and decreased right parietal lobe activity. Similar studies showed that when speaking in tongues activity in the frontal lobes and left temporal lobe decreased significantly.

From these studies, among others they mention, Brown and Strawn make the following conclusions: acts of rationality are "based on" and are "an outcome of" patterns of brain activity;[65] "interpersonal empathy is based on mirroring the emotional experience of the other's pain within your own brain";[66] morality is based on brain activity; and "religious states are associated with identifiable changes in the distribution of brain activity."[67] Taken together, Brown and Strawn conclude that rationality, relationality, morality, and religiousness are outcomes of the functioning of our bodies and brains, not a soul or mind.

Much can be said in reply to these kinds of arguments. First, these studies only show close correlations between specific mental states and localized brain states. This correlation may be evidence of a causal interaction between mental states and brain states. However, the direction of causation is by no means always clear; in some cases, it looks like the causal relation may be from the mental state to the brain state. However, in cases where mental states do seem dependent on brain states, there is no reason for a dualist to resist such claims. So, when Brown and Strawn speak of certain mental states as "based on" or "an outcome of" certain brain states, we can only take them to mean that there is a causal or dependence relation between these mental states and these brain states. However, almost every dualist affirms this kind of dependence and interaction. Although neuroscience has helped us understand *how* the mind depends on the brain in some cases, *that* biology plays a role in our thoughts and behavior was known by the ancient Hebrews and first-century Christians.[68]

Brown and Strawn seem to assume that if dualism is true then the mind should not depend on the brain in any way. However, minimal dualism accepts the possibility that such dependence may be pervasive. However, the fact that mental states may depend on brain states does not show that they are identical. Nor does it show that there is no dependence in the other direction. No discoveries in neuroscience show that mental states play no important causal role in our lives. In fact, if neuroscience did show anything like that, it would undermine the kind of "nonreductive physicalism" Christian physicalists typically affirm. If mental states are completely explicable in terms of brain activity, then it is hard to see how one could resist a reductive form of physicalism.

Physicalists often fail to recognize the logical relations that hold between self-conscious beings and their bodies. For example, it does not follow from any neuroscientific findings that because the brain is used to do certain things that the brain is what does those things. As Roderick Chisholm observes,

> Many have assumed—quite obviously incorrectly—that from the fact that one thinks by *means of* the brain, it follows logically that it is *the brain* that thinks. We walk by means of our feet, but our feet do not walk in the sense we do (if they did, then they would have feet).[69]

Even ardent antidualist Nancey Murphy admits that current neurological evidence does not rule out dualism.[70] The dualist can always interpret such studies as showing that the nature of the soul is such, that while embodied, it is dependent on the brain in a variety of ways. Significant minimal dualism is completely open to whatever causal dependence is supported by the evidence. The only way neuroscience could disprove this kind of dualism would be to prove epiphenomenalism. But as one of us has argued elsewhere, the discovery that the mind is epiphenomenal would imply that our experience of ourselves as conscious agents is illusory. However, this cannot be the case, as all of science, including neuroscience, depends on our self-understanding as conscious agents.[71]

PROBLEMS FOR CHRISTIAN PHYSICALISM AND SPIRITUAL FORMATION

So far, we have shown that each objection from Brown and Strawn fails to undermine dualism in favor of Christian physicalism. In the following section, we demonstrate how the Christian physicalism of Brown and Strawn is incompatible with certain theological and philosophical preconditions of spiritual formation theory and practice.

First, an observation: it isn't clear how the view of spiritual formation that Brown and Strawn present is distinctively Christian, or Christian at all. For example, their view lacks a robust role for the Holy Spirit's active role in sanctification. The clear teaching of scripture is that the Holy Spirit is the empowering/transformational agent of ongoing sanctification (e.g., 1 Cor. 6:11; Gal. 5:16–24; 2 Thess. 2:13; 1 Peter 1:2). Hence, any account of spiritual formation must be grounded in the sanctifying activity of the Holy Spirit. Yet, Brown and Strawn fail to attribute any activity to the Holy Spirit. All the work is done by church bodies functioning as self-forming systems, networks of communication and interaction between persons who imitate those who imitate Christ. However, it is in partaking of and participating in the divine nature—not the mere moral influence of a church body—that the believer is sanctified (2 Peter 1:4; see also 1 Cor. 1:9). It is entirely possible for their account of sanctification that God does not even exist. Of course, this needn't be the case for all Christian physicalist accounts. It seems to be produced by an overemphasis on a purely scientific, rather than a scientifically informed, account of spiritual formation.

Christian Physicalism's Fragmented Persons

According to Brown and Strawn, Christian physicalism holds that, as bodies, we have a single, unified nature.[72] However, we will argue that their ontology of the human person is neither holistic, unified, nor substantial. Rather, it implies we are biological aggregates. Consequently, their view cannot ground central features of human persons like persistence, agency, and the unity of consciousness, each of which is necessary for any account of Christian spiritual formation.

Their account in the chapter titled "How Bodies Become Persons" is often unclear, moving between talk of the person, the brain, and the human mind, each of which are characterized as having a self-organizing nature.[73] Elsewhere, they state that a person is "a uniquely organized pattern that is dynamic in its developmental process of self-organization"[74] and that, "we human beings are also complex dynamical systems."[75] They seem to reject the self as a substance, a genuine unified entity, and identify the self as a function or process.

Brown and Strawn offer two pictures of the self. They maintain that we are wholly physical bodies. Yet, they also assert that we are a function or process. It isn't clear how both descriptions can be correct. But suppose we assume that the human person is a wholly physical body and that such a body is just a collection of complex processes and functions. It is not clear how such a view can explain how humans can be subjects of consciousness and agents. Nor it is clear how it can explain how humans can undergo psychological and spiritual transformation. These facts pose significant problems for physicalists.[76]

As a biological organism, the human body undergoes an unrelenting process of part replacement. Moment by moment your body absorbs new parts and expels old parts. This takes place through respiration and metabolic processes, among others. Strictly speaking, the body you had twenty seconds ago is not exactly the same body you have now. It is similar, but not identical. This is true for the same reason that the body you have now is not identical to the body you had when you were an infant. If your body is nothing more than a wholly physical biological organism comprised of various complex processes and functions, then your body does not exist from one moment to the next. Your body five minutes ago does not have all the parts that your body has right now. It is fairly obvious that they are not identical. However, if you are identical to your body, a wholly physical biological organism comprised of various complex processes and functions, then you do not persist through part replacement either. That is, the person that existed five minutes ago is not the person you identify as yourself right now.

Like other Christian physicalists, Brown and Strawn hold that there are features of your consciousness, agency, and psychology that are emergent and thus not reducible to your body. So perhaps they can escape this objection by holding that although the body to which you are identical does not persist, the emergent properties that are a part of what comprises you do persist, and so in some sense you do as well. But why think this is possible? If it is possible, this implies that the persisting entity is not wholly physical, since it has nonphysical emergent properties that seem essential to it.

Consciousness, agency, and psychological change are features of an individual person. Consciousness does not exist without a subject of consciousness. The same is true for agency and psychological change. However, if the body does not persist and the body is the person, then the consciousness and agency of that person, that biological organism, does not persist. Likewise, if the body I am identical to does not persist then there is literally nothing that undergoes psychological change. Hence, emergent properties are not sufficient to ground the persistence of a human person.

Consequently, Christian physicalism, at least the version of Brown and Strawn, makes the notion of spiritual formation incoherent. Spiritual formation is a process that an individual person goes through. When a self grows in patience or peace that self must persist through that change. If some other thing replaces the self, then the initial self does not develop but passes out of existence. A self that does not persist cannot undergo any transformation at all. Ironically, like many Christian physicalists, Brown and Strawn present Christian physicalism as a holistic and unified view of human persons, and claim that dualists must reject this. However, as we have shown, the opposite is true.

Interpersonal Knowledge, Phenomenal Consciousness, and Christian Physicalism

Eternal life, and by extension spiritual formation, is characterized by Jesus as knowledge of God (John 17:3). As one of us has argued elsewhere, the kind of knowledge Jesus refers to here is an *interpersonal knowledge*, which is a species of knowledge by acquaintance.[77] To see this, consider the following propositions:

a. Laura and Jan know that Jesus is the smartest person to have ever lived.
b. Laura and Jan know Jesus.

These two propositions express different kinds of knowledge. In (a) what is known is a proposition about Jesus, that he is the smartest person to have ever lived. However, in (b) what is known is not a proposition, but a person, Jesus. Here is another way to understand how these two kinds of knowledge are distinct. Suppose that Jan knows everything there is to know about Laura, even though they've never met. Consider what happens when Jan spends the day with Laura. Clearly Jan "gets to know" Laura in a way different from all the facts that Jan knows about Laura. Jan gains interpersonal knowledge of Laura in virtue of her experience of Laura, her knowledge by acquaintance of Laura. This knowledge couldn't have come from any more propositional knowledge about Laura.

This kind of interpersonal knowledge by acquaintance is present in instances of shared attention and interpersonal attachment between individuals, both of which Brown and Strawn recognize as of great developmental importance.[78] Moreover, the kind of knowledge present in many spiritual formation practices is a type of *intrapersonal self-knowledge*, which is also a species of knowledge by acquaintance. Knowing the truth that anger keeps me from unity with God is vastly different than my experiential knowledge of a lack of unity with God when I am angry. Consequently, knowledge by acquaintance is a central feature of spiritual formation. However, we argue that Christian physicalism is at odds with interpersonal knowledge by acquaintance.

A prominent thought experiment many take seriously to undermine physicalism can be adapted for our purposes here.[79] In Eleonore Stump's version, we are invited to consider Mary, a neuroscientist who is omniscient of the scientific facts about interpersonal knowledge. However, Mary has never met another person before. That is, Mary has never experienced interpersonal knowledge. Imagine one day Mary is united with her biological mother who loves her very much. For the first time, Mary will come to know what it is like to be loved by another. Stump writes,

> And this will be new for her, even if in her isolated state she had as complete a scientific description as possible of what a human being feels like when she senses that she is loved by someone else . . . Mary will also come to know what it is like to be touched by someone else, to be surprised by someone else, to ascertain someone else's mood, to detect affect in the melody of someone else's voice, to match thought for thought in conversation, and so on.[80]

Mary will also come to know her mother—have knowledge of her mother—in addition to knowing what it is like to know and experience her mother.

Cases like this have proven extremely difficult for physicalism. We argue this difficulty extends to Christian physicalism as well. If physicalism is true then the physical facts about the world should exhaust all the facts about the world. Hence, if one knows all the physical facts about interpersonal knowledge, then there are no further facts one can know regarding interpersonal knowledge. However, this is not what happens in Mary-type thought experiments. Mary knows all the physical facts about interpersonal knowledge, however, she still comes to know something new when she meets a person, her mother, for the first time. That is, the physical facts are not the only facts. Hence, physicalism is false.

Consider again the intrapersonal self-knowledge mentioned earlier. This kind of knowledge is necessarily first-person and cannot be known through third-person scientific inquiry. I can read in the Bible or learn from a friend that my anger keeps me from full unity with God, but that is not sufficient or even necessary for me to attend to the phenomenology of feeling God's distance from me in my anger. That knowledge I gain in my first-person experience, not through third-person propositional knowledge. But the kind of knowledge that Brown and Strawn focus on and ground their view of spiritual formation in is third-person scientific knowledge. That is, their account of spiritual formation does not have, and can't seem to gain, the recourses to account for the kind of self-knowledge involved in important aspects of spiritual formation.

Christian Physicalism and the Nature of Emotions

Lastly, we wish to draw attention to the metaphysics of emotion that Brown and Strawn advance. This is significant, as emotions have been taken as a significant aspect of spiritual formation. Jesus, for example, begins his Sermon on the Mount with a profound treatment of anger and contempt. He does so as these complex mental states, including emotions, are at the ground floor of what needs to be transformed in us. Although Brown and Strawn recognize that emotions are an important part of spiritual formation, their account of emotions faces difficult problems.

According to Brown and Strawn, "Emotions are continuous brain-body adjustments and attunements to our current situation, most particularly our social situation."[81] They continue with the following: "they [emotions] are by-products of automatic bodily adjustments to the situation that, when experienced consciously, provide information about the nature of our current relationship to the social surrounding.[82] In a later chapter, Brown and Strawn state,

> [W]e are not saying that subjective, inner experiences and emotions are not important in the Christian life. Rather, emotions and feelings are bodily reactions that serve the purpose of giving us information about the significance of the events, including religious events, that we are involved in, physically or in our imaginations.[83]

The view as stated is at best unclear and at worst obviously incoherent. Brown and Strawn first say that emotions are "by-products of bodily adjustments," but then go on to imply that emotions are "subjective, inner experiences." But it is not clear how both can be true. Emotions, as subjective inner experiences with motivational and epistemological components, cannot be identical to bodily reactions.[84] Identity is a necessary relation. A thing must be identical to itself. So, if emotions are identical to bodily reactions or brain states, then there cannot be an instance of an emotion that is not a bodily reaction or brain state. However, the Christian physicalist is faced with the following counterexample: biblically, God has emotions, but does not have a body, brain states, or any physical features.[85] Therefore, emotions cannot be identical to any physical thing, bodily reactions, brain states, or otherwise.

Perhaps Brown and Strawn mean only that human emotions are identical to bodily reactions or brain states. This would escape our counterexample, although this move seems ad hoc. Regardless, this view faces a number of problems. First, notice that on their account, phenomenal consciousness is irrelevant to emotion. An individual can have the brain state or bodily reaction of anger although that person does not have the phenomenal experience of anger. The phenomenal experience of anger is not identical to the bodily reaction of anger, which is why one can appear to be angry, yet not actually be angry. Likewise, one can exhibit the bodily responses of fear, such as increased pulse rate, perspiration, and trembling without having any fear at all. For example, someone might tremble from excitement while entering a hot room expecting a surprise.[86] But an account of emotions that leaves out the phenomenal quality, the "what-its-like-to-experience" feature of emotions, has simply eliminated the fundamental feature of emotion. To *feel* anger just is to *be* angry. So, the account of Brown and Strawn does not provide a sufficient condition for what it is for one to be in an emotional state. Moreover, because people can behave as if they are afraid, yet not actually be afraid, their account also does not give a necessary condition for emotions.

BODILY SOULS AND SPIRITUAL FORMATION

Now that we have responded to the objections to dualism from Brown and Strawn and offered some problems for their version of Christian physicalism, we wish to make a positive contribution to the discussion. A main theme of this chapter has been that dualism is often misunderstood by its critics. In order to help remedy this problem we now present a specific version of dualism we have defended elsewhere.[87] We call this form of dualism the bodily soul view.

The Bodily Soul View

We agree with Christian physicalists that there is biblical emphasis on the value of the body. We retain this by borrowing from Augustine and Aquinas the insight that we are the kinds of souls that require bodies. Augustine, like Aristotle and Aquinas, considers the soul the very life of the body.[88] The body does not exist on its own, but subsists through the soul.[89] My body lives through,[90] and is vivified by my soul.[91] We are selves to be sure, but bodily selves that cannot function properly and be all they are intended to be without bodies. We might say that we are *bodily souls*, souls that exist in a bodily form or bodily manner.

Paradoxically, thinking of my soul as identical to myself rather than a part of myself allows for a more ontologically intimate relation between body and soul. It allows me to think of the body not as a part of myself, but my actual manner of being as a whole. I am a soul, but I am not a pure spirit, like an angel, but rather an incarnate or bodily self or soul. The relation between soul (or self) and body can be as intimate as you like. One might believe that the self cannot exist at all without a body. Or, perhaps more wisely, following Augustine and Aquinas, we could hold that the self cannot exist in the fullest and richest sense without a body. The soul can exist between death and the resurrection but cannot carry out all its functions if it does not exist in a bodily form. Thus, human salvation without a resurrected body would be incomplete.[92]

One might ask why, if self and body are so intimately related, we should not simply identify a person with his or her body. Why not opt for Christian physicalism, rather than dualism? The answer is that a person as a self must be distinguished from his or her body. Identity is a necessary relation. If I am identical to my body, then it is necessary that what is true of my body is also true of me and vice versa. However, because a person has some characteristics *qua self* that the person does not have *qua body*, it is not logically possible to identify a person with his or her body.

In our view, the human body plays a dual role. The self is a bodily *self*, and thus *my* body is not simply another object in the world. It is rather the form in which I exercise my agency. If I move from point A to point B, I do so by walking or biking or otherwise moving my body. However, the body is also experienced as an object in the world. It can and does exhibit the same indifference and recalcitrance as the rest of the physical world. If my legs are trapped under a car, I will not be able to move from point A to point B. If a brain tumor invades the region of my brain that controls my motor functions, I will similarly be unable to walk and move.

I thus find myself necessarily thinking of my body in two distinct ways: both as the locus of my agency; the form in which I exist as a conscious self, and as an object in the world; a physical entity that, like other physical entities, follows the laws of nature and does not always act as I want it to act. When we think of the body in this second way, we naturally think of it as something distinct from our self; we think of the body as if it were merely another object in the world, an entity whose characteristics I must take account of when I act. And when I think of my body as a material object in the world, it is natural and in fact valuable to objectify it, to study it scientifically as one might study any other object in the world.

When I think of my body as the form in which I exist as a self, it is not a mere object, but myself. When I think of my body in this second, objectified manner, however, it is natural to think of it, not as myself, but as something that the self must take into account in its agency, a part of the physical world. When I think of the body in this objectified way, it is natural to think of it as something distinct from the self. Hence, the language of body and soul as two distinct entities is not only appropriate because of the possibility of life after death; it is also appropriate insofar as we conceive of the body in this objectified manner.

Christians should continue to affirm the traditional Christian view that human persons are souls or selves, and that souls are not identical with any physical objects. However, we should not think of our souls as ghostly entities that live inside us. Strictly speaking we do not have souls; we are souls. However, on a Christian view this in no way diminishes the importance of the body, because we are embodied, incarnate souls. I am at the same time wholly soul and yet fully bodily. Wittgenstein says that, "The human body is the best picture of the human soul."[93] That seems right from a Christian perspective.

Bodily Souls and Embodied Spiritual Formation

In conclusion, we would like to offer brief statements as to how our bodily soul view explains some bodily aspects of Christian spiritual formation. We offer the following for consideration.

1. Because our body is the primary manner in which we manifest our presence in the world, our body must be at the center of our sanctification.
2. Because our body is the primary manner in which we manifest our presence in the world, we must pay attention to how we make our selves known through our bodies and also how we can hide our selves by concealing our bodies.
3. Because our body has both private and social dimensions our sanctification will also have private and social dimensions. One cannot flourish without the other.
4. Because there are intimate interactions between body and soul, what happens to my body significantly shapes my sanctification. Hence, my environment will always contribute to my spiritual formation.

CONCLUSION

The history of psychology, psychiatry, and by extension neuroscience, is one of increasing reductionism, some of which was the product of political motivations, rather than empirical discovery. Somogy Varga explains that various changes to the definition of what qualifies as a mental disorder in the DSM-III and DSM-IV were produced, not by scientific discovery, but sociological pressure to legitimize psychiatry as a science. This was done by redefining mental disorders in biological terms, and by eliminating any kind of talk that might imply dualism.[94] Of course, not all reductions come about this way. But what this shows is that academic communities have in the past rejected dualism and embraced physicalism for illegitimate reasons. This is true, or so we have argued, of the antidualism and Christian physicalism of Brown and Strawn.

NOTES

1. See Steve L. Porter, "Sanctification in a New Key: Relieving Evangelical Anxieties over Spiritual Formation," *Journal of Spiritual Formation & Soul Care* 1, no. 2 (2008): 129–128.

2. See, 2 Corinthians 3:18; Romans 5:1–5; 12:2; Ephesians 4:14–16; Colossians 3:4–17; 2 Peter 1:2–11 and 3:18.

3. See Dallas Willard, "Spiritual Formation as a Natural Part of Salvation," in Dallas Willard, *Renewing the Christian Mind: Essays, Interviews, and Talks*, ed. Gary Black Jr. (New York: HarperOne, 2016), 301–319.

4. See for example, Dallas Willard, *Knowing Christ Today: Why We Can Trust Spiritual Knowledge* (New York: HarperOne, 2009) and Dallas Willard, *The Disappearance of Moral Knowledge* (forthcoming).

5. C. Stephen Evans, "Separable Souls: A Defense of 'Minimal Dualism,'" *Southern Journal of Philosophy* 19, no. 3 (1981): 313–332.

6. C. Stephen Evans, "Separable Souls: Dualism, Selfhood, and the Possibility of Life after Death," *Christian Scholar's Review* 34, no. 3 (2005): 327–340.

7. See, for example, Lynne Rudder Baker, "Need a Christian Be a Mind/Body Dualist?" *Faith and Philosophy* 12, no. 4 (1995): 498–504, and "Christian Materialism in a Scientific Age," *International Journal of Philosophy of Religion* 70, no. 1 (2011): 47–59; Trenton Merricks, *Objects and Persons* (Ithaca, NY: Oxford University Press, 2001), chapters 4 and 5; Peter van Inwagen, *Material Beings* (Ithaca: Cornell University Press, 1990), chapter 9; and Kevin Corcoran, *Rethinking Human Nature* (Grand Rapids, MI: Baker Academic, 2006), chapter 3.

8. Warren S. Brown and Brad D. Strawn, *The Physical Nature of Christian Life: Neuroscience, Psychology, and the Church* (New York: Cambridge University Press, 2012).

9. See for example, Lewis Ayers, "The Soul and the Reading of Scripture: A Note on Henri De Lubac," *Scottish Journal of Theology* 61, no. 2 (2008): 173–190; Dallas Willard, "Spiritual Disciples, Spiritual Formation, and the Restoration of the Soul," *Journal of Psychology & Theology* 26 (1998): 101–109; and J.P. Moreland, "Restoring the Substance of the Soul to Psychology," *Journal of Psychology and Theology* 26, no. 1 (1998): 29–43.

10. Brown and Strawn, 69.

11. For a reply to the conservation of energy objection see, Robin Collins, "The Energy of the Soul," in *The Soul Hypothesis: Investigations into the Existence of the Soul*, ed. Mark C. Baker and Stewart Goetz (New York: Continuum, 2011), 123–137.

12. For a detailed defense of this claim, see Dennis Bielfeldt, "Can Western Monotheism Avoid Substance Dualism," *Zygon* 36, no. 1 (2001): 153–177.

13. Raymond E. Brown, *An Introduction to the New Testament* (New York: Doubleday, 1997), 92, cited in Brown and Strawn, 22.

14. Brown and Strawn, 23. This argument is reiterated through the book, especially chapters 1–3.

15. Dallas Willard, *Renovation of the Heart: Putting on the Character of Christ* (Colorado Springs, CO: NavPress, 2002), 17.

16. Willard, *Renovation of the Heart*, 45.

17. Willard, *Renovation of the Heart*, chapter 9.

18. Brown and Strawn, 163.

19. Charles Taliaferro, "The Virtues of Embodiment," *Philosophy* 76 (2001): 111–125.

20. Richard, Swinburne, "What's So Good About Having a Body?" in *Comparative Theology: Essays for Keith Ward*, ed. Timothy Walter Bartel (London: SPCK, 2003), 137.

21. See Howard Robinson, "A Dualist Perspective on Psychological Development," in *Philosophical Perspectives on Developmental Psychology*, ed. J. A. Russell (Oxford: Basil Blackwell, 1987), 119–139, and "A Dualist Theory of Embodiment," in *The Case for Dualism*, ed. John R. Smithies and John Beloff (Charlottesville: University of Virginia Press, 1989), 43–58.

22. See for example, Gordon Barnes, "Is Dualism Religiously and Morally Pernicious?" *American Catholic Philosophical Quarterly* 78, no. 1 (2004): 103.

23. Augustine, *Confessions*, III.6.

24. See C. Stephen Evans, "Who is the Other in *The Sickness unto Death*? God and Human Relations in the Constitution of the Self," and "Kierkegaard's View of the Unconscious," both in C. Stephen Evans, *Kierkegaard on Faith and Self: Collected Essays* (Waco, TX: Baylor University Press, 2006).

25. For example, Boyd writes, "The word *self* has replaced the term *soul* in popular culture, with the effect that people tend to think of themselves without thinking that God is important to their self-concept. We live in a pre-Copernican age where God, if God is thought to exist at all, is understood as being in orbit around the self, strengthening self-esteem or weakening the self through guilt feelings. The center of focus in our time is on the self, on the individual and the individual's need for autonomy, self-determination, fulfillment, happiness, and self-sufficiency" (Jeffrey H. Boyd, "Losing Soul: How and Why Theologians Created the Mental Health Movement," *Calvin Theological Journal* 30 [1995]: 473).

26. Dallas Willard, *The Spirit of the Disciples: Understanding How God Changes Lives* (New York: Harper & Row, 1988), 33–34.

27. Dallas Willard, *The Spirit of the Disciples*, 31.

28. Brown and Strawn, 32.

29. Brown and Strawn, 47.

30. In fact, his fascination with anatomy and physiology prompted one suggestion that "if Descartes were alive today, he would be in charge of the CAT and PET scan machines in a major research hospital" (Richard Watson, *Cogito Ergo Sum: The Life of René Descartes*, rev. ed. [Boston: David Godine, 2007], 15).

31. For contemporary work on dualism and self-awareness, see J. P. Moreland, "Substance Dualism and the Argument from Self-Awareness" *Philosophia Christi* 13, no. 1 (2011): 21–34. See also, David Barnett, "The Simplicity Intuition and Its Hidden Influence on the Philosophy of Mind," *Noûs* 42 (2008): 308–355; and "You are Simple" in *The Waning of Materialism*, eds. Robert Koons and George Bealer (Oxford: Oxford University Press, 2010), 161–174.

32. Stewart Goetz and Charles Taliaferro, *A Brief History of the Soul* (Oxford: Wiley-Blackwell, 2011), 155.

33. See, for example, Paul Bloom, *Descartes' Baby: How the Science of Child Development Explains What Makes Us Human* (Cambridge, MA: Basic Books, 2004), chapter 7.

34. Bloom, *Descartes' Baby*, xii.

35. See for example, Karl R. Popper and John C. Eccles, *The Self and Its Brain*, 2nd ed. (New York: Springer-Verlag, 1985); John C. Eccles, *The Human Psyche* (New York: Springer, 1980), lectures 1 and 2; *Evolution of The Brain: Creation of The Self* (London: Routledge, 1989), chapter 9; and *How the Self Controls Its Brain* (Berlin: Springer-Verlag, 1994), chapters 2 and 10.

36. Wigner argued that quantum mechanics requires a commitment to a strong variety of mind-body dualism. For example, Wigner writes, "Until not many years ago, the 'existence' of a mind or soul would have been passionately denied by most physical scientists. . . . There are [however] several reasons for the return, on the part of most physical scientists, to the Spirit of Descartes' 'Cogito ergo sum' . . . When the province of physical theory was extended to encompass microscopic phenomena, through the creation of quantum mechanics, the concept of consciousness came to the fore again: it was not possible to formulate the laws of quantum mechanics in a consistent way without reference to consciousness" (Eugene Paul Wigner, "Remarks on the Mind–Body Question," originally published in *The Scientist Speculates*, ed. I. J. Good [London: Heinemann, 1961], 284–302, reprinted in *Quantum Theory and Measurement*, ed. J. A. Wheeler, and W. H. Zurek [Princeton, NJ: Princeton University Press, 1983], 168–169).

37. See Wilder Penfield, *The Mystery of Mind* (Princeton: Princeton University Press, 1975).

38. Stanford states, for example, "God has created us as embodied spirits, having physical and spiritual aspects to our being" (Matthew S. Stanford, *The Biology of Sin: Grace, Hope, and Healing for Those Who Feel Trapped* [Downers Grove, IL: InterVarsity, 2010], 12).

39. Jeffrey Schwartz, "A Role for Volition and Attention in the Generation of New Brain Circuitry: Toward a Neurobiology of Mental Force," *Journal of Consciousness Studies*, 6, nos. 8–9 (1999): 115–142; and Jeffrey Schwartz and Sharon Begley, *The Mind and the Brain: Neuroplasticity and the Power of Mental Force* (New York: HarperCollins, 2002), 54–95.

40. Nancy S. Duvall, "From Soul to Self and Back Again," *Journal of Psychology and Theology* 26, no. 1 (1998): 6–15.

41. Todd W. Hall, "The Soul or Substantive Self as Experiencer, Actualizer, and Representative in Psychoanalytic Theory," *Journal of Psychology and Theology* 26, no. 1 (1990): 55–65.

42. See for example, Jeffrey H. Boyd, *Reclaiming the Soul: The Search for Meaning in a Self-Centered Culture* (Cleveland, OH: The Pilgrim Press, 1996); "The Soul as Seen Through Evangelical Eyes, Part I: Mental Health Professionals and 'The Soul,'" *Journal of Psychology and Theology* 23, no. 3 (1995): 151–160; "The Soul as Seen Through Evangelical Eyes, Part II: On Use of the Term 'Soul,'" *Journal of Psychology and Theology* 23, no. 3 (1995): 161–170; and "A History of the Concept of the Soul during the 20th Century," *Journal of Psychology and Theology* 26, no. 1 (1998): 66–82.

43. Eric L. Johnson, *Foundations of Soul Care: A Christian Psychology Proposal* (Downers Grace, IL: IVP Academic, 2007). Johnson holds that the soul is an immaterial substance (16–17). See also Johnson, "Whatever Happened to the Human Soul? A Brief Christian Genealogy of a Psychological Term," *Journal of Psychology and Theology* 26, no. 1 (1998): 16–28.

44. Sherwood O. Cole, "Don't Disembody Me Just Yet! A Christian Perspective on our Biological Nature," *Journal of Psychology and Christianity* 21, no. 2 (2002): 15–60.

45. Greggo argues that compassionate clinical care is enriched by dualism, whereby we have "the increased awareness that persons as living souls are formed by a creative convergence of both human and divine nature and nurture" (Stephen P. Greggo, "Soul Origin: Revisiting Creationist and Traducianist Theological Perspectives in Light of Current Trends in Developmental Psychology," *Journal of Psychology and Theology* 33, no. 4 [2005]: 266).

46. See for example, Riccardo Manzotti and Paolo Moderato, "Neuroscience: Dualism in Disguise," in *Contemporary Dualism: A Defense*, eds. Andrea Lavazza and Howard Robinson (New York: Routledge, 2014), 81–97; Alessandro Antonietti, "Must Psychologists Be Dualists?" in *Psycho-Physical Dualism Today: An Interdisciplinary Approach*, eds. A. Antonietti, A. Corradini, and E. J. Lowe (Lanham, MD: Rowman & Littlefield, 2008), 37–67; and Mario Beauregard and Denyse O'Leary, *The Spiritual Brain* (New York: HarperCollins, 2007), chapters 5 and 6.

47. See, M. R. Bennett and P. M. S. Hacker, *Philosophical Foundations of Neuroscience* (Malden, MA: Blackwell, 2003); W. Teed Rockwell, *Neither Ghost nor Brain* (Cambridge, MA: MIT Press, 2005); William R. Uttal, *The New Phrenology: The Limits of Localizing Cognitive Processes in the Brain* (Cambridge, MA: The MIT Press, 2001), and William R. Uttal, *Dualism: The Original Sin of Cognitivism* (Mahwah, NJ: Lawrence Erlbaum Associates, 2004).

48. See for example, Eric LaRock, "Neuroscience and the Hard Problem of Consciousness," in *Neuroscience and the Soul: The Human Person in Philosophy, Science, and Theology*, eds. Thomas M. Crisp, Steven L. Porter, and Gregg A. Ten Elshof (Grand Rapids, MI: Eerdmans, 2016), 151–180; Eric LaRock and Robin Collins, "Saving Our Souls from Materialism," in *Neuroscience and the Soul,* 137–146; Eric LaRock, "Is Consciousness Really a Brain Process?" *International Philosophical*

Quarterly 48, no. 2 (2008): 201–222; and J. P. Moreland, "Christianity, Neuroscience, and Dualism," in *The Blackwell Companion to Science and Christianity*, eds. J. B. Stump and Alan Pagget (Oxford: Blackwell, 2012), 467–479; Jeffrey M. Schwartz and Sharon Begley, *The Mind and the Brain*; Mihretu P. Guta, "Neuroscience or Neuroscientism?" *Perspectives on Science and Christian Faith* 63, no. 1 (2011): 69–70; and Terence Horgan, "Nonreductive Materialism and the Explanatory Autonomy of Psychology," in *Naturalism: A Critical Appraisal*, eds. Steven Wagner and Richard Warner (Notre Dame, IN: University of Notre Dame Press, 1993), 313–314.

49. See for example, Paolo Legrenzi and Carlo Umiltà, *Neuromania: On the Limits of Brain Science*, trans. Frances Anderson (Oxford: Oxford University Press, 2011); Robert G. Shulman, *Brain Imaging: What it Can (and Cannot) Tell Us about Consciousness* (Oxford: Oxford University Press, 2013); Sally Satel and Scott Lilienfel, *Brainwashed: The Seductive Appeal of Mindless Neuroscience* (New York: Basic Books, 2013); and Raymond Tallis, *Aping Mankind: Neuromania, Darwinitis and the Misrepresentation of Humanity* (New York: Routledge, 2011).

50. For a defense of this claim see, Moreland, "Christianity, Neuroscience, and Dualism."

51. Joseph Levine, "Materialism and Qualia: The Explanatory Gap," *Pacific Philosophical Quarterly* 64 (1983): 354–361.

52. David Chalmers, *The Conscious Mind: In Search of a Fundamental Theory* (Oxford: Oxford University Press, 1996), 107.

53. Owen Flanagan, *The Science of the Mind*, 2nd ed. (Cambridge, MA: MIT Press, 1991), 313. The new mysterians include Colin McGinn, *The Mysterious Flame: Conscious Minds in a Material World* (New York: Basic Books, 1999), chapter 2; and Thomas Nagel, *Mind and Cosmos: Why the Materialist Neo-Darwinian Conception of Nature Is Almost Certainly False* (Oxford: Oxford University Press, 2012), 58–65.

54. Brown and Strawn, 87.

55. Peter Brown, *The Body and Society: Men, Women and Sexual Renunciation in Early Christianity* (New York: Columbia University Press, 1988), 235–236.

56. According to scripture, we present our *bodies* to God as a living sacrifice (Rom. 12:1), with repeated *bodily exercise* (1 Cor. 9:24–27; 1 Tim. 4:7–8) involving specific *body parts* (Rom. 6:11–13,19), resulting in putting to death our bad habits (Col. 3:5).

57. J. P. Moreland, *Kingdom Triangle: Recover the Christian Mind, Renovate the Soul, Restore the Spirit's Power* (Grand Rapids, MI: Zondervan, 2007), 152.

58. Willard, *The Spirit of the Disciplines*, 158–190.

59. Willard, *The Spirit of the Disciplines*, 152.

60. They write, "For some reason, our dualist presuppositions about persons created a disconnect between our understanding of Christian formation in children and our comprehension of the forces at work in adult Christian life. It is not true that the impact on human development of all of these processes of ongoing reciprocal interaction with one's social environment comes to an end somewhere in later childhood or early adolescence. Rather, this developmental process is ongoing, allowing for continuing development, formation, and change as adults" (Brown and Strawn, 70).

61. Brown and Strawn, 30–46.

62. For an insightful treatment of the gradual nature of sanctification by a dualist, see, Steven L. Porter, "The Gradual Nature of Sanctification," *Themelios* 39, no. 3 (2014): 470–483.

63. For a detailed dualist account of this, see Robinson, "A Dualist Perspective on Psychological Development."

64. For such an account, see John Coe and Todd Hall, *Psychology in the Spirit: Contours of a Transformational Psychology* (Downers Grove, IL: IVP Academic, 2010), section 3.

65. Brown and Strawn, 32.

66. Brown and Strawn, 35.

67. Brown and Strawn, 46.

68. Stanford observes, "The ancient Hebrews and the first-century Christians were unaware of how the brain and nervous system function. They were not unaware, however, that biology played a significant role in thoughts and behavior (e.g., Jeremiah 17:10, Psalm 26:2, 73:21–22)" (Stanford, *Biology of Sin*, 135). For more, see, R. Shane Tubbs, et al., "Roots of neuroanatomy, neurology, and neurosurgery as found in the Bible and Talmud," *Neurosurgery* 63 (2008): 156–162.

69. Roderick Chisholm, "Mind," in *Handbook of Metaphysics and Ontology*, eds. Hand Burckhardt and Barry Smith (Munich: Philosopia Veril, 1991), 544.

70. Nancey Murphy, *Bodies and Souls, or Spirited Bodies?* (Cambridge, UK: Cambridge University Press, 2006), 112.

71. Evans, "Separable Souls: Dualism, Selfhood, and the Possibility of Life After Death," 335.

72. Brown and Strawn, 5.

73. Brown and Strawn, 54.

74. Brown and Strawn, 125.

75. Brown and Strawn, 75.

76. This section is influenced by the following papers: J. P. Moreland, "Restoring the Substance of the Soul to Psychology"; and J. P. Moreland, "Spiritual Formation and the Nature of the Soul," *Christian Education Journal* (2000): 25–43. Although Moreland raises different objections than ours, his approach to analyzing substances in contrast to aggregates or property things inspirited our approach in this section.

77. Brandon Rickabaugh, "Eternal Life as Knowledge of God: An Epistemology of Knowledge by Acquaintance and Spiritual Formation," *Journal of Spiritual Formation and Soul Care* 6, no. 2 (2013): 204–228.

78. Brown and Strawn, 58–63.

79. The original thought experiment was introduced in Frank Jackson, "Epiphenomenal Qualia," *Philosophical Quarterly* 32 (1982): 127–136; and more fully developed in, "What Mary Didn't Know," *Journal of Philosophy* 83, no. 5 (1986): 291–295.

80. Eleonore Stump, *Wandering in Darkness: Narrative and the Problem of Suffering* (Oxford: Oxford University Press, 2010), 52.

81. Brown and Strawn, 148.

82. Brown and Strawn, 149.

83. Brown and Strawn, 162.

84. On the nature of emotions see, Robert C. Roberts, *Emotions: An Essay in Aid of Moral Psychology* (Cambridge, UK: Cambridge University Press, 2003), chapter 2; and Michael S. Brady, *Emotional Insight: The Epistemic Role of Emotional Experience* (Oxford: Oxford University Press, 2013), chapter 2.

85. We are aware that there is a theological tradition that holds that God is "impassible" and thus does not have emotions. We cannot argue for our view here, but simply want to affirm that it is hard to see how the view that God has no emotions can be consistent with the biblical picture of God.

86. This example is from M. R. Bennett and P. M. S. Hacker, *Philosophical Foundation of Neuroscience* (Oxford: Blackwell, 2003), 209.

87. C. Stephen Evans and Brandon Rickabaugh, "What Does It Mean to Be a Bodily Soul?" *Philosophia Christi* 17, no. 2 (2015): 315–330.

88. Augustine, *On Freedom of the Will*, II, XVI. 41; *The Trinity*, IV.I.3; *Confessions*, II.6.

89. Augustine, *The Immortality of the Soul*, An. XV.24.

90. Augustine, *Tractates on the Gospel of John*, XXVII.6.

91. Augustine, *Tractates on the Gospel of John*, XXVII.6.

92. For a comprehensive defense of bodily resurrection from a dualist view, see Brandon Rickabaugh, "Dismantling Bodily Resurrection Objections to Mind-Body Dualism" (chapter 16).

93. Ludwig Wittgenstein, *Philosophical Investigations*, trans. G. E. Anscombe, 3rd ed. (New York: Macmillan, 1953), 178.

94. Somogy Varga, *Naturalism, Interpretation, and Mental Disorder* (Oxford: Oxford University Press, 2015), 141.

SUGGESTED FURTHER READING

Antonietti, Alessandro. "Must Psychologists Be Dualists?" In *Psycho-Physical Dualism Today: An Interdisciplinary Approach*, edited by A. Antonietti, A. Corradini, and E. J. Lowe. Lanham, MD: Rowman & Littlefield, 2008.

Crisp, Thomas M., Steven Porter, Gregg A. Ten Elshof, eds. *Neuroscience and the Soul: The Human Person in Philosophy, Science, and Theology*. Grand Rapids, MI: Eerdmans, 2016.

Evans, Stephen C., and Brandon Rickabaugh. "What Does it Mean to Be a Bodily Soul?" *Philosophia Christi* 17, no. 2 (2015): 315–330.

Johnson, Eric L. *Foundations for Soul Care: A Christian Psychology Proposal*. Downers Grove, IL: InterVarsity Press, 2007.

LaRock, Eric. "Is Consciousness Really a Brain Process?" *International Philosophical Quarterly* 48, no. 2 (2008): 201–229.

Manzotti, Riccardo, and Paolo Moderato. "Neuroscience: Dualism in Disguise." In *Contemporary Dualism: A Defense*, edited by Andrea Lavazza and Howard Robinson. New York: Routledge, 2014.

Moreland, J. P. "Restoring the Substance of the Soul to Psychology." *Journal of Psychology and Theology* 26, no. 1 (1998): 29–43.

———. "Spiritual Formation and the Nature of the Soul." *Christian Education Journal* 4, no. 2 (2000): 25–43.

Robinson, Howard. "A Dualist Perspective on Psychological Development." In *Philosophical Perspectives on Developmental Psychology*, edited by James Russell. Oxford, UK: Basil Blackwell, 1987.

———. "A Dualist Account of Embodiment." In *The Case for Dualism*, edited by John R. Smythies and John Beloff. Charlottesville: University of Virginia Press, 1989.

Swinburne, Richard. "What's So Good About Having a Body?" In *Comparative Theology: Essays for Keith Ward*, edited by Timothy Walter. London: SPCK, 2003.

———. *Mind, Brain, and Free Will.* Oxford, UK: Oxford University Press, 2013.

Willard, Dallas. *Renovation of the Heart: Putting on the Character of Christ.* Colorado Springs, CO: NavPress, 2002.

Chapter 13

Hope for Christian Materialism?

Problems of Too Many Thinkers

Jonathan J. Loose

The question of whether a materialist view of human persons is consistent with the Christian doctrine of the resurrection of the body has been described by Hud Hudson as a "centerpiece" in the dispute over Christian materialism.[1] Hudson develops his own materialist metaphysic of the human person involving the controversial ontology of temporal parts and arguing that the associated account of the possibility of resurrection addresses issues that have not been successfully dealt with by others, including animalists or proponents of the Constitution View. After giving a rough sketch of Hudson's views, I argue that this account of resurrection suffers from a theological difficulty that renders it no less problematic than its materialist rivals. This difficulty turns out to be rooted in a more general problem that strongly suggests a nonmaterialist view of human persons.

Of the various reasons to adopt a temporal parts view of material objects (also termed "four-dimensionalism"),[2] Hudson limits his motivation to just one: the resolution of paradoxical puzzles that seem to result from a commonsense understanding of the composition of material objects. Hudson highlights two such paradoxes, including the problem of fission that I will discuss. In order to solve a third, related problem, Hudson adds to four-dimensionalism a counterpart theory of *de re* modal relations.[3] Thus, his metaphysic contains highly controversial elements, but these have generated much interest and some support. We begin, then, by briefly outlining the temporal parts view, before considering the resulting account of resurrection, its problematic theological consequence, and then the larger problem for materialism.

The idea that events have temporal parts is uncontroversial (a soccer match is indeed a game of two halves), but the "temporal parts view" controversially holds that objects have temporal parts too. Hudson explains the principal idea that

> necessarily, for each way of exhaustively dividing the lifetime of any object, x, into two parts, there is a corresponding way of dividing x itself into two parts, each of which is present throughout, but not outside of, the corresponding part of x's lifetime.[4]

Thus, persons do not endure as a whole from moment to moment but have a location and extension in time by having temporal parts spread out across time in the same way that we have location and extension in space by having parts spread out across space. A temporal part incorporates *all* of the person's other parts for as long as that temporal part exists. Thus, persons often are visualized as space-time worms (e.g., with vertical-spatial extension and horizontal-temporal extension). Those who adopt the temporal parts view typically agree with Hudson that we are wholly material; that extended temporal parts are fusions of momentary ones; and that Universalism is true. The universalist accepts that, as David Lewis put it, "any old class of things has a mereological sum. Whenever there are some things, no matter how disparate and unrelated, there is something composed of just those things."[5]

What, then, is the problem of fission that Hudson takes to be one of the important motivators for four-dimensionalism and which will have a direct consequence for his account of resurrection? First, imagine that my brain is removed from my skull and fissioned (the hemispheres are separated). My body is then destroyed. Now compare two fission scenarios: in the *non-branching* scenario, just one hemisphere is successfully transplanted into a waiting, brainless body while the other hemisphere is destroyed. One living human person results. In the *branching* scenario, both of the hemispheres are transplanted into different brainless bodies and two living human persons result. The question is what happens to *me* in each case?

Given non-branching fission the common and confident response is that I survive. However, branching fission is more difficult since it includes *two* instances of what previously seemed to be a survivable operation, yet the transitivity and necessity of identity dictate it would be incoherent to claim that I survive as *two* beings. The scenario offers no way to distinguish the two transplants to suggest that I survive as one rather than the other. Thus the common conclusion is that *I fission out of existence*.[6] However, the four-dimensionalist is able to avoid this conclusion without either denying classical identity or making arbitrary decisions about which fission product I am. This is because the temporal parts view allows the claim that there were *two persons present all along*.

Recall that on the temporal parts view, objects do not endure from moment to moment as wholes but are space-time worms, spread across time and composed of temporal parts. Thus, we can say that two objects *share* a temporal part, remaining distinct objects at all times while being indistinguishable during the period in time across which that shared part is located. Hence branching fission reveals that I have a temporal part that is shared with another person, and that part is temporally located across the period from the moment I began to exist until the moment of fission. The other person and I each have later temporal parts that we do not share, and so at later moments we are observable as the distinct individuals that in fact we are at all times. Visualized as space-time worms, the two persons are clearly distinct objects that share parts at one point, just as two different railway lines might share a single piece of track for part of their length.

Turning to the possibility of the resurrection, Hudson finds fault with existing materialist accounts. He accepts that a materialist metaphysic must explain how it is that the same person could be present at different times and how resurrection is possible,[7] and he argues that views that accept temporal parts can do this in a way that avoids the difficulties of others that do not.

Difficulties arise for both traditional divine reassembly views and currently popular constitution views,[8] but of most interest is animalism. Van Inwagen's animalist account of resurrection holds that God preserves corpses for the last day by instantaneous body-switching at the moment of each person's death such that what is buried is not a corpse but a simulacrum. However, the consequent systematic deception of the bereaved renders God a deceiver, and so an alternative is offered by Zimmerman that seeks to avoid this consequence.[9] He suggests that the simples which compose a body might have the power to *fission* at the last moment of earthly life such that one fission product leaps the temporal gap to the resurrection while leaving the other (which is truly one's corpse) on earth.[10] However, this model is undermined by its requirement for a closest-continuer account of personal identity. Hudson thus presents his four-dimensional account as a way to avoid divine deception and closest-continuer theories, as well as the need for reassembly and the belief that constitution is not identity.

The four-dimensionalist's account of resurrection follows from the branching fission problem, as he "simply applies his solution to standard fission cases by recognizing overlapping (but non-co-located) continuants."[11] Resurrection is possible since a human person can be understood to be "an extended (earlier) temporal part which mereologically overlaps a human animal and an extended (later) temporal part which, in the words of St. Paul, is a new and imperishable spiritual body."[12] In other words, there is an earthly, temporal part of me that is also a temporal part of a particular human organism (call this part "PERISHABLE"). PERISHABLE is a *living* human organism. The larger human organism that has PERISHABLE as a part also has another temporal part that extends from the moment of my death throughout its period of existence as a corpse. It is because PERISHABLE is a temporal part of a larger human organism that I am a *human* person. I also have another heavenly temporal part (call this part "IMPERISHABLE"), which is a new and imperishable spiritual body that extends eternally from the last day forward. I am thus composed of temporally scattered parts, something that is unproblematic given a universalist view of composition.

The result of all this is something like Zimmerman's fissioning account, but without a problematic closest-continuer theory (because there is no need of any theory of diachronic *identity*). To hold that PERISHABLE and IMPERISHABLE are parts of me we need only establish that they are linked by a psychological gen-identity relation in the way that temporal parts should be if we are to understand them to compose persons.[13] The difficulties of other views also fail to apply. Reassembly of the same thing at a later time is meaningless on the temporal parts view, and given the possibility of shared temporal parts we can explain coextensive entities at a time while holding that constitution *is* identity. There is no deceptive body-snatching, and the matter that remains on earth is in a meaningful sense *my* corpse, since it is part of a human animal that shares a temporal part with me. Hudson believes the view stands "head and shoulders above" the others.[14]

However, without rehearsing either the serious objections to four-dimensionalist metaphysics and counterpart theory[15] or the inadequacy of Hudson's theological account of the intermediate state and resurrection in comparison to the most authoritative treatment,[16] his view of resurrection has a serious theological problem of its own. A satisfactory theological anthropology will show not only how the nature of human persons is consistent with what Christian doctrine has to say about us, but also that what a human person is capable of knowing includes what Christian doctrine tells us we can know. This includes knowing that we have hope for *our own* postmortem futures.

The New Testament concept of hope has been summarized as "trust in God, patient waiting and confidence in God's future."[17] Four-dimensionalism does not entail that I cannot experience such hope, having acquired it in theologically acceptable ways such as through observation of God's prior activity, comprehension of revealed promises, or participation in an eschatological community within which the Holy Spirit provides a basis in present experience for hope in what is to come. However, despite this, a very unusual anthropological question remains. We must ask whether we can be sure that this acquired attitude amounts to anything more than a mere *quasihope*, where quasihope is an experience of hope in a future that belongs *not to the experient but to another.* Can I know whether or not the object of my hope is my own future or whether it is the future belonging to someone else? This unusual problem is a serious one for the friend of temporal parts.

The four-dimensional account that purports to demonstrate that it is *possible* that I will stand again on the last day also renders me incapable of knowing if it will be *me* who will do so. My confidence in God's future is as likely to be a *quasihope* as it is to be a hope that is rightly mine. To understand why this is the case, first consider again the puzzle of branching fission. Given four-dimensionalism I know prior to fission that I will later be one of the fission products (and that I cannot be both), but I do not know *which* of the fission products I will be and thus which of the two persons I presently am. This is because *I am entirely indistinguishable from the other person during the period in which we share a temporal part.* During this period neither I nor anyone else can know if I am Jonathan or someone else. This matters greatly if the futures of the two persons are to be significantly different postfission. For example, if Jonathan is to be rewarded while the other is to be tortured, then it will be a matter of great concern to me to know who I am. The reason I cannot know this is clarified by the illustration of two railway lines that share a piece of track for part of their length. While it is on the shared *track*, we have no idea on which *line* an unmarked train is traveling (and thus what its destination will be). For that information we must wait until it reaches a location at which the lines are once again on separate tracks.

Next consider the resurrection case, noting that the two objects of which PERISHABLE is a temporal part are both *thinkers*. If PERISHABLE thinks and is a temporal part shared by both a human organism and a person (Jonathan), then both Jonathan and the human organism think.[18] Furthermore, the futures of Jonathan and the human organism could not be more different. The human organism will become a corpse, while Jonathan will go on to resurrection life. So, it is a matter of some concern to me as I write these words to be able to answer this question: am I Jonathan or the thinking human organism with which Jonathan currently shares a temporal part? Given the ontology of temporal parts, I simply cannot know and thus cannot know whether I experience hope that will not disappoint (because I am Jonathan) or quasihope that will (because I am the human organism). I cannot know the answer to this troubling question until I am located temporally at the point at which the human organism and Jonathan do not have an overlapping temporal part. By then, if I am the human organism, I will know nothing at all since I will be a corpse.

So, it does indeed seem that the very four-dimensional metaphysic introduced to demonstrate the possibility of my standing again at the last day renders me incapable on this day of knowing whether it will be me who does so. This is because there are at least two thinkers in my chair where there seems to be but one human body. The situation for the friend of temporal parts seems, quite literally, hopeless, and this view of resurrection is at least as problematic as the other materialist views that Hudson canvases, albeit for a different reason.

Here, then, is the underlying problem: materialist views have a habit of problematically generating too many thinkers. This problem arises in different ways. One important instance is a general problem described by Peter Unger that causes him to reject a physicalist view of human persons. I will describe this problem and Unger's resolution.

Unger introduced a dilemma for the common-sense view that a perception of a single physical object of a certain type—be it a cloud in an otherwise clear sky, a salt shaker on the kitchen table, or a person sitting in a chair—reflects the fact that there is indeed a single object of that type present. He argued instead that in such cases either there is no object present or else there are many millions of objects present, hence "the Problem of the Many or the None."[19] Since Unger took the idea of the many to be intolerable, the dilemma becomes, simply, the Problem of the Many.

Why is there a Problem of the Many? It is rarely, if ever, the case that a given situation contains just *one* precisely defined group of entities that could together compose a concrete complex of a given type (such as a cloud). The perception that a distant cloud has a sharp boundary might encourage us to believe it must be one such complex constituted by a precisely defined, sufficiently numerous, and not too widely scattered group of water droplets. However, if we were to fly into the cloud and inspect more closely, we would discover that the cloud's boundary is unclear and there are very many water droplets for which we cannot say with certainty that they are or are not constituents of it. If we decide to make an initial selection of the droplets that we consider constitute the cloud then many, highly similar and almost entirely overlapping groups will be rejected, each involving a difference of just a few droplets more or less, here or there. The differences between these many groups would not be so great as to have any impact on whether they qualify for cloudhood. Thus, if these many "overlappers" are indeed equally qualified then:

> either *all* of *them* make it or else *nothing* does; in this real situation, either there are many clouds or else there really are no clouds at all. This dilemma presents our problem of the many.[20]

While the Problem of the Many is illustrated clearly by thinking about clouds and recognizing that, if there is one cloud, then there are millions of clouds, it also applies much more widely and most likely to all complex physical objects (including salt shakers, human bodies, and much else besides). For example, Unger notes that even the surface of a stone will include many atoms or molecules "whose status with regard to our typical stone, nature has left unclear,"[21] and thus the problem applies even here.[22]

Hudson took Unger's problem to be important in the initial development of his materialist metaphysic of the human person. He applies it (for clarity's sake without making use of four-dimensionalism) to the case of "Legion":

a wholly material object that is a human person.[23] At any time, Legion is the fusion of a very large number of material simples.[24] The set of simples that jointly compose him is named "the Primary Set" and the fusion of these simples, "Tweedledee." He next considers a further set of simples that overlaps with the Primary Set almost perfectly, apart from the exclusion of just one material simple that the Primary Set includes and the inclusion of just one material simple that the Primary Set excludes. This overlapping set he names "the Secondary Set" and the fusion of its members, "Tweedledum."

Given the almost total overlap between Tweedledee and Tweedledum, it seems either that both satisfy the conditions for composing a person or that neither does, and herein lies the problem. Where we thought we had one person (Legion), we must have at least two (Tweedledee and Tweedledum), but—of course—there is a vast number of sets of material simples that differ from the Primary Set in just the way that the Secondary Set does, and so it seems there is a vast number of persons where we thought there was only one.

Both Unger and Hudson believe the Problem of the Many presents a serious issue, the resolution of which will have important consequences for our understanding of the metaphysics of human persons. They disagree, however, about the way in which the dilemma should be resolved and thus about what those anthropological consequences *are*. Unger takes the existence of many thinkers to be the *greater* evil, initially adopting a broadly nihilist response as a result. Contrastingly, Hudson considers the existence of many thinkers to be the *least* among evils, being preferable not only to the nihilist alternative (which he rejects firmly as incoherent)[25] but also to various alternative solutions (excluding his own "Partist" view) that he surveys under the heading "The Many Problematic Solutions to the Problem of the Many":[26]

> if I were without recourse to the Partist solution . . . I think I would have to recommend the many-persons approach as the least embarrassing of the currently available alternatives open to materialists who are committed (as I am) to eschewing bruteness where possible.[27]

Hudson's interest in the Problem of the Many arises because it seems to threaten the coherence of a set of assumptions, including materialism, that enjoy wide acceptance and are regarded as sensible by a broad section of philosophers. He thus seems primarily concerned to establish which of the materialist views on offer is most able to deal with the problem and secondarily to reject nonmaterialistic views, including dualist and idealist views within his survey of "Many Problematic Solutions." However, it is not clear why dualism and idealism offer problematic solutions. Dualism, he suggests, can be developed to accommodate the problem (though he describes this development as a "perversion of their theory"),[28] and so it is unrelated traditional objections to dualism that are his greatest concern. Idealism is on even stronger ground, as Hudson notes that if there are no human organisms or material objects then our problem simply disappears, and this is the strongest argument for idealism that he has seen, even if he cannot consider it sound because when it comes to idealism, he "cannot believe a word of it."[29] Whatever the seriousness of the problems traditionally pressed against dualism and idealism, Hudson is not adding to their woes by presenting the Problem of the Many, and discussion of these views seems out of place in the list of "Problematic Solutions."[30]

The development of the dualist theory to which Hudson refers is that we "count just one human person when we have some immaterial person connected to many substantially overlapping human organisms."[31] Very interestingly, this is the view ultimately expressed by Unger as the one to which the Problem of the Many inexorably leads.[32] Unger acknowledges dualism even in his initial presentation of the problem. He argues that persons are concrete individuals and, if *physical*, then plausibly they must also be complex, in which case The Problem applies to them with devastating effect. He writes,

> At first blush, then, our problem would give comfort to dualistic views regarding the mental and the physical, I am a concrete entity that is *only* mental or spiritual in nature and *not* physical or material.[33]

Unger then rejected this possibility because he believed that it undermined the common practice of identifying persons by distinguishing one body from others. For example, when I identify my wife I assume that the body I identify is "the body of a single person and this person has no body but this one."[34] However, this is a poor reason to reject the view if the many human organisms to which my wife is connected overlap to such a great extent. In that case, I can rest content that the number of *persons* remains as it seems to be, even if the number of human organisms is very many more and that whichever of the millions of overlapping human organisms I identify when I believe myself to be recognizing my wife, I can be sure that I am successfully identifying *her*.

In later work, Unger no longer offered this objection. Rather, he came to believe that even though the problem has universally *counterintuitive* results, those results are not always *intolerable*.[35] Consider the power to digest. It is merely a physical propensity that is derived from the powers of the physical entities that compose a complex physical being. Thus, the consequence of recognizing the problem of the many *digestings* amounts to little more than the introduction of a new way to speak about it. We might choose to speak *loosely*, saying that in the place where I am located there is but one process of digesting going on; or we might speak strictly, which given the Problem of the Many means we must say that there are many very similar overlapping digestings going on. While the latter is less intuitive than the former, little else rides on the difference and so the strict claim that there are many digestings is tolerable.

This contrasts sharply with the power to *experience.* The situation here is different. Experiencing is not a physical propensity, but a mental one. The power to experience is not obviously derived from the powers of the physical entities that compose a complex physical being, and so it is at the least radically emergent. The consequences of recognizing this "Experiential Problem of the Many" are much more significant. For example, they include the fact that each of the overlapping experiencers will be wrong in his firm and fundamental beliefs that he is alone in his situation and that he is having experiences that are unique, private, and not communicated to millions of others. Unger writes, "Really, now, can anything like that possibly be right? The very suggestion is, I think a terribly disturbing idea."[36] Such consequences cannot be safely ignored by speaking *loosely* of just one thinker. Here, then, is the version of the Problem of the Many that generates a troubling and serious philosophical problem that leads Unger to *substance* dualism.

If we try to explain the power to experience in terms of an emergent *property* dualism (what Unger calls "Nonentity Emergentism") then the experiencer is the complex physical being (the body), and therefore we fall victim to the Experiential Problem of the Many with its intolerably counterintuitive consequences. So Unger instead turns back to the "first blush" solution. He explores the possibility of an emergent *substance* dualism on which the immaterial soul that I am causally interacts directly and equally with each of the many highly overlapping complex bodies (each of my bodies). It does not seem obviously problematic to hold that each of the many bodies is one of my bodies and none of them are the bodies of anyone else. There would, of course, be many neural systems that would need to promote just a single experiencer, but neural systems are like digestive ones in being nothing more than physical propensities of a physically complex being that derive from its physical constituents, and so fundamental material particles could jointly possess a propensity to promote a *single* experiencer when appropriately configured with many others. In that way, all of these overlapping bodies and neural processes would promote only that single experiencer and only the experiencing that belongs to it. I may speak strictly of my many bodies or, without seriously problematic loss, speak loosely of "my body," but in each case I speak only of myself. Here, then, is Unger's dualist solution to the Problem of the Many.

I noted earlier that Hudson also takes this Problem of the Many to be serious and draws a different conclusion from it than Unger. Hudson offers his own *Partist* solution: an original and sophisticated metaphysical account in which parthood is understood relative to a spatial region such that we can think of my parts at different regions in the way that a three-dimensionalist would think of my presence at different times. On this view, I can be numerically identical across multiple simultaneous regions even if my composition at those regions differs. Despite the originality and sophistication of this creative response, I dare say that many will agree that such a view only goes to show just how far the materialist has to stretch credulity to be able to accommodate these fundamental difficulties. Even with Partism, a different problem of too many thinkers then arises in the account of resurrection, as I have argued. Thus, the Problem of the Many presents a general and powerful example of problems of "too many thinkers" that plague materialist but not nonmaterialist views, and their power is such as to persuade a nontheistic philosopher such as Unger to adopt a dualist view despite the lack of fit to his broader worldview and his acknowledgment that, "Nowadays, it's hard for academically respectable philosophers to believe in mentally powerful nonphysical beings."[37]

Whatever hope remains for materialist theories of the possibility of resurrection (and new theories continue to proliferate, not least from Hudson himself), it is clear that a four-dimensionalist account that parallels the solution to the problem of fission will not suffice as the centerpiece of the materialist's case, and this because it suffers from a problem of too many thinkers that is a significant general problem for materialism.

NOTES

1. Hud Hudson, *A Materialist Metaphysics of the Human Person* (Ithaca, NY: Cornell University Press, 2001), 148.

2. For a helpful introduction to this view see, for example, Eric T Olson, *What Are We? A Study in Personal Ontology* (Oxford: Oxford University Press, 2007), chapter 5.

3. For the problem of fission, see Derek Parfit, "Personal Identity," in *Personal Identity*, ed. John Perry (Berkeley: University of California Press, 1975), 199–223; David Lewis, "Survival and Identity and Postscripts," in *Philosophical Papers,* vol. 1 (Oxford: Oxford University Press, 1983), 55–77. The other important puzzle resolved by the doctrine of temporal parts is Wiggins's puzzle of Tibbles and Tib. See David Wiggins, "On Being in the Same Place at the Same Time," *The Philosophical Review* 77, no. 1 (1968): 90–95. If two objects cannot be co-located without being identical then consider Tib—a proper part of a cat, Tibbles, minus her tail. If Tibbles loses her tail, Tibbles and Tib are now co-located and both seem to survive. Are Tibbles and Tib identical after all? The four-dimensionalist says that Tibbles and Tib are four-dimensional continuants that overlap by sharing a temporal part that begins at the point that the tail is lost. It is Gibbard's related puzzle of the lump of clay that are

perfectly coincident throughout the course of their existence that requires an additional commitment to counterpart theory. See Allan Gibbard, "Contingent Identity," *Journal of Philosophical Logic* 4 (1975): 187–221.

4. Hudson, *A Materialist Metaphysics of the Human Person*, 58.

5. David Lewis, *The Plurality of Worlds* (Oxford: Blackwell, 1986), 211.

6. Even this conclusion is not trouble free. When combined with the claim that I survive in the single transplant scenario we conclude that my persistence has nothing to do with the physical thing that I am before or after the transplant, or with the operation itself. Rather, my persistence depends on the fate of the *other* hemisphere; a seemingly absurd consequence. See Harold W Noonan, *Personal Identity*, 2nd ed. (London: Routledge, 2003).

7. See Trenton Merricks, "There Are No Criteria of Identity Over Time," *Noûs* 32 (1998): 106–24; Dean W Zimmerman, "Criteria of Identity and the 'Identity Mystics,'" *Erkenntnis* 48 (1998): 281–301.

8. If the same matter is shared by successive individuals, most strikingly by the cannibal and his victim, then this ensures that the raw materials are unavailable for God to reassemble everyone on the last day. The Constitution View is "insufficiently motivated, its commitment to co-location an impossibility, and its constitution relation a mystery" (Hud Hudson, "Multiple Location and Single Location Resurrection," in *Personal Identity and Resurrection: How Do We Survive Our Death?*, ed. Georg Gasser [Farnham, UK: Ashgate, 2010], 87–101); see also Lynne Rudder Baker, "Persons and the Metaphysics of Resurrection," *Religious Studies* 43, no. 3 (2007): 333–48.

9. See Peter van Inwagen, "The Possibility of Resurrection," in *The Oxford Handbook of Philosophical Theology*, vol. 2, ed. Michael C Rea (Oxford: Oxford University Press, 2009), 321–27; Dean W Zimmerman, "Materialism and Survival: The Falling Elevator Model," *Faith and Philosophy* 16 (1999): 194–212. I review these accounts in detail in Jonathan J. Loose, "Materialism Most Miserable: The Prospects for Dualist and Physicalist Accounts of Resurrection," in *The Blackwell Companion to Substance Dualism*, ed. Jonathan J. Loose, Angus J. L. Menuge, and J. P. Moreland (Oxford: Wiley-Blackwell, Forthcoming, 2018).

10. For the claim that Zimmerman's model merely changes the method of divine deception rather than removing it, see William Hasker, "Materialism and the Resurrection: Are the Prospects Improving?" *European Journal for Philosophy of Religion* 3, no. 1 (2011): 83–103.

11. Hudson, *A Materialist Metaphysics of the Human Person*, 189.

12. Hud Hudson, "The Resurrection and Hypertime," in *Paradise Understood: New Philosophical Essays About Heaven*, ed. T. Ryan Byerly and Eric J Silverman (Oxford: Oxford University Press, 2017), 263–73; see also Hudson, "Multiple Location and Single Location Resurrection," 94–95; Hudson, *A Materialist Metaphysics of the Human Person*, chapter 7.

13. See Hudson, "Multiple Location and Single Location Resurrection," 94–95.

14. Hudson, *A Materialist Metaphysics of the Human Person*, 189.

15. Hudson notes that the theory of temporal parts has been charged with "incoherency, declared unmotivated, and criticized for the company it keeps (i.e., for its close

association with counterpart theory)," (Hudson, "The Resurrection and Hypertime," 266–67).

16. See Hudson, *A Materialist Metaphysics of the Human Person,* chapter 7; cf. John W Cooper, *Body, Soul and the Life Everlasting: Biblical Anthropology and the Monism-Dualism Debate* (Grand Rapids, MI: Eerdmans, 2000).

17. J. M. Everts, "Hope," in *Dictionary of Paul and His Letters*, ed. Ralph P. Martin, G. F. Hawthorne and Daniel G. Reid (Downers Grove, IL: IVP Academic, 1993), 415–17.

18. Even if our concept of human person does not include the human animal (as Hudson's does not), the problem rests only on the claim that there are two thinkers present, and it is clear that the human organism is at least a thinking nonperson.

19. Peter Unger, "The Problem of the Many," *Midwest Studies in Philosophy* 5 (1980): 412.

20. Unger, "The Problem of the Many," 415.

21. Unger, "The Problem of the Many," 430.

22. Such everyday objects, he argues, are actually *better* demonstrations of the dilemma since we are more confident that these are *concrete* entities in the first place and so we are more confident that if they have a range of possible boundaries (or possible groups of constituents) then the problem applies: if there is one such object, there must be (at least) millions of such objects.

23. Hudson, *A Materialist Metaphysics of the Human Person*, 11–14.

24. The term "material simple" refers to the lowest level uncuttable particles, whatever they turn out to be.

25. Hudson, *A Materialist Metaphysics*, 17–18.

26. Hudson, *A Materialist Metaphysics*, chapter 1.

27. Hudson, *A Materialist Metaphysics*, 38.

28. Hudson, *A Materialist Metaphysics*, 21.

29. Hudson, *A Materialist Metaphysics*, 21.

30. Proponents of all views must face typical objections, but the common dismissal of dualism via certain common objections looks increasingly suspect. See, for example, William G Lycan, "Redressing Dualism," in *The Blackwell Companion to Substance Dualism.*

31. Hudson, *A Materialist Metaphysics of the Human Person*, 20.

32. Peter Unger, *All the Power in the World* (Oxford: Oxford University Press, 2006), chapters 6–7.

33. Unger, "The Problem of the Many," 461.

34. Unger, "The Problem of the Many," 461.

35. See Unger, *All the Power in the World*, 376–81.

36. Unger, *All the Power in the World*, 376–77.

37. Unger, *All the Power in the World*, 381.

SUGGESTED FURTHER READING

Cooper, John W. *Body, Soul and the Life Everlasting: Biblical Anthropology and the Monism-Dualism Debate*. Grand Rapids, MI: Eerdmans, 2000.

Hudson, Hud. *A Materialist Metaphysics of the Human Person*. Ithaca, NY: Cornell University Press, 2001.

———. "Multiple Location and Single Location Resurrection." In *Personal Identity and Resurrection: How Do We Survive Our Death?*, edited by Georg Gasser. Farnham, UK: Ashgate, 2010.

Lewis, David. *The Plurality of Worlds*. Oxford, UK: Blackwell, 1986.

Loose, Jonathan J. "Materialism Most Miserable: The Prospects for Dualist and Physicalist Accounts of Resurrection." In *The Blackwell Companion to Substance Dualism*, edited by Jonathan J. Loose, Angus J. L. Menuge, and J. P. Moreland. Hoboken, NJ: Wiley-Blackwell, forthcoming, 2018.

Lycan, William G. "Redressing Dualism." In *The Blackwell Companion to Substance Dualism*, edited by Jonathan J. Loose, Angus J. L. Menuge, and J. P. Moreland. Hoboken, NJ: Wiley-Blackwell, forthcoming, 2018.

Olson, Eric T. *What Are We? A Study in Personal Ontology*. Oxford, UK: Oxford University Press, 2007.

Parfit, Derek. "Personal Identity." In *Personal Identity*, edited by John Perry. Berkeley: University of California Press, 1975.

Unger, Peter. *All the Power in the World*. Oxford, UK: Oxford University Press, 2006.

———. "The Problem of the Many." *Midwest Studies in Philosophy* 5 (1980): 411–67.

Chapter 14

How to Lose the Intermediate State without Losing Your Soul

James T. Turner, Jr.

Elsewhere in print, I have offered a number of reasons for thinking that Christians should deny the intermediate state.[1] By the "intermediate state" I mean the purportedly disembodied postmortem preresurrection plane of existence in which redeemed humans (at least) exist in a heavenly/paradisiacal sort of condition.[2] For the purposes of this chapter, I am going to assume that those arguments go through. From that position, I aim to offer a constructive proposal that acts as a "how to" guide for positing an immediate resurrection view. But, unlike some immediate resurrection accounts that are often (rightly, in my view) disparaged, my view denies the following: "soul sleep," "gappy existence" (i.e., that a human ceases to exist after her death and then pops back into existence at her resurrection), and that there's no general resurrection. In other words, I offer an immediate resurrection view that affirms that all humans who are to be resurrected will be resurrected at the same time, that is, at the time of Christ's return. I also affirm that dead redeemed humans *immediately* are in the presence of the incarnate Christ. I take it that this is an immediate resurrection view that is unlike any that's been offered.[3]

I begin constructing my proposal for an immediate resurrection view in the first section by filling out some further biblical theological reasons for motivating an immediate resurrection theory. In the second section, I sketch a model of time such that theologians can affirm that a dead redeemed human person *immediately* will find herself at the general resurrection, in the presence of Christ in Paradise. Lastly, in the third section, I provide a nonphysicalist account of human beings that works with the model of time I provide in the second section. I do this because it is sometimes taken to be the case that immediate resurrection views are the property of physicalist anthropologies. I argue this is not the case; in fact, a particular sort of dualism best provides for an immediate resurrection.[4]

BIBLICAL THEOLOGICAL DATA

One well-worn philosophical idea is that part of what it is to know what a thing is is to know its purpose, its *telos*. For example, if one wants to know what a hammer is, one helpful way of explaining it is by explaining what it is used for, that is, smashing metal spikes into substances through which one cannot, by one's bare hand, push the spikes. If one wanted to design a hammer, one would begin with the end in mind: smashing metal spikes into hard substances.

Now, when it comes to thinking about our current topic, namely, life after death and—with respect to the volume writ large—the human person, one might think of the *telos* of the human person as a way to figure out what sort of thing a human is. And theologians have done so from various angles. Some say that humans are made primarily for relationships, particularly *with God*, an immaterial being; some say that humans are made to reflect the *mental* attributes of God—a mind greater than which none can be conceived.[5] Some say that the purpose of a human is to co-rule the cosmos with God, its creator.[6]

Now, I agree with this last statement. But, as it stands, it doesn't provide any direct bearing on the ontological constitution of the human person. That is to say: that humans are meant to co-rule the cosmos doesn't tell us anything about whether physicalism or dualism about the human person is correct. So, instead of beginning with a discussion of the human person, I wish to begin thinking about the human being by reviewing some biblical theological data concerning the purpose of *the entire created order*. This is fitting because there are long held philosophical and theological reasons for thinking that human beings are pictures of the cosmos in miniature: *microcosms*. If that's correct, then one helpful way to understand what sort of thing a human being is is to understand what sort of thing the cosmos is. To do that we need to know the cosmos's purpose.

Here I wish to follow much of the recent biblical theological work advanced by G. K. Beale, J. Richard Middleton, Jon D. Levenson, Benjamin L. Gladd and Matthew S. Harmon, and John H. Walton (to name but a few). By their corporate lights, the purpose of the cosmos is to be a *temple* in which Yahweh dwells with His creation.[7] They suggest that this is one of the main themes in the Old and New Testaments, from the creation narrative in Genesis to the closing of the Book of Revelation: God's dwelling place is meant to be *with* man in a cosmos where God's will is done on earth as it is in Heaven (Matt. 6:9–10). In other words, the goal is that Heaven and earth will be united (cf. Rev. 21 and 22). Or, in the language of the Psalms and Prophets, God's glory is meant to cover the earth "as the waters cover the sea" (cf. Ps. 57:5, 57:11, 72:19, 108:4–5; Isa. 6:3, 11:9).[8]

If this is correct, if it's the case that the material creation has been, from the beginning, purposed to be a cosmic temple in which Yahweh dwells, then this may say something about the purpose of human beings. That is, it might say something about what it is for humans to have been, as the Scriptures declare, made "in the image of God." Unsurprisingly, these biblical scholars think it does. And, the purpose it reveals isn't what many systematicians and philosophers think. Rather, in the words of Middleton:

> *a virtual consensus* has been building since the beginning of the twentieth century among Old Testament scholars concerning the meaning of *imago Dei* in Genesis, and this view is quite distinct from the typical proposals found among systematic theologians. This virtual consensus is based [in part] on exegesis of Genesis 1:1–23, the textual unit that forms the immediate literary context of 1:26–27 ["let us make man in our own image"] . . . Such exegesis notes the predominantly royal flavor of the text, beginning with the close linkage of image with the mandate to rule and subdue the earth and its creatures in 1:26 and 28 (typically royal functions). But beyond this royal mandate, the God in whose image and likeness humans are created is depicted as sovereign over the cosmos, ruling by royal decree ("let there be") and even addressing the divine council or heavenly court of angelic beings with "let us make humanity in our image," an address that parallel's God's question to the seraphim at the call of Isaiah [Isaiah 6:8] . . . Just as Isaiah saw YHWH "seated on a throne, high and exalted" (6:1), so the writer of Genesis 1 portrays God as king presiding over "heaven and earth," an ordered and harmonious realm in which each creature manifests the will of the creator and is thus declared "good." Humanity is created *like* this God, with the special role of representing or imaging God's rule in the world.[9]

Being made "in the image of God," then, is a *royal title* with a royal function and purpose. What it is to image Yahweh just is to be a royal representative, a vice-regent of his sovereign rule.[10]

Moreover, against the backdrop of the Ancient Near East, and its popular theology within which the cult image "is a precisely localized, visible, corporeal representation of the divine," we have good reason to think that the Genesis creation story presents *humankind* as the *embodied* and *visible* expression of Yahweh's rule, in contradistinction to the various wooden, stone, and what have you, cult images of competing Ancient Near Eastern deities.[11] Yahweh's image is crafted by Yahweh Himself, rather than by human hands.[12] If one supposes that the property "being a localized, visible, and corporeal representation of Yahweh" is an essential property of human beings, then a human cannot—so long as she exists—fail to be localized, visible, and corporeal. If that's right, humankind lacks the ability to exist in a disembodied state.

I'm no biblical scholar. I'm a theologian. So, I'll let the exegetes fill in the previously mentioned work. The reason I've offered this cursory examination of (a plausible view of) the cosmos's purpose, and the purpose of humans within it, serves one basic end: to explain why I am inclined to posit an immediate bodily resurrection view. How does the preceding discussion relate to immediate resurrection? Well, given what I have said, if the redeemed are to find themselves immediately in the presence of the incarnate Christ following their deaths, it is not going to be in a disembodied state (*ex hypothesi*, they're essentially visible and localizable, i.e., embodied). Since Christians think the resurrection of the dead occurs when Christ returns (a future event), we need some way of accounting for how such a thing could follow immediately after a human's death—an event presumably that's temporally well before the *parousia*.

One problem with disembodied notions of immediate presence with Christ, as I see it, is that most explications of such views do not say anything about fulfilling the peculiarly royal and *earthly* purpose that exegetes say humans are created to accomplish, that is, live and work in the cosmos to expand the glory and worship of the Yahweh in His cosmic temple (though this isn't to say they *couldn't* do this). Problematically, many explanations of, for example, the beatific vision, appear to take no account of the Ancient Near Eastern background to the Genesis creation story, the narrative that frames the rest of the story of Scripture, not least the story's *telos* (New Creation) (e.g., it seems entirely missing in Thomas Aquinas's noteworthy and seminal account of the beatific vision).[13]

Perhaps many theologians won't be bothered by the idea that, on many theological accounts of the human *telos*, the Ancient Near Eastern setting for the story of humankind's creation isn't in view. Instead, such theologians might rest comfortably on the history of interpretation through the tradition. Admittedly, that's not a weak position. But if the exegetes I've mentioned (and the purportedly consensual deliverances of the last century of Old Testament exegetes) are correct about what the Bible means concerning the purpose of the cosmos and the purpose of God's image bearers, personal eschatology is *not* best viewed as the individually experienced "heavenly" state, nor is the biblical story about a human's individually experiencing-being united to God in a beatific vision. Rather, the story vis-à-vis humanity moves forward to a particular and divinely ordained climax: being bodily resurrected into the New Creation (better: Renewed Creation) to live and work as a corporate people in God's cosmic temple as redeemed and glorified images; that is, images that won't rebel. I grant these musings don't rule out a disembodied intermediate state. I highlight them merely to help explain why I take it that embodiment is critically important to the human species and God's purpose for them. As I say above, I have sustained arguments against the intermediate state elsewhere.

Helpfully, this way of thinking about the creation's purpose does strengthen a historically well-established theological position about resurrection: that the human body that dies is numerically the same human body that rises from the grave. Though this position recently has been reduced in prominence, there seem to be *very* strong exegetical and theological reasons for defending it. Joshua Mugg and I have reviewed and defended a number of these important reasons elsewhere.[14] I'll not reiterate those arguments here. What I will offer, though, are some of the arguments' conclusions: (1) "Resurrection" just means that the body that died is the body that rises (hence: resurrection of *the* body). Acquiring a numerically distinct body is reincarnation, *not* resurrection. (2) Jesus's resurrection is a foretaste of the redemption of the cosmos writ large *and* our bodies (i.e., the microcosms). What this tells us is that, just as in the gospels' resurrection narratives Jesus's body is *not annihilated*, the numerically same body that goes into the tomb comes out (this is the *explanans* for the empty tomb in the gospel witnesses).

Moreover, because Jesus's resurrection is a model for the redemption of the cosmos, we can infer that the redeemed cosmos (the New Creation) will be numerically identical to *this* cosmos (see also 1 Cor. 15 wherein Christ is described as the "first-fruits" of the resurrection).[15] I take it that this is the obvious point of Romans 8:19–23: "For *the* creation waits with eager longing for the revealing of the sons of God . . . that *the* creation itself will be *set free from its bondage to corruption and obtain the freedom* of the glory of the children of God . . . And not only the creation, but we ourselves . . . wait eagerly for . . . the redemption of our bodies" (so that we can fulfill our intended purpose as described above).[16] We can sum up the lesson from Jesus's resurrection and the Romans 8 passage with the following conditionals: (3) If Jesus's resurrected body is numerically identical to his premortem body, then the New Creation is numerically identical to the pre-eschatological cosmos. (4) If the New Creation is numerically identical to the pre-eschatological cosmos, then a resurrected human's body will be numerically identical to her premortem body. (5) If Jesus's resurrected body is numerically identical to his premortem body, then every resurrected human body will be numerically identical to its premortem body.

I think these conditionals are true. I further think that the biblical theological data I've provided give good reasons for thinking that humans are essentially embodied. Now, a physicalist might think that what I've said so far is an argument in physicalism's favor. Not so! As we'll see in the third section, I think that physicalism struggles to account for the numerical identity between premortem bodies and resurrection bodies. In my view, if physicalism cannot account for this numerical identity, that would be a fatal theological problem for physicalism.

Following from and rounding out this bit of this discussion, allow me to put forward a list of theological affirmations that drive me to provide a model of immediate resurrection:

a. Bodily resurrection is in *no way* superfluous to a human being's life after death.
b. We have good exegetical reasons for thinking that redeemed humans *immediately* reach Paradise and the presence of the incarnate Christ the moment following their biological deaths.
c. Bodily resurrection *requires* that the body that dies is the body that rises *again* to everlasting life.
d. The general resurrection of the dead happens at Christ's return.
e. The purpose of the human is to be the localizable, visible, and corporeal representative of Yahweh's rule in His cosmic temple.
f. The teleologically fulfilled and eschatological state of the cosmos is such that Heaven and earth are joined into a cosmic temple filled with Yahweh's glory.

Now, it's likely that some of these theological affirmations are controversial, at least among certain systematicians. My point in offering them, and the preceding discussion, is not fully to convince the systematician that this position is correct (though I do think I've given some good, if brief, reasons to think so). Rather, my point is to fill out my purpose for *this* chapter. This is a *how-to* chapter, after all. As with many "how" questions, there precedes a "why" question. Take the above as providing a set of answers to the "why" question. Moreover, in the interest of this volume, my purposes for providing a model of immediate resurrection is *not* motivated by philosophical anthropology. Nor is it motivated by supposed scientific findings vis-à-vis human evolution and neuroscience.[17] Rather, my motivations for rejecting a disembodied intermediate state are biblical and theological. And these are my same motivations for attempting to construct an immediate resurrection model *and* to reject physicalism.

For the remainder of the chapter, assume (for argument) that (a)–(f) are true. The model of immediate resurrection I tentatively offer means to accommodate, at least, claims (a)–(f). It's to that task I now turn.

SKETCHING A MODEL OF THE NOW AND NOT-YET: "ESCHATOLOGICAL PRESENTISM"

With a healthy portion of the Christian tradition, I affirm claim (b). That is, there seem to me good scriptural reasons for affirming what most Christians in the tradition affirm: at least some redeemed humans go immediately to Paradise, in the presence of the incarnate Christ, following their biological deaths. As a Protestant, I want to affirm something further: there are good scriptural reasons for thinking that *all* redeemed humans go immediately to Paradise following their biological deaths.[18] Again, I don't have the space to defend this, but such a claim seems consistent with the *Westminster Confession of Faith* chapter 32; Answer 57 of the Heidelberg Catechism; and so on. One biblical passage that provides prima facie affirmation of immediate postmortem presence in Paradise is Luke 23:43, Jesus's promise to the robber on the cross. As such, I want to sketch a way to make sense of this promise, "today you will be with me in Paradise." But, I want to put "today"—that is, immediacy—in the context of the *eschatological* resurrection.

To do so requires a model of the metaphysics of time within which it's coherent to say that one's *eschatological* (that is, at Christ's return) bodily resurrection follows immediately after one's biological death. Providing such a model isn't at all easy. Even so, I want to outline such a model so that the common critiques of "immediate resurrection" (in particular, that it requires individuals popping up in the resurrection world at different times) might be put to pasture. Before I do that, let me briefly mention models of time that I don't use.

I don't affirm eternalism or four-dimensionalism. Eternalism is a thesis about time that says all times are ontologically on par; they all exist, even though they don't all exist at the same time. On eternalism, the year 1981 is just as real as 2017. Four-dimensionalism is a thesis about the way things persist *through* time. It posits that substances are "spread out" over time, in much the way that objects are spread out over space, and are divided up into "time slices," "temporal parts," or "person stages," such that it's the cumulative temporal parts of any given thing—taken as a whole—that *is* the substance.[19] On such a view, Jones, a human person, is not wholly present at any one given time. All that exists at one particular time is a "time-slice" (or "temporal part" or "person stage") of Jones. Jones, the substance, is the whole "space-time worm" (think of the segments of a worm as analogous to time-slices/temporal parts). This way of persisting through time is called *perdurance*.[20] Again, for lack of space, I cannot fully spell out my rejection of these views, but I reject them for theological reasons. Briefly, it seems to me that an implication of these views is that temporal parts of the entire creation and individual created things (e.g., humans) are never redeemed. They are eternally "stuck" within the ravages of sin and evil.[21] I could say more about this, but I need to press on.

For the purposes of sketching my model of immediate resurrection, I deny "presentism" too. Presentism is the model of time that's seen as something of a contradiction to eternalism. This is a view of time that seeks to provide a metaphysic for what intuitively seems to be the case: all that actually exists is whatever's present. The past no longer exists and the future does not yet exist. If it's time T_1, T_1 (and all things at T_1) is all that exists. If it's time T_2, T_2 (and all things at T_2) is all that exists. (Compare: on eternalism, supposing that T_1 and T_2 denote different moments in time, both T_1 and T_2 eternally exist.) On presentism, if things persist through time, they do so by *enduring*. That is, things that persist through time are *wholly* present at each individual moment in time. Again, all too briefly, though I am sympathetic to this account of time, I ultimately reject it because it cannot provide a way for a biologically dead human being immediately to enter the time of the eschatological bodily resurrection—a future point in time, a time that, given presentism, doesn't exist. As you can see here, my theology is driving my metaphysics.

So, then, what model *do* I use? The model I put forward is a model of time I call "Eschatological Presentism" (EP). It's an amended version of what philosopher Barry Dainton calls "Compound Presentism."[22] EP is a view that borrows from four-dimensionalist/eternalist views *and* presentism.

To get an idea of how EP works, consider Figure 14.1[23]

On the x-axis are temporal co-ordinates (viz., T_1, T_2, T_3, T_4). On the y-axis are sum totals of reality denoted by S_1, S_2, S_3, and S_4. T_Ω denotes the time of the eschatological resurrection, the return of Christ. Moreover, assume that (borrowing Dainton's words), "the sum total of reality consists of two coexisting but non-simultaneous very brief reality-slices (each spatially three-dimensional)."[24] These reality slices are denoted in Figure 14.1 by the letters

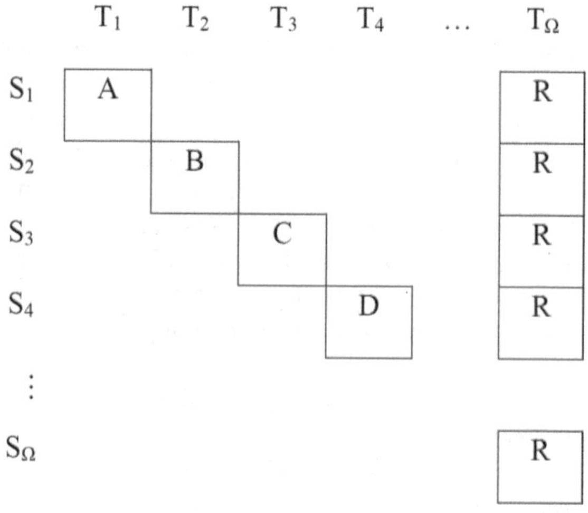

Figure 14.1 Eschatological Presentism. *Source*: Graphic courtesy of the author.

A, B, C, D, and R. By deploying a notion of reality slices—and that there are two moments in time that nonsimultaneously coexist in the sum total of what exists—I borrow from four-dimensionalism. That is, the (capital "P") Present is "spread out" over two times, as are the *things within* the Present. The sum total of what exists are two temporal moments and two reality slices (e.g., T_1+T_Ω and A+R). But, borrowing from presentism, when one pre-T_Ω moment is (small "p") present, that's the only time and reality slice in the pre-T_Ω temporal series that exists. When T_1 is present and A exists, there is no T_2/B that exists alongside it. When T_2 is present and B exists, there is no T_3/C that exists alongside it (nor T_1/A before it). And so on. Let us say that a time is (small "p") *present* if and only if there is a reality slice at the intersection of a T- and S-coordinate. So, we can conceive of the timeline $T_1 \ldots T_{\Omega\text{-}1}$ the way a presentist would. The only time that exists in the $T_1 \ldots T_{\Omega-1}$ time series is whatever time is present.

On EP, the temporal moment that *does* exist alongside each pre-T_Ω moment is T_Ω, the time of Christ's return and the resurrection of the dead. The sum total of reality when T_1 is present is S_1, and S_1 consists of T_1, T_Ω, and their corresponding reality slices, namely, A and R. And so on for each pre-T_Ω moment. When a given pre-T_Ω moment is present and its corresponding reality slice exists, the only other temporal moment that exists is T_Ω and its reality slice, R. The Present is two temporal moments compounded together. The sum total of reality is two reality slices compounded together (e.g., A and R when T_1 and T_Ω are Present); reality is "spread out" over two (small "p") present moments.

I mention above that this view of time also requires that the things in the Present be spread out across two temporal moments (e.g., T_1 and T_Ω). For space reasons, I'll bracket out discussion of all the possible and actual things that do or could exist. Instead, and to the purposes of this chapter, I'll focus on humans. A human being, on this view, is such that she is "spread out" over the present pre-T_Ω temporal moment *and* T_Ω. For example, if Jones, a candidate human, exists when T_1 is present, she exists in the Present T_1+T_Ω. Moreover, if Jones exists when T_1 is present, she has a reality slice in A and, because T_Ω is in the Present with T_1, Jones also has a reality slice in R. If Jones is still alive when T_2 is present, she exists in the Present T_2+T_Ω (existing as a member of the corresponding reality slices, viz., B and R). On this way of thinking, given that Jones is "spread out" across two times and composed of two reality slices, she can be thought of as a "space-time pill bug."[25] (Compare: the "space-time worm" of four-dimensionalism where, supposing for example that Jones existed for ten temporal moments, she'd be composed of ten reality slices ["temporal parts"]. On EP, Jones only ever is [at most] composed of two reality slices.)

While there is pre-eschatological time (i.e., while there are temporal moments that exist pre-T_Ω), a human being is a space-time pill bug composed of two reality slices (or temporal parts, or whatever descriptor one likes) and spread out over two present moments in the Present—the Eschatological Present. It should be obvious by now why I label it the Eschatological Present: the eschaton is in the Present with each pre-T_Ω temporal moment. This is, no doubt, a bizarre thesis.[26] But, if it's coherent, it carries some important consequences.

First, it's obvious that there's no intermediate state, nor is there a need for any sort of temporally interim state between one's death and one's bodily resurrection. To see this, suppose that (looking at Figure 14.1), at T_1, Jones is alive, but, at T_2, she's dead. When T_2 is present, where (or when) is Jones? The answer is clear: she's at T_Ω as a part of the R reality slice, the return of Christ and resurrection of the dead. When Jones dies, she's *immediately* at the eschatological resurrection of the dead. Moreover, she's there *with everyone else who is to resurrect* because *every human* who is to resurrect is spread out over some pre-T_Ω reality slice and T_Ω/R.

Second, Eschatological Presentism is either consistent with or helps safeguard some of the theological affirmations I list in the second section. I present them again here:

a. Bodily resurrection is in *no way* superfluous to a human being's life after death.
b. We have good exegetical reasons for thinking that redeemed humans *immediately* reach Paradise and the presence of the incarnate Christ the moment following their biological deaths.
c. Bodily resurrection *requires* that the body that dies is numerically the same body that rises *again* to everlasting life.
d. The general resurrection of the dead happens at Christ's return.
e. The purpose of the human is to be the localizable, visible, and corporeal representative of Yahweh's rule in His cosmic temple.
f. The teleologically fulfilled and eschatological state of the cosmos is such that Heaven and earth are joined into a cosmic temple filled with Yahweh's glory.

In particular, it's either consistent with or safeguards five of the six of these. The one exception is (c). It's not yet clear how EP might help secure that bodily resurrection *requires* that the body that dies is numerically the same body that rises *again* to everlasting life.

The reason that it's not yet clear how EP is consistent with (c) is because EP, by itself, is not enough. There's an additional thesis that needs, even if briefly, to be offered: a particular thesis about human beings: hylemorphic dualism.

HYLEMORPHIC DUALISM AND ESCHATOLOGICAL PRESENTISM

Let's take stock of our progress. In an attempt to provide a way for dead human beings immediately to reach the eschatological bodily resurrection, I've put forward Eschatological Presentism, a thesis both about the nature of time and the nature of persistence through time. The EP model posits that time and (at least some) things in time are spread out between two temporal moments, one of which is (at least, when there's pre-eschatological time) the pre-eschatological "present" and the other is the temporal moment of Christ's return. This latter temporal moment I've labeled: T_{Ω}. And, so I say, doing so provides an account of resurrection that denies an intermediate state; it denies "soul sleep" (or anything relevantly similar); and it denies that resurrected humans are popping up one by one at different temporal moments rather than simultaneously at the general resurrection. In other words, EP allows us to suggest that, when a human dies, she immediately reaches the eschaton, but not without the rest of us.[27]

The preceding ruminations account only for (a), (b), (d), (e), and (f) in the above list. I still need to account for (c). And, at least prima facie, it's not obvious how the numerical identity of the *body* is secured on EP. Perhaps, at this point in the chapter, the best I can say is that EP is consistent with (c). But, it's also—at least, so far—seemingly consistent with body-switching, so long as the *person* is spread out over the pre-T_{Ω} temporal moment and T_{Ω}.

In my view, the best metaphysic on offer to explain how a human body persists through time (no matter the theory of persistence one uses), through death and immediate resurrection is what I'll call a broadly Thomistic hylemorphic account of human beings. By "broadly Thomistic," I mean to suggest that I borrow my understanding of hylemorphism from Thomas Aquinas and his modern-day expositors. I use the qualifier "broadly" to suggest that my preferred version of hylemorphism takes a number of exceptions to Aquinas's own outworking, the first and foremost of which is that I deny—*contra* most (all?) Thomists—that human souls can exist independent of a human body. Fortunately, because I'm constructing a metaphysic for skipping a disembodied intermediate state, this issue, and my disagreement with Thomas on this score, needn't detain us. Nor, for that matter, should any discussion of the current debates between "survivalism" and "corruptionism" in the Thomist camps.[28] Given, then, the list of theological affirmations I've listed above, and that the point of this chapter is to provide an account of immediate bodily resurrection, allow me to give a brief overview of my hylemorphic account of human beings and the identity conditions it provides for the numerical identity between the body that dies and the body that resurrects at Christ's return.

Hylemorphism is a metaphysic concerning things that exist. For my purposes, obviously enough, the things on which I wish to focus are human beings. On hylemorphism generally, and my own account specifically, human beings are *substances*. That is, they are genuinely *one* thing, not a collection or aggregation of things (like puzzles and Lego sets). And, on my way of thinking, human substances are such that the following claims hold (more about some of the terminology anon):

6. A human being[29] = A human body
7. A human body = a human form/prime matter composite
8. A human body ≠ prime matter
9. A human being ≠ a human form
10. A human being ≠ prime matter

In (6)–(10), note that by "=" I mean numerical/strict identity.[30] Take our candidate human, Jones. On my account, Jones is numerically identical to her body. Jones just is her body. So, the very obvious reason why I take:

(a) Bodily resurrection is in *no way* superfluous to a human being's life after death to be the case is precisely because Jones has no hope of life after death without her bodily resurrection. The only way she *could* live again is if her body does, because she and her body *is* one thing.[31] This also provides further reason for why I think (c) is true, that is, that the numerical identity between one's dead body and one's resurrection body must obtain. For, if not, then one doesn't resurrect.

Further, on my view, a human is a very specific sort of entity, namely an *organism*, a living biological substance. To further label my broadly Thomistic hylemorphism, I'll adopt the (recently deployed) moniker "hylemorphic animalism," wherein the term "animalism" denotes a philosophical thesis that posits that human persons/beings are identical to human organisms.[32] Why this distinction is important will become apparent shortly.

To make good on the promissory note I wrote just before listing (6)–(10), I'll further explain certain, nonobvious, terms in the list. In particular, I'll flesh out, albeit briefly, the terms "form" and "prime matter." This will help explain how my animalism is consistent with *dualism*. To do this, I first need to address—again, briefly—one of the driving questions in hylemorphic metaphysics: what causes what exists potentially to exist actually? To answer this question, hylemorphists (generally) posit four causes: an efficient (or agent) cause, a formal cause, a material cause, and a final cause. These explain the move from potential to actual.

Here's a rough and ready example of how this works. Consider Michelangelo's statue, David. The efficient cause of David's existence (i.e., one cause for why it doesn't remain a concept in the artist's mind) is Michelangelo's hammering, sculpting, and the like. The formal cause, the *form*, of David's existence—in that it's *David* and not some other statue—is its design, structure, and shape (at least). David's material cause, its *matter*, is the marble out of which David is made; the matter marks David out as a concrete individual and provides the potential ways in which David could be shaped and structured. David's final cause is David's *purpose* or *telos*; it tells us *why* David was created, namely to be a piece of art that functions in such and such a way. Why did David move from potential thing to actual thing? Because Michelangelo had (a) particular *reason(s)* for David's existence; David has a *purpose*. That's the final cause for why David actually exists.

Now, I say this is a rough and ready example because statues and human beings are, on hylemorphism, quite different sorts of things. At the very least, on my version of hylemorphism, statues aren't substances. They are, instead, aggregates. They are one thing, but only in a derivative sense, a sense that *must* appeal to the bits of material that make it up. Humans, on the other hand are substances, one thing. You might think (rightly) that a human has parts. But, contrary to aggregates, a human is not the sum of her parts. This is for very many reasons, reasons that I haven't the room to explain here. So, for the foregoing, let's grant that humans are substances and statues are not.

Here's why this is important. That a human is a substance and a statue isn't tells us something about the *kind* of form that is the formal cause of a human and the formal cause of a statue. The formal causes of all substances are called "substantial forms." Note that this doesn't suggest that the forms *are* substances. Rather, it suggests that they are forms that account for why the particular things they inform are in fact substances. The formal causes of nonsubstances are called "accidental forms." Now, substances have accidental forms, too. They allow for substances to have certain accidents predicated of them (e.g., Socrates sitting). But the idea is that nonsubstances do *not* have, as their formal cause, a substantial form. Humans, as substances, do.

Further, for living substances, the substantial form provides not just its formal cause but also its efficient and final causes. The explanation for how a living thing comes to be what it is designed to be—how it grows and develops—is filled out by its substantial form. Substantial forms provide the *essence* of a given substance; they explain why a particular thing is the thing it is and not something else. Hylemorphists call the substantial form of living things "soul."[33] So, the soul of a human being is her substantial form. It explains why and how she's a *human* and not some other thing.

I say in (7) that a human body (i.e., a human being) is identical to a form/prime matter composite. With this way of phrasing things, I take a departure from (at least, on my reading) a number of Thomists. For, on many occasions in hylemorphic literature, a human is said to be a form/*matter* composite or a composite of body and soul.[34] But, again, that needn't detain us here. Much of the debate surrounding this particular issue is wrapped up in whether a body and soul can come apart such that the soul survives. Since I'm eschewing that discussion for the reasons given above, I needn't explain why I think a hylemorphist should deny that a hylemorphic compound is a compound between a soul and a body.

I've said a bit about what a form is, particularly a human soul. What, then, is prime matter? Prime matter is *not* a body; it is *not* visible/corporeal matter. On my view, human souls explain why a particular body is *human*. What explains why a human is a material thing is prime matter, the principle of pure potency. Now, I grant that prime matter is a deeply mysterious sort of entity. On the hylemorphic scheme, it is completely nonactual, nonabstract, and not nothing.[35] In fact, according to hylemorphism, there just never is an instance where prime matter exists without some form or other informing it. The same is true (barring what most Thomists, at least, say about the human substantial form) for substantial forms: they never exist apart from informing prime matter. One comes to know of these two metaphysical constituents via abstraction; they explain the act/potency of a given concrete entity. One doesn't come to know about them because one observes them existing "out there" apart from concrete hylemorphic compounds.

On my version of hylemorphism, there's a perfectly good reason for this (aside from the obvious incoherence of saying that a thing that's purely potential actually exists in pure potentiality): substances are *not* aggregates of metaphysical parts. Substances are *one thing* (in Aquinas's terminology *unum simpliciter*). And what accounts for this essential unity is the inseparability of the formal and material causes. If the substantial form of a thing could exist apart from any material principle, *it*, the form, would be the substance; the physical being—the hylemorphic compound—would be an aggregate, its matter being an add-on to the substance, the form. The same holds for the human body. It, too, is not separable from a human form. A human form, after all, just is the thing that explains why a human body is an actual (and living!) human body. A body without a human soul is not a human body (more than that: bodies just are actualized prime matter). This is just the sort of thing I mean to imply by deploying the "=" symbol in (7).

Much more could be said about this.[36] But I hope this gives some insight into the way I think about hylemorphism and what I take human souls to be. Since I take it that there's no disembodied state, that an existing human soul is a sufficient condition for the existence of a human body, and that a human body is identical to a human being, I think I can offer (borrowing from Christopher Brown) the following set of identity conditions for a human's persistence through time:

> (HI): For any material substances x and y, x at time, T_1, is numerically identical to y at time, T_2 (where T_2 is any time later than T_1), if and only if the substantial form of x at time, T_1, is numerically identical to the substantial form of y at time, T_2.[37]

With this, we can look back to Figure 14.1. Suppose that Jones, our human, is in reality slice A at time T_1. Suppose further that there's a human in reality slice R at time T_Ω that has the numerically same substantial form as Jones. By (HI), that human at T_Ω is identical to Jones. Further still, suppose that, at time T_3, Jones is dead (say she's dying at T_2). If the numerically same substantial form persists through T_1 and T_2, then Jones does, too (as a human organism). And if the substantial form of the human at T_Ω is numerically identical to the substantial form of Jones at T_1 and T_2, then the human being at T_Ω is Jones (a human organism). So, even if Jones is dead at T_3, the human being at T_Ω is still Jones.

At this point, what I've said concerning (HI) and resurrection appears consistent with a position that might violate a principle many metaphysicians affirm, a principle I call "Locke's Axiom." According to this axiom, what begins to exist and then fails to exist cannot begin to exist again. Elsewhere this axiom has been termed the "principle of non-repeatability."[38] If one is concerned not to violate this sort of principle, it's helpful that EP gets around potential worries about humans popping into and out of existence. For on EP a human being is spread out across two temporal moments. Prior to and at the eschaton, one of those moments is the eschatological moment, T_Ω. What this means is that, during the pre-eschatological period in which she's alive, Jones, a form/prime matter composite (living organism), is spread out over one pre-eschatological temporal moment and the time of the general bodily resurrection, T_Ω. Neither Jones, nor her soul, goes completely out of existence at her death. Instead, upon her pre-eschatological death, she immediately finds herself at the general resurrection, still existing in the sum of all reality, with all the rest of the redeemed.

Notice here how central the human soul is for this account. Given (HI) and EP, it's precisely the soul that is a necessary and sufficient condition for the existence of a living human organism. It's what explains how an organism is actual and it's what provides for an organism's numerical identity at all times at which it exists, in this life and in the next. The bits of informed matter that come to compose the human organism do not. And though prime matter is a necessary metaphysical cause of the human being, it is not sufficient for a human's existence. It *must* be informed by a human soul to result in an actual human being.

By contrast, I think physicalist accounts of the human organism struggle to provide criteria for an organism's diachronic numerical identity. Here, I'll understand "physicalism" to be the thesis that human beings are identical to things that have physical properties essentially, and that are composed only of physical parts.[39] It should be clear, with this understanding of "physicalism," why hylemorphism isn't physicalist. For, though my hylemorphic animalism suggests that humans have physical properties and parts essentially, they're not composed *only* of physical parts; rather, they are dualistic form/prime matter composites. Now, Trenton Merricks, a Christian and physicalist, highlights the physicalist's struggle to provide criteria for a human's diachronic identity. And, he bites the bullet. By his lights, anticriterialism—the view that there just are no *informative* necessary and sufficient conditions for personal identity over time—is the best way forward for physicalist accounts of identity over time and through resurrection.[40]

It's hard to disagree with Merricks on this point. For, at least with respect to Christian thinkers doing work in this area—particularly with respect to accounts of resurrection—the best physicalist view on offer for providing criteria for an organism's identity is in the work of Peter van Inwagen. To see that this is so, just check the sorts of identity conditions assumed in recent physicalist literature on resurrection. The "Falling Elevator Model" used by Dean Zimmerman and Kevin Corcoran and van Inwagen's own "Simulacrum Thesis," for example, each assume van Inwagen's account of organism identity over time. For space reasons, I cannot explicate fully van Inwagen's view, but I can provide a brief synopsis of a problem his account faces—a problem that generalizes over the previously mentioned models. At first blush, van Inwagen successfully provides criteria of diachronic identity for an organism; and one might think this is enough (given that he's an animalist). But there's a problem: by his own admission, his account of an organism's diachronic identity presupposes that the *life* of the organism remains numerically identical over time. What's central to organism identity is *life* identity. But, again by his own admission, he cannot provide criteria of identity for a *life*.[41] Now, if lives, on his account, were primitive and unanalyzable, then no problem. But, on his account, they're neither primitive nor unanalyzable. So, taking his position to its logical conclusion ends up doing just what Merricks claims all physicalists must do: assume some uninformative necessary and sufficient conditions for object identity over time; that is, deny that there are criteria of identity.

For those of us whose intuitions beg for criteria of identity, anticriterialism is a nonstarter. In any case, though I do think van Inwagen gives a nice and coherent account of what an organism is, in the end, his view fails to provide informative necessary and sufficient conditions for an organism's persistence through time. And, *if* that's right, then views borrowing from his account of an organism's persistence cannot provide criteria for the numerical identity of the body in the resurrection. Moreover, if it's correct that there are informative necessary and sufficient conditions for the numerical identity of the human organism over time, and if it's correct that physicalism cannot provide them, then physicalism cannot provide an account for a human person's life after death. This is because, *ex hypothesi*, on the sorts of physicalist accounts I have in mind, a human person and her organism are identical. Here's the upshot: I take it that it's not too much of a gamble to suggest that any view of the human being that cannot account for a human's life after death is incompatible with Christian theology. Thus, physicalism appears incompatible with Christian theology. However, with EP, my account of hylemorphism can account for the life after death and bodily resurrection of a human person; thus, it *is* compatible with Christian theology.

CONCLUSION

My fundamental goal in this chapter is modest: to show that there is (at least) one model of time and human beings that might allow for an immediate resurrection account that satisfies the spirit, rather than the letter, of believing in the intermediate state. That is to say, I aim to show that physicalism is neither the sole owner of immediate resurrection theories nor their best fit. To help motivate my position, I've offered some biblical theological reasons for thinking that humans are essentially embodied beings. With this in hand, and affirming immediate postmortem presence in Paradise, I offer Eschatological Presentism and a hylemorphic account of human beings. With these two things, I provide a model of immediate eschatological resurrection *and* provide an account of necessarily embodied human beings that denies physicalism. Indeed, I hope I've provided some reason to think that one can lose the intermediate state without losing one's soul. In fact, it may be the only way to do so![42]

NOTES

1. James T. Turner, "On the Horns of a Dilemma: Bodily Resurrection or Disembodied Paradise?" *International Journal of Philosophy and Theology* 75, no. 5 (2015): 406–421; James T. Turner, "On Two Reasons Christian Theologians Should Reject the Intermediate State," *Journal of Reformed Theology* 11 (2017): 121–139.

2. I'm using "the intermediate state" in the narrow sense defined above, not as a term to denote a general disembodied state that precedes bodily resurrection. I do deny that there are such states, but that's an argument to be made elsewhere. I also deny all accounts of Purgatory. See my arguments in James T. Turner, "Purgatory Puzzles: Moral Perfection and the Parousia," *Journal of Analytic Theology* 5 (2017): 197–219. On the narrow sort of view about "the intermediate state" in Christian tradition, see Wayne Grudem, *Systematic Theology* (Leicester: IVP, 1994); N. T. Wright, *Surprised by Hope* (New York: HarperOne, 2008); Joshua R. Farris, *The Soul of Theological Anthropology* (London: Routledge, 2017), 131–150; Matthew Levering, *Jesus and the Demise of Death: Resurrection, Afterlife, and the Fate of the Christian* (Waco, TX: Baylor University Press, 2012), 1–22. In his *Benedictus Deus* of 1336, Pope Benedict XII taught a view consistent with the intermediate state. See Peter C. Phan, "Roman Catholic Theology," in *The Oxford Handbook of Eschatology*, ed. Jerry

L. Walls (Oxford: Oxford University Press, 2008), 216. On the Apostolic Fathers and Apologists, see Brian Daley, *The Hope of the Early Church* (New York: Cambridge University Press, 1991), 14–24. For a theological titan, see also Daley, *The Hope of the Early Church*, 141–146 for an overview of Augustine's affirmation of the intermediate state. Or see Augustine, "Enchiridion," in Vol. 3 of *Nicene and Post-Nicene Fathers*, ed. Philip Schaff (Peabody, MA: Hendrickson Publishers, 2004), 272; Augustine, "On the Soul and Its Origin," in Vol. 5 of *Nicene and Post-Nicene Fathers*, ed. Philip Schaff, (Peabody, MA: Hendrickson Publishers, 2004), 2.14, 337; Augustine, *City of God*, 433–445, 20.23; Augustine, "Reply to Faustus the Manichaean," in Vol. 4 of *Nicene and Post-Nicene Fathers*, ed. Philip Schaff (Peabody, MA: Hendrickson Publishers, 2004), 341. 33.1. For Aquinas see, for example, Thomas Aquinas, *The Soul*, trans. John Patrick Rowan (London: B. Herder Book Co., 1951), article 1. For Calvin see John Calvin, *Institutes of the Christian Religion*, III.xxv.6, etc. Many more could be cited.

3. For the sorts of views that are disparaged, see, for example, John Hick, *Death and Eternal Life* (Louisville, KY: Westminster John Knox Press, 1994), 278–296; John Hick, "Resurrection Worlds and Bodies," *Mind* 82, no. 327 (July 1973): 409–412. Hick in effect denies a general resurrection because, on his account, people individually and at different moments pop into the resurrection world. John Cooper's complaint against *all* immediate resurrection views is that they deny a literal general resurrection. This is a worry I aim to get around. See John W. Cooper, *Body, Soul, and Life Everlasting* (Grand Rapids, MI: Eerdmans, 2000), 106. In the philosophical literature, there's the "Falling Elevator Model" of postmortem survival, a view posited by Dean Zimmerman, Kevin Corcoran, Timothy O'Connor, and Jonathan Jacobs (perhaps others) that suggests that, at the moment of death (or just prior), a human's body fissions leaving both a fissioned off corpse and sending a fissioned off body to the afterlife. One might count such a view as consistent with an "immediate resurrection" view, too. None of them say in print (so far as I'm aware), just *when* the fissioned human being "jumps" to. So, it's not clear whether these views deny a general resurrection or not. See Dean W. Zimmerman, "The Compatibility of Materialism and Survival: The 'Falling Elevator' Model," *Faith and Philosophy* 16, no. 2 (April, 1999): 194–212; Kevin Corcoran, "Physical Persons and Postmortem Survival," in *Soul, Body, and Survival: Essays on the Metaphysics of Human Persons*, ed. Kevin Corcoran (Ithaca, NY: Cornell University Press, 2001), 201–217; Timothy O'Connor and Jonathan Jacobs, "Emergent Individuals and the Resurrection," *European Journal for Philosophy of Religion* 2 (2010): 69–88.

4. See, for example, Cooper's complaints against immediate resurrection views in *Body, Soul, and Life Everlasting*, 166–167. In my view, Cooper does not anticipate well the sorts of physicalist accounts that crop up around the beginning of the twenty-first century.

5. Consider what Aquinas borrows from Damascene, namely, that because man is made in God's image it must be the case that man is intelligent and is the fundamental principle of his actions (i.e., free willed) (*Summa Theologiae*, IIa., prologue). Jaroslav Pelikan outlines a Patristic development of this sort of thought in *Christianity and Classical Culture* (New Haven, CT: Yale University Press, 1993), 121–135.

For "communal" sorts of emphases, see, for example, Stanley J. Grenz, *The Social God and the Relational Self* (Louisville, KY: Westminster, 2001).

6. Some recent examples: G. K. Beale, *The Temple and the Church's Mission* (Downers Grove, IL: InterVarsity Press, 2004), 70ff; throughout J. Richard Middleton, *The Liberating Image* (Grand Rapids, MI: Brazos, 2005); Willem Vangemeren, *The Progress of Redemption* (Grand Rapids, MI: Baker Books, 1988), 52ff; James K. A. Smith, *You Are What You Love* (Grand Rapids, MI: Brazos, 2016), 172–173.

7. Cf. G. K. Beale, *The Temple and the Church's Mission,* 163–166, 401. In this work, Beale argues that the tabernacle and temple of the OT were, themselves, microcosms of the entire created order; Jon D. Levenson, "The Temple and the World," *The Journal of Religion* 64/3 (July 1984): 275–298; Benjamin L. Gladd and Matthew S. Harmon, *Making All Things New* (Grand Rapids, MI: Baker Academic 2016); J. Richard Middleton, *A New Heaven and a New Earth* (Grand Rapids, MI: Baker Academic, 2014), 46–49, 164ff; J. Richard. Middleton, *The Liberating Image*, 81ff; John H. Walton, *The Lost World of Genesis One* (Downers Grove, IL: IVP Academic, 2009). For a critique of this view, see Daniel I. Block, "Eden: A Temple? A Reassessment of the Biblical Evidence," in *From Creation to New Creation: Essays in Honor of G. K. Beale*, ed. Daniel M. Gurtner and Benjamin L. Gladd (Peabody, MA: Hendrickson Publishers, 2013), 3–29.

8. Cf. Beale, *The Temple and the Church's Mission,* 163–166, 401; Gladd and Harmon, *Making All Things New*, 135, 149; Middleton, *A New Heaven and a New Earth*, 107.

9. Middleton, *The Liberating Image*, 25–26. My emphasis and insert. Perhaps Middleton's point is slightly overstated, for there are systematicians tracking with this line of thought. See David Fergusson, "Creation," in *The Oxford Handbook of Systematic Theology*, 74–75.

10. Importantly, but not of central relevance to my point in this chapter, imaging God is not *merely* royal; it's also priestly. That is, given that the thing over which and in which God rules—the cosmos—is a *temple*, helping to rule and sustain the cosmos is a function of tending and keeping a temple, a priestly duty. Middleton, *A New Heaven and a New Earth*, 45ff; Middleton, *The Liberating Image*, 89ff; Beale, *The Temple and the Church's Mission*, 68ff; Walton, *The Lost World of Adam and Eve*, 104–127. I cite what I do here not to suggest that there are no competing views, but simply to get my project up and running. One theologian that works with opposing conclusions to my reading of the biblical text is Joshua Farris. He seems to think that this functional account implies some further substantive account of what sort of thing a human is. While I agree that—given the precise functions Yahweh desires of his images—that some further substantive account falls out of the functional account, Farris and I disagree on the sort of substantive model we think is implied. He's a substance dualist; I am not, as you see in what follows. Cf. Farris, *The Soul of Theological Anthropology*, 35.

11. Middleton, *The Liberating Image*, 25.

12. On the importance of the cosmos and Yahweh's image not being crafted by human hands, see Joel B. Green, "Eschatology and the Nature of Humans: A Reconsideration of Pertinent Biblical Evidence," *Science and Christian Belief* 14/1 (April

2002): 48–49; Middleton, *The Liberating Image*, 121–130; Beale, *The Temple and the Church's Mission*, 152.

13. Aquinas, *Summa Theologiae,* IIa Q.5 A.3–5. This is not to say that everyone who posits a classical beatific vision is ignorant of the Ancient Near Eastern background of the creation narratives. Matthew Levering, for example, seems well aware. See his *Jesus and the Demise of Death.*

14. Joshua Mugg and James T. Turner, Jr., "Why a Bodily Resurrection? The Bodily Resurrection and the Mind/Body Relation," *Journal of Analytic Theology* 5 (2017): 121–144.

15. Cf. David Fergusson, *Creation* (Grand Rapids, MI: Eerdmans, 2014), 4–9; Richard Bauckham, *Bible and Ecology: Rediscovering the Community of Creation* (London: Darton, Longman, & Todd, 2010), 157; W. Waite Willis, Jr., "A Theology of Resurrection: Its Meaning for Jesus, Us, and God," in *Resurrection: The Origin and Future of a Biblical Doctrine*, ed. James H. Charlesworth (London: T&T Clark, 2006), 212; Middleton, *A New Heaven and New Earth*, 131–175; Richard Bauckham, "Eschatology," in *The Oxford Handbook of Systematic Theology*, 308–311; Mugg and Turner, "Why a Bodily Resurrection," 122–128.

16. My emphasis.

17. See John Cooper's worries in his, "Whose Interpretation? Which Anthropology? Biblical Hermeneutics, Scientific Naturalism, and the Body-Soul Debate," in *Neuroscience and the Soul*, eds. Thomas M. Crisp, Steven L. Porter, and Gregg A. Ten Elshof (Grand Rapids, MI: Eerdmans, 2016), 238–257.

18. There is now a growing body of literature for a view of Purgatory that's purportedly consistent with Protestant soteriology. See, especially, Jerry L. Walls, *Heaven, Hell, and Purgatory* (Grand Rapids, MI: Brazos Press, 2015); Jerry L. Walls, *Purgatory: The Logic of Total Transformation* (Oxford: Oxford University Press, 2012); David Vander Laan, "The Sanctification Argument for Purgatory," *Faith and Philosophy* 24, no. 3 (2007): 331–339; Justin D. Barnard, "Purgatory and the Dilemma of Sanctification," *Faith and Philosophy* 24, no. 3 (2007): 311–330; Neal Judisch, "Sanctification, Satisfaction, and the Purpose of Purgatory," *Faith and Philosophy* 26, no. 2 (2009): 167–185. I've addressed these versions of Purgatory (and offer an argument against the very notion) in James T. Turner, "Purgatory Puzzles: Moral Perfection and the Parousia."

19. Which term one wants to use to explain four-dimensionalism will depend on which flavor of four-dimensionalism one espouses. For a nuanced overview, see Michael Rea, "Four-Dimensionalism," in *The Oxford Handbook of Metaphysics*, eds. Michael J. Loux and Dean W. Zimmerman (Oxford: Oxford University Press, 2003), 246–280. For a "person stage" account, see Theodore Sider, *Four-Dimensionalism* (Oxford: Oxford University Press, 2001).

20. Though Sally Haslanger argues that stage-theoretic accounts are theories of "exdurance" rather than "perdurance." Sally Haslanger, "Persistence Through Time," in *The Oxford Handbook of Metaphysics*, 319. For more on the endurance/perdurance distinction, see Neil McKinnon, "The Endurance/Perdurance Distinction," *Australasian Journal of Philosophy* 80, no. 3 (September 2002): 288–306.

21. Suppose that the substance Susie is in pain at time T1. On four-dimensionalism, she's eternally in pain at that moment. That pain never goes away. Moreover, because

Susie's temporal part at T1—the part that's in pain—is *essential* to Susie, God cannot eliminate the pain. Multiply this sort of evil over all the humans that have existed and make many of them horrendous, and one can see why this is intuitively problematic. God can never redeem—that is, renew—those parts of his eternally existing cosmos.

22. You can find an explanation of Compound Presentism in Barry Dainton, *Time and Space*, 2nd ed. (London: Acumen, 2010), 95–101.

23. Figure idea borrowed (though amended for my purposes) from Dainton, *Time and Space*, 96.

24. Dainton, *Time and Space*, 95–96. Note that, as Dainton warns with his own similar figure, "temporal dynamism cannot be fully captured in a static diagram." Figure 14.1 *appears* to show that all the times coexist. But that is not correct. Think, rather, that when there is a vacancy at the S and T co-ordinates, that time does not exist. For example, at T3, T1 does not exist—there is nothing at the intersection of T1 and an S coordinate when T3 is populated.

25. I owe this expression to Thom Atkinson who coined the phrase when I explained EP in personal conversation.

26. Among the hostages my view gives to fortune is that time, on Eschatological Presentism, is discrete (i.e., finitely divisible). This is quite against the "received" view in the philosophy of time, namely, that time is continuous (i.e., infinitely divisible). It requires a discrete view because I need discrete temporal moments that can be compounded together in the Eschatological Present. Moreover, my view requires that humans, for example, have a temporal part that *endures* through the pre-T_{Ω} time series. Normally, temporal parts are eternally existing and static (on four-dimensionalism) or else one *wholly* moves through individually present moments. Additionally, I don't have any way of spelling out how time and things in time progress after the T_{Ω} moment. This last problem, though, isn't unique to me. I don't know of any thinker who presumes to know with all clarity just what sort of existence awaits us in the New Creation or how time (if there is time) will work in that cosmos. I'm inclined to think that, if a compound presentist theory of time is correct, that it continues into the New Creation. But arguing for this is a project for another day.

27. It's possible that at least one person in the tradition thinks in a nearly identical way. The Luther scholar, Paul Althaus, suggests that Luther's personal eschatology is the following: "For those who have died, the Last Day comes very soon after their death—even 'immediately' when they die. 'Each of us has his own Last Day when he dies.' Therefore we arrive at the end of the world and the Last Day at the moment of our death. And yet it comes no sooner to the departed than to us and to all generations after us until the temporal end of the world" (Paul Althaus, *The Theology of Martin Luther*, trans. Robert C. Schultz [Philadelphia, PA: Fortress Press, 1966], 416).

28. Survivalism is the idea that, at the death of the body, the human person survives as a soul (either identical to it or constituted by it). Corruptionism suggests that, though the soul survives, the human person doesn't. I get these terms from Patrick Toner, "On Hylemorphism and Personal Identity," *European Journal of Philosophy* 19/3 (September 2011): 454–473. But Toner originally labels survivalism the "alternative [Thomist] view" and corruptionism the "standard [Thomist] view." Patrick Toner, "Personhood and Death in St. Thomas Aquinas," *History of Philosophy Quarterly* 26/2 (April 2009): 121.

29. For Christological reasons, in what follows I deploy the term "human being" rather than "human person." For, on my account, all human beings are persons. But not all human beings are *human* persons (Jesus, for example, is a human being but not a human person; he's a divine Person, viz., God the Son).

30. Harold Noonan and Ben Curtis, "Identity," *The Stanford Encyclopedia of Philosophy* (Spring 2017 Edition), edited by Edward N. Zalta. https://plato.stanford.edu/archives/spr2017/entries/identity/. Accessed May 26, 2017.

31. Notice the copula is purposefully in the singular.

32. See Eric T. Olson, *The Human Animal* (Oxford: Oxford University Press, 1999); Andrew M. Bailey, "Animalism," *Philosophy Compass* 10, no. 12 (2015): 867–883; Patrick Toner, "Hylemorphic Animalism," *Philosophical Studies* 155, no. 1 (2011): 65–81; Mugg and Turner, "Why a Bodily Resurrection?" 138ff.

33. In the Aristotelian and Thomist literature, at any rate, there are three types of soul: vegetative, animal, and human. Each of these describes a particular sort of set of boundary conditions that mark out the sorts of capacities available to a living entity in each class. For space reasons, I'll not get into that here.

34. For example, Eleonore Stump, *Aquinas* (London: Routledge, 2003), 203. Robert Pasnau spells out Aquinas's view in this way in, *Thomas Aquinas on Human Nature* (Cambridge, UK: Cambridge University Press, 2002), 8ff; Jason T. Eberl, "Aquinas on the Nature of Human Beings," *Review of Metaphysics* 58, no. 2 (December 2004): 333–365. It's tricky trying to interpret Aquinas on this score. As Christopher Hughes rightly notes, Aquinas often seems to equivocate on the term "matter." Christopher Hughes, "Matter and Actuality in Aquinas," in *Thomas Aquinas: Contemporary Philosophical Perspectives*, ed. Brian Davies (Oxford: Oxford University Press, 2002), 70.

35. Jeffrey Brower and John Haldane are Thomists that think of prime matter as a "stuff of no kind." John Haldane, "A Return to Form in the Philosophy of Mind," *Ratio* 11 no. 3 (December 1998): 263; Jeffrey E. Brower, "Matter, Form, and Individuation," in *The Oxford Handbook of Aquinas*, eds. Brian Davies and Eleonore Stump (Oxford: Oxford University Press, 2012), 93.

36. I address all of this more fully in Turner, "We Look: PhD diss." But a heavily revised version of my arguments and explanation is now in the monograph *We Look for the Resurrection of the Dead*, under review.

37. Borrowed and modified from Christopher M. Brown, *Aquinas and the Ship of Theseus* (London: Continuum, 2005), 119.

38. Christina Van Dyke, "Human Identity, Immanent Causal Relations, and the Principle of Non-Repeatability: Thomas Aquinas on the Bodily Resurrection," *Religious Studies* 43, no. 4 (December 2007): 373–394.

39. Another broad way of understanding "physicalism" is to couch it in terms of a thesis such that human beings can be explained exhaustively by physics. I take it that the way I've expressed physicalism is consistent with this.

40. Trenton Merricks, "There Are No Criteria of Identity Over Time," *Nous* 32, no. 1 (March 1998): 106–124; Trenton Merricks, "How to Live Forever without Saving Your Soul," in *Soul, Body, and Survival: Essays on the Metaphysics of Human Persons*, ed. Kevin Corcoran (Ithaca, NY: Cornell University Press, 2001).

41. For his account of human beings and personal identity, see Peter van Inwagen, *Material Beings* (Ithaca, NY: Cornell University Press, 1990), particularly 85ff. For his statements concerning the inability to provide an account for the identity of a life, see, pages 157–158.

42. Thanks to Joshua Farris and Jordan Wessling for providing helpful comments on earlier drafts of this chapter.

SUGGESTED FURTHER READING

Beale, G. K. *The Temple and the Church's Mission.* Downers Grove, IL: InterVarsity Press, 2004.

Corcoran, Kevin, ed. *Soul, Body, and Survival: Essays on the Metaphysics of Human Persons.* Ithaca, NY: Cornell University Press, 2001.

Dainton, Barry. *Time and Space.* 2nd ed. Durham, UK: Acumen, 2010.

Farris, Joshua R. *The Soul of Theological Anthropology.* London, UK: Routledge, 2017.

Middleton, J. Richard. *A New Heaven and a New Earth.* Grand Rapids, MI: Baker Academic, 2014.

Mugg, Joshua, and James T. Turner, Jr. "Why a Bodily Resurrection? The Bodily Resurrection and the Mind/Body Relation." *Journal of Analytic Theology* 5 (2017): 121–144.

Toner, Patrick. "Hylemorphic Animalism." *Philosophical Studies* 155, no. 1 (2011): 65–81.

Turner, James T., Jr. "On the Horns of a Dilemma: Bodily Resurrection or Disembodied Paradise?" *International Journal of Philosophy and Theology* 75, no. 5 (2015): 406–421.

Walton, John H. *The Lost World of Genesis One.* Downers Grove, IL: IVP Academic, 2009.

Wright, N. T. *The Resurrection of the Son of God.* Minneapolis, MN: Fortress, 2003.

Chapter 15

Dismantling Bodily Resurrection Objections to Mind-Body Dualism

Brandon Rickabaugh

"My only comfort in life and death is that I belong, body and soul, to my faithful Savior Jesus Christ."[1] This confession is grounded in two historic and prevalent Christian beliefs:[2]

> *Bodily Resurrection*: We, human persons, will exist in the life to come with a glorified and incorruptible resurrected body.
> *Mind-Body Dualism*: We, human persons, are not identical to any purely physical thing, but are or have an embodied immaterial soul.

Although most Christians throughout the history of the church have maintained both beliefs, some Christian materialists argue that these two doctrines are in conflict. Some argue that bodily resurrection is trivialized by substance dualism (here after, dualism), that dualism makes expiations of why bodily resurrection is truly difficult, or that dualism should be rejected as bodily resurrection is better accounted for by Christian physicalism. Let's call such arguments resurrection objections.

These criticisms are somewhat understandable. Dualism is often stated with little to no mention of the body. Regarding the core commitments of dualism, Dean Zimmerman observes,

> (a) they believe that, for every person who thinks or has experiences, there is a thing—a soul or spiritual substance—that lacks many of the physical properties the body shares with unthinking material objects; and (b) they believe that this extra thing is essential to the person, and in one way or another responsible for the person's mental life.[3]

The emphasis is on the soul, with only passing mention of the body. This too is somewhat understandable as arguments for dualism are often framed in debate with physicalism. Christian dualists agree that embodiment is a crucial aspect of human persons. Alvin Plantinga observes that

> on the traditional Christian view, God has designed human beings to have bodies; they function properly only if embodied; and of course, Christians look forward to the resurrection of the body. My body is crucial to my well-being and I can flourish only if embodied.[4]

Likewise, for Dallas Willard, "*the body lies at the center of the spiritual life*."[5] and is "an essential part of who we are and no redemption that omits it is full redemption."[6]

Nevertheless, contemporary dualist accounts of embodiment or bodily resurrection are scarce. Kevin Corcoran observes:

> Yet it is plausible to believe that it is precisely that doctrine that needs to be addressed by Christian Dualists, for none of the ecumenical creeds of the church confesses belief in a doctrine of soul survival. The Christian doctrine has been understood as the doctrine of bodily *resurrection*.[7]

It is worth briefly explaining that Corcoran's interpretation of the creeds is problematic. The creed makers, like most at the time, assumed dualism of one kind or another.[8] With no need to defend dualism, their goal was to emphasize the uniquely Christian claim of bodily resurrection. There is a fundamental assumption of dualism evident in the Apollinarian controversy surrounding Chalcedon and neo-Chalcedonian Christology. The central debate was whether or not the Son needed a soul *in addition to* him or his mind.[9] Chalcedon explicitly rejects the Apollinarian and Arian "God with a body" Christology, yet affirms that the Son has a rational soul, meaning Christ assumed a soul and a body. The Cappadocian fathers reject Apolinarius's teaching and Arian Christology, yet affirm that the Logos must assume a soul and a body. Further, the Cappadocians continually argue that the Logos's relationship to his human nature is just like the relationship between the body and the soul. Moreover, the rejection of Origenism at the Fifth Ecumenical Council is telling. On one version of Origenism, possibly affirmed by Gregory of Nyssa, a soul lives on without a body. In response, the council sought to affirm the body, but did not deny the soul.[10] Moreover, the Fourth Council of Constantinople states: "the old and new Testament teach that a man or woman has one rational and intellectual soul, and all the fathers and doctors of the church, who are spokesmen of God, express the same opinion."[11] Hence, Corcoran's use of the early creeds is misleading.[12]

Still, Corcoran is right about the lack of dualist work on embodiment. Although, Corcoran, and every other Christian physicalist I am aware of, overlooks the works of Charles Taliaferro,[13] Howard Robinson,[14] and Richard Swinburne.[15] Still, inattention to embodiment is a weakness in contemporary dualism.[16] There are, of course, other theological objections to dualism, but those have received substantive replies.[17] While dualists[18] and Christian materialists[19] have raised resurrection objections against various forms of Christian materialism, Christian materialists have responded.[20] The same cannot be said of dualists.[21] This chapter is intended to help move this conversation forward.

BODILY RESURRECTION, WHAT'S THE PROBLEM?

According to Lynne Rudder Baker, Christian views of the resurrection must account for three doctrines.

> EMBODIMENT: Resurrection requires some kind of bodily life after death.
> IDENTITY: The very same person who exists on earth is to exist in an afterlife.
> MIRACLE: Life after death, according to Christian doctrine, is a gift from God.[22]

These doctrines are largely uncontroversial.[23] Trenton Merricks, however, defends a moderately controversial criterion.

> BODILY IDENTITY: An individual's resurrected body must be numerically identical to their preresurrection body.[24]

Taken together, this four-part desideratum rouses the strongest resurrection objections to dualism.

Frequently, the objection is that dualism cannot account for at least one of the resurrection desiderata, while Christian materialism can. At other times the stronger objection is that dualism makes explaining these criteria impossible. But how so? According to Baker,

> The best that metaphysics can do is to show how resurrection is metaphysically possible. That is, any candidate for a metaphysics of resurrection must conceive of human persons in such a way that it is metaphysically possible (even if physically impossible) that one and the same person, whose earthly body is corruptible, may also exist with a post-mortem body that is incorruptible.[25]

If this is the best that metaphysics can do, then resurrection objections should attempt to show that the conjunction of dualism and bodily resurrection is metaphysically impossible. In reply, dualists need only show that this conjunction is metaphysically possible.

While I find Baker's criterion undeniable, I am skeptical of BODILY IDENTITY. Yet there is, I will argue, no reason to think that dualism is at odds with BODILY IDENTITY. My goal is to show that not one of these criteria provides a definitive or even serious problem for dualism, which in turn undermines a common motivation for Christian physicalism.

RESURRECTION OBJECTIONS FROM EMBODIMENT

According to EMBODIMENT, resurrection requires a bodily afterlife. Those in Christ will be raised by God with a physical, glorious, incorruptible, powerful, and immortal body (cf. 1 Cor. 15:42–43, 53–54) like the resurrected body of Jesus (Phil. 3:21). Paul expected this to happen not at death, but at the advent (1 Cor. 15:20–24, 51–54; 1 Thess. 4:14–17), as part of Christ's renewal of all things (Phil. 3:20–21). So, how is EMBODIMENT a problem for dualism? Merricks and Baker offer slightly different objections. Let us take them in turn.

Merricks: EMBODIMENT is Not Necessary for Dualism

Merricks's objection from EMBODIMENT begins with a question: "But if dualism were true, it is hard to see why our resurrection would be a big deal."[26] He then argues,

> Now the dualist might object that a soul in Heaven without a body is somehow mutilated or incomplete, and so the dualist might, therefore, insist that resurrection is a blessing. But it is hard to know just how much stress should she put on the value of resurrection, since stress on what we gain in resurrection is by its very nature, stress on what we lack before resurrection. Pre-resurrection existence united with God in Heaven is not supposed to be too bad; indeed, it is supposed to be very good.[27]

Merricks assumes that souls in the intermediate state are conscious. That isn't something dualism entails, and those who hold that souls "sleep" until the body is resurrected easily avoid this problem. Secondly, the claim is not that resurrection cannot be a great good given dualism. Merricks's claim is much weaker: given dualism, it is difficult to assess how valuable resurrection is because embodiment is not needed to enjoy the greatest good of being united to God. But what follows from this cannot be that embodiment is not a great good! It does not follow from the fact that *x* is a great good, that *x* in conjunction with *y* is not an even greater good.

Merricks makes the further point as to what the dualist cannot argue.

> And however the dualist might deal with this problem, one thing is certain: The dualist cannot say that resurrection is necessary for eternal life . . . one cannot maintain both that life after death occurs *before* resurrection and also that life after death *requires* resurrection.[28]

This needn't worry the dualist. First, Merricks shifts between talk of eternal life and life after death. These notions, though, are not equivalent. For example, if annihilationism is true, then one can have a life after death without an eternal life. Likewise, if the doctrine of eternal hell is true, one can have an eternal life in terms of duration, but not in terms of quality. *What this means is that eternal life is not reducible to life after death or the persistence of identity*. Once we distinguish Merricks's conflation of these two doctrines, his objection is less plausible.

Eternal life in the biblical sense is much more than unending postmortem existence. Eternal life is resurrection, as N. T. Wright notes:

> The meaning of "resurrection" as "life after 'life after death'" cannot be overemphasized, not least because much modern writing continues to use "resurrection" as a virtual synonym for "life after death" in the popular sense.[29]

Eternal life is one overarching event with present and future aspects. In the present, eternal life makes available a renewed or resurrected life, the sign of which is to trust and be permeated by agape love.[30] Death begins now, as does life in the Spirit.[31] According to Jesus, those who believe have eternal life *now*.[32] The future aspect of eternal life includes a distinctive kind of survival of death which includes the righting and overcoming of sin and its consequences, touching the body before and after death.[33] This process of glorification starts before death.[34] So, eternal life refers both to duration as well as quality of life. According to several New Testament authors, eternal life, in terms of quality, can begin in this life. "The new life," Wright notes, "which will be consummated in the resurrection itself works backwards into the present, and is already doing so in the ministry of Jesus."[35] Resurrection is something that has become available now. Wright observes,

> Resurrection in John continues to be both present and future, and we should resist attempts to flatten this out by marginalizing the "future" emphasis of overemphasizing the "realized eschatology." It is, of course, true for John, as for Paul, that "eternal life" is not simply future, but already to be enjoyed in the present.[36]

What this means is that eternal life is not mere postmortem existence, but requires resurrection, a part of which is bodily resurrection. So Merricks is mistaken. One can maintain both that eternal life begins before bodily resurrection and that eternal life, in the qualitative sense, requires bodily resurrection.

Consequently, the dualist can hold that bodily resurrection is necessary for eternal life, when we understand that an eternal kind of life is necessarily a bodily resurrected life. Furthermore, we have no reason to think that disembodiment is anything other than a natural consequence of sin, just as death is (Rom. 6:23). Hence, a disembodied life is a soteriologically incomplete life. It is a great good, but not the greatest good. Moreover, resurrection, the righting and overcoming of sin and its consequences, demands re-embodiment. As such, bodily resurrection is needed for the defeat of sin and death. Without it, God's mission of resurrection in the full sense is not fulfilled.

Baker: Disembodiment Excludes an Explanation of EMBODIMENT

According to Baker, "Mind-body dualism would provide no obvious explanation of why resurrection should be bodily (since, according to mind-body dualism, we can exist unembodied)."[37] This assumes that if a theory holds that human persons can exist unembodied, then that theory provides no obvious explanation for EMBODIMENT. What seems to motivate this is a reductive theory of bodily resurrection. Like Merricks, Baker reduces bodily resurrection to the persistence of personal identity. Her argument assumes that if we get an explanation of resurrection as persistence after death that does not require EMBODIMENT, then that explanation offers no obvious explanation of EMBODIMENT. But this assumes that bodily resurrection is reducible to postmortem persistence. However, I have shown that is false. Consequently, Baker's objection from EMBODIMENT fails.

Medieval and Contemporary Arguments for EMBODIMENT

Often Christian physicalists proffer less of a criticism and more a shifting of the burden of proof. They claim that dualists have failed to offer a reason for EMBODIMENT. This claim is easily rejected once we recognize the extensive medieval literature on why the human soul needs a body. My point here is not to develop or defend these accounts beyond chapters 1 and 2 of this volume.[38] I simply wish to show that the burden-shifting move is illegitimate as there are many arguments for why a soul needs a body. In turn, the burden is on the Christian physicalist to show how these accounts fail.

Not all medieval thinkers held the same ontology of the human person, although they uniformly rejected physicalism, arguing that thinking things must be immaterial.[39] Many adopted a neo-Aristotelian metaphysics making use of hylomorphism, a view not obviously at odds with dualism. So, while not every argument from the medieval era would aid dualism, at least the following can:

1. *Appetite Satisfaction and Perfect Happiness*: Bonaventure, Aquinas, and Scotus recognize that our desire for our body is so powerful that we would not be perfectly happy without, not just any body, but our individual body.
2. *Metaphysical Completeness*: According to Bonaventure, the human soul perfects the body, and is, therefore, naturally inclined to be joined to its body. This inclination is frustrated when the soul is disembodied. Therefore, in virtue of the fact that resurrection is fundamentally about bringing creation into perfection, the resurrected person must be an embodied soul.
3. *Metaphysical Perfection*: Aquinas argues that embodiment is part of God's soteriological plan. Being disembodied is not a perfected state, but a punishment for the fall. However, Christ's passion merits the permanent restoration to God's original intent for human persons as embodied souls. If that is how God made us, then that is how he will perfect us.

These are only a few of the arguments made by Bonaventure, Aquinas, and Scotus, among others. The claim that dualists lack reasons for EMBODIMENT is just historically naïve.

While most of the medieval arguments appeal to our desire and inclination to have a body, they do not obviously explain why it is good that we should have such a desire and inclination. Contemporary dualists offer such explanations. Taliaferro argues that being an embodied person consists in the exercise of six types of virtue: sensory, agency, constitutional, epistemic, structural, and affective.[40] Swinburne argues that having a body makes possible great goods, including the ability of free choice between good and evil and the ability to influence others and the inanimate world.[41]

Gordon Barnes argues that the telos of the human soul, as created by God, is embodiment. Consequently,

> If we take this telos of a particular human soul to be constitutive of its very identity, then its embodiment in a particular parcel of matter is also constitutive of its identity, even if the soul and that parcel of matter are really distinct and separable.[42]

Maximus the Confessor offers alternative teleological arguments for dualism.[43] God, argues Maximus, created us as embodied souls as an ontological preparation for the eschatological mystery of the incarnation of Christ. Our ontology is set up, as it were, for the incarnation. Maximus also argues that God creates us as embodied souls so that our ontology reflects and aids us in serving as a mediator in relation to God and his creation. Lastly, Maximus argues that we are a microcosm reflecting elements of the entire world, in body and soul. Without being comprised of both body and soul, human persons would not truly reflect the world in its relationship to God. We would not be a true microcosm or mediator.

Again, my point is not to defend these arguments, but to point out that these arguments have not been addressed by those defending resurrection objections against dualism. The common claim that dualism has no account for EMBODIMENT is simply false. There are many accounts. Taken together they help explain why dualists embrace EMBODIMENT.

BODILY RESURRECTION OBJECTIONS FROM IDENTITY

According to IDENTITY, the very same person who exists on earth is to exist in the afterlife. This is far from controversial. However, Christian physicalists argue that the conjunction of IDENTITY and dualism is somehow problematic. Baker offers both a diachronic and a synchronic version of problem.

Baker's Diachronic IDENTITY Problem

Here is Baker's diachronic IDENTITY objection:

> There is a metaphysical problem with immaterialism: in virtue of what is a soul the same soul both before and after death? Perhaps the best answer is that souls are individuated by having a "thisness" or haecceity. This is an intriguing suggestion that I cannot pursue here. A haecceity view, if otherwise satisfactory, may well be suitable as a metaphysics of resurrection—if it did not leave dangling the question of why resurrection should be bodily.[44]

Note that the first sentence is not an objection, but merely a question. Posing a question does not by itself produce a problem. What we need is a reason to think that dualism cannot answer the question. Baker does not provide one. In fact, she admits that souls could be individuated by having a "thisness" or haecceity. However, she faults such an account as it presumes another doctrine that is problematic for dualism: EMBODIMENT. But faulting an account of persistence for leaving open the question of EMBODIMENT is not an objection from IDENTITY. It certainly cannot be the case that if a theory satisfies IDENTITY it must also satisfy EMBODIMENT unless one assumes Christian physicalism. Baker has, by her own admission, simply stated that even if dualism can satisfy IDENTITY, the problem of EMBODIMENT remains. However, as we saw in the previous section, the objection from EMBODIMENT fails.

Baker later observes that a soul must be subject to change in virtue of the fact that religious practice involves conversion.[45] From this she argues:

> Consider Augustine before and after his conversion—at t1 and t2, respectively. In virtue of what was the soul at t1 the same soul as the soul at t2? The only answer that I can think of is that the soul at t1 and the soul at t2 were both Augustine's soul. But, of course, that answer is untenable inasmuch as it presupposes sameness of person over time, and sameness of person over time is what we need a criterion of sameness of soul over time to account for. So, it seems that the identity of a person over time cannot be the identity of a soul over time.[46]

The dualist has several responses.[47] First, for independent reasons, one might deny there is such a thing as criteria of diachronic identity. Merricks defends such a view.[48] Following Lowe, one might hold that persistence is "primitive or ungrounded, in that it can consist neither in relationships between non-persisting things nor in the persistence of other sorts of things."[49] On such views, Baker's demand for criteria in virtue of which a soul at t_1 is the same soul as the soul at t_2 is in principle impossible regardless of one's ontology of the human person.

Secondly, Baker suggests that a haecceity view might work. According to this view, a soul has a nonqualitative property which is responsible for its individuation and identity. I will offer another account in section 3.4.2. There are, as it turns out, many ways to avoid Baker's diachronic IDENTITY objection.

BAKER'S SYNCHRONIC IDENTITY PROBLEM

Additionally, Baker offers a synchronic IDENTITY objection. Here the idea is that without a body the individual person cannot satisfy IDENTITY. Baker argues,

> In virtue of what is there one soul or two? If souls are embodied, the bodies individuate. There is one soul per body. But if souls are separated from bodies—existing on their own, apart from bodies—then there is apparently no difference between there being one soul with some thoughts and two souls with half as many thoughts. If there is no difference between there being one soul and two, then there are no souls. So, it seems that the concept of a soul is incoherent.[50]

This seems right. If the body is the only thing that can individuate the soul, then a soul without a body cannot be individuated. What she is mistaken about is that this objection renders the concept of a soul incoherent. The most obvious way out of this problem is to reject the notion that a soul is individuated by a body. Baker has already given the dualist a way out by admitting that the soul could be individuated by a haecceity. Consequently, it is difficult to see this objection as having much force.

BODILY RESURRECTION OBJECTIONS FROM MIRACLE

According to MIRACLE, life after death must be understood as a miraculous gift from God (cf. 1 Cor. 15:38). One is hard pressed to find an objection from miracle among contemporary Christian philosophers, although some theologians press this objection. Baker mentions such an objection regarding the dualism of the ancient Greeks.[51]

First, I am unaware of any contemporary Christian dualist who holds that the soul is naturally immortal.[52] Richard Swinburne, for example, rejects this thesis and considers arguments to that conclusion fallacious.[53] He recognizes that such a view is "out of line with the Christian emphasis on the embodiness of men as their normal and divinely intended state."[54] According to Swinburne, neither philosophical nor scientific arguments support the immortality of the soul *on its own powers*. He does, however, think that Scripture and the Creeds evidence the continued existence of the soul after death *due to divine act*.[55] Other dualists, like Karl Popper, are skeptical of the mind's existence after death,[56] while Robert Audi thinks it cannot be guaranteed or ruled out.[57]

Secondly, objections from MIRACLE rest on a false assumption. "The possibility of immortality," says Reichenbach, "should not be confused with the actuality of it."[58] Likewise, Swinburne argues, "even if the soul is simple and separable from the body, it does not follow that it will continue to exist after death, let alone exist forever with a mental life, with thoughts, feelings, and sensations."[59] That a soul continues to exist after its body dies does not mean there are no other conditions under which a soul could cease to exist. Why not think that the death of the body would, save for God's miraculous intervention, result in the death of the soul? The metaphysical possibility of disembodied existence may very well be made actual only by the miraculous activity of God.

Finally, suppose the dualist cannot cite the continued existence of the soul after death as a miracle. It does not follow that life after death, in the full biblical sense of resurrection, is not a miraculous gift from God. That one receives a glorified, incorruptible body is not a consequence of dualism. Hence, one will receive a resurrected body only if God makes it so. These arguments taken together undermine the objection from MIRACLE.

BODILY RESURRECTION OBJECTIONS FROM BODILY IDENTITY

The final resurrection objection is made in terms of numerical identity. According to BODILY IDENTITY, an individual's resurrected body must be numerically identical to their preresurrection body. There are at least two ways one could reply: reject BODILY IDENTITY or show that it is not incompatible with dualism.

Some Christian philosophers do not believe that strict philosophical identity is taught in Scripture or required to preserve important Christian teaching about resurrected persons.[60] In fact, many Christian materialists reject BODILY IDENTITY,[61] as do most contemporary theologians.[62] So, we may have good reasons to reject BODILY IDENTITY. However, we needn't make this move to avoid Merricks's objection. In what follows, I will simply assume BODILY IDENTITY but argue that it is not inconsistent with and can be accounted for by dualism.[63]

Merricks: No Parthood, No BODILY IDENTITY

Merricks's objection to dualism from BODILY IDENTITY presses the following dilemma.

> Some might suggest that my current body will be identical with whatever resurrection body has the same (substantial) soul as is had by my current body. But a soul is not part of a body. And I doubt that the identity of one physical object (such as a body) might be entirely a matter of the identity of a second object (such as a soul) when that second object is not itself a part of the first object. In this regard, taking a soul to be the guarantor of bodily identity is less plausible than taking the bone from the base of the spinal cord to be that guarantor. For at least that bone is a part of the relevant body.[64]

On the one hand, says Merricks, the dualist may argue that (i) if a soul is a part of the body then perhaps the body could persist if the soul persists. However, Merricks points out that the soul is not a part of the body, so the body is not the only or best means of accounting for BODILY IDENTITY.[65] On the other hand, the dualist may argue (ii) that a body can persist in virtue of a soul's persisting. However, Merricks rejects this claim by appealing to a thesis I state as follows.

> *Part Identity*: the identity of one physical thing, B_1, at some time, t_1, cannot be identical with a physical thing, B_2, at another time, t_2, in virtue of some further thing, S_1's, persisting between t_1 and t_2 (if S_1, in our case the soul, is not a proper part of B_1 or B_2).

It is unclear what physicality is doing in this principle other than ensuring it does not entail that it is impossible for God to guarantee the identity of anything that is not a proper part of God. Still, I have a hard time seeing how Part Identity could be defended. Regardless, considering how the soul can guarantee the identity of the body across time even though the soul is not a part of the body will be enough to answer Merricks. Although it is beyond the scope of this chapter to defend a full dualist account of bodily persistence, I offer the following sketch as a plausible view.

Excursus: Bodily Souls, the Body as a Mode of the Soul

C. Stephen Evans and I have defended what we call the bodily soul view.[66] On this view, the human person is identical to an immaterial substance: the soul. However, the person, as embodied, is a bodily soul, where the soul is in a sense the form of the body. As Edmund Husserl says, "the soul . . . besouls the Body."[67] Hence, the body is not merely another object in the world, but the mode in which we manifest our presence in the world and exercise our agency and relationality. "To live as a person," says Husserl, "is to posit oneself as a person, to find oneself in, and to bring oneself into, conscious relations with a 'surrounding world.'"[68]

I suggest this view be infused with a robust neo-Aristotelian metaphysics of substances and modes, especially that of the late medieval Aristotelians[69], and the work of E. J. Lowe,[70] J. P. Moreland,[71] and others.[72] Although I find this view extremely interesting, philosophically fruitful, and underexplored, I offer it only as a plausible view of mind-body dualism. There are other ways a dualist could answer Merricks. I simply offer this as one possible and interesting dualist model that can answer Merricks's objection.

What is most important for my reply to Merricks is the essence of the soul and its relation to the body and the body's persistence. On the proposed view, the essence of the soul is ontologically fundamental, such that facts about the essence of the soul determine, among other things, the soul's natural kind. In the terms of late medieval Aristotelians, the essence of the soul is a thin particular, which includes the essence/form, the nexus of exemplification, and prime matter.[73] On Moreland's view, it is a bare particular, not prime matter, that individuates the soul.[74] Like Augustine,[75] Aquinas,[76] and Suárez,[77] this view takes from Aristotle the notion that the soul is "the cause and source of the living body."[78] The essence of the soul contains, as a primitive unity, powers for developing the body. The essence of the soul is both the internal efficient cause of and teleological guide for the internal structure and development of the body.[79] That is, the essence of the soul is both the first efficient cause of the body's development, as well as the final cause of its functions and structure.[80] Consequently, the body is an ensouled physical structure, not a mere physical machine or aggregate of separable parts standing in external relations. The body is merely a physical thing but has both physical and non-physical aspects. [81] The body is a complex structural mode of the soul.

Back to the Objection

With the previous model in mind, it is plausible, contrary to Merricks's claim, that facts about the soul ground facts about BODILY IDENTITY. Because the body is a mode of the soul, fundamental facts about the body obtain in virtue of facts about the soul.[82] Without the soul, there is no body. Hence, the body persists just in case the soul persists. Of course, one might reject the view I've sketched. However, this would not undermine my reply to Merricks's objection from BODILY IDENTITY. I have offered this bodily soul view only as a possible model for dualism, and I remain open to other types of mind-body dualism. However, all I need to reply to Merricks is a metaphysically possible account that can explain BODILY IDENTITY. To that end, this bodily soul view succeeds.

Dead Souls Cannot Be Resurrected?

Merricks offers a further objection to dualism from BODILY IDENTITY:

> What if we were not identical with our bodies? Then it would be hard, if not impossible, to make sense of the idea that dead people will be resurrected. Moreover, the importance of the doctrine that, on the Day of Resurrection, one gets a body identical to the body one had in this life would be difficult to explain. Indeed, I cannot think of any plausible explanation at all, much less one that rivals the very straightforward and absolutely compelling explanation that flows directly from the claim that each of us is identical with his or her body.[83]

The fact that dead people will be resurrected, says Merricks, is explained much better if we are identical to our body, such that when our body is resurrected we are resurrected. But if we are not identical to our body, then we will not be resurrected; only our body will be resurrected. So, the fact that we are resurrected can only be explained (or, at least, is much better explained) if we are identical to our body.[84]

Again, the hidden assumption is that resurrection is nothing more than postmortem survival, which I have shown is false. In terms of God's overall project of resurrection, the dualist should hold that a soul undergoes its own kind of resurrection. In fact, on the bodily soul view sketched above, the resurrection of the body, in terms of restoration, will include the soul, as the body is a mode of the soul. Resurrection will include restoring the soul-body relation. Certain deficiencies in the soul, as well as the soul's relation to the body, will be transformed, recovered from death, and made alive. Re-embodiment does not leave the soul unchanged. Contrary to Merricks's assumption, the dualist can argue that the whole person, not merely their body, is resurrected.

Still, it is unclear what Merricks is actually arguing here. In what follows I raise objections for various interpretations of Merricks's argument. Perhaps he is arguing something like the following. We must account for this fact: it is very important that on the day of resurrection one gets the body that is identical to their premortem body. It is very important on the view that one is identical with one's body because one's premortem body is needed for one to persist after death. Any reason the dualist gives for BODILY IDENTITY will not be as important as the Christian physicalist's reason. Bodily resurrection is a matter of existence given Christian materialism, but not for dualism. Understood this way, Merricks's argument is that the value of resurrection is higher on physicalism than it is on dualism.[85]

But what follows from this argument is not that dualism is inconsistent with or cannot account for BODILY IDENTITY. What follows is that the materialist account has greater value. But that one account is more valuable than another certainly does not mean that the more valuable account is the correct or more justified account. Of course, that I exist is very important, at least to me! However, it is not important enough. Many have this intuition about eternal hell or Sisyphus. The value of resurrection is not merely that I exist, but that I exist in a resurrected state where the damages of sin are overcome. Mere existence does not get us resurrection. Resurrection requires much more. This point seems to be lost on Patrick Lee and Robert George, who write,

> If I just were a soul, even though I had a natural orientation to union with my body, then the nonexistence of the resurrection might be disappointing, but it is hard to see how it would render the faith futile (as St. Paul argues). And it would be difficult to explain why bodily resurrection would be at the center, rather than, say, "icing on the cake," for the central teaching about life with Christ.[86]

Far from "icing on the cake," the cornerstone of resurrection is God's redemption of creation by restoring the conditions under which it flourishes, including our body (Rom. 8:18–25). I see no reason why the dualist cannot account for their continued existence as a miraculous act of God that is partly constitutive of the resurrection. True, their sustained existence does not require BODILY IDENTITY. However, as Merricks admits, being present before God is a great good. This great good is missing for the Christian materialist who holds, as Merricks seems to, that resurrection requires that one go out of existence. While the Christian materialist does not exist and can enjoy nothing, the dualist enjoys the great good of being in the presence of God. So, how do we evaluate which view entails that resurrection is more valuable? Merricks's view works on the presumption that dualism cannot include existence as a great good of resurrection. But this is false. Perhaps the dualist can argue that BODILY IDENTITY is part of the conditions under which we flourish. Some might think the medieval arguments mentioned earlier might bolster this view. Metaphysical perfection, for example, might require BODILY IDENTITY. Likewise, it may well be that the God-given telos of human persons includes embodiment.

> The human soul is created by God for embodiment in a particular parcel of matter . . . It is constitutive of the human soul, per se, that it is naturally directed towards embodiment. Thus, part of what it is to be a human soul is to have this telos. Thus . . . each and every individual human soul is naturally directed towards embodiment in a particular parcel of matter.[87]

This alone gives the dualist reason to think that BODILY IDENTITY is true, provided this teleological fact is true. Such a teleological fact can be disputed. But that is not the point. What this shows is that dualism is not at odds with BODILY IDENTITY, but has a possible reason for thinking it is true.

Corcoran: Reassembly and Gappy Bodies

Corcoran presses another issue, arguing that BODILY IDENTITY poses a difficulty for both dualism and Christian materialism.[88] He asks,

> How can a physical object that exists in the hereafter be numerically identical with a physical object that has either radically decayed or passed out of existence under more gruesome circumstances?[89]

Corcoran uses van Inwagen's reassembly argument to show that sameness of parts is not sufficient among the persistence conditions of bodies. According to van Inwagen's argument, the reassembly of your body cannot ground personal identity over time, because God could reassemble all the material particles of my five-year-old body alongside all the material particles of my 25-year-old body. But clearly, these two bodies cannot be the numerically same body. Therefore, reassembly of parts does not give us BODILY IDENTITY.[90]

Perhaps this is, as Corcoran claims, a genuine problem for Christian materialism, but it needn't be for dualism. Recall the bodily soul view sketched above, where a body is not merely an aggregate of material parts standing in external relations. So, God couldn't reassemble my five-year-old body alongside my 25-year-old body, because my body is necessarily ensouled by me. My soul is what makes my body a body. Hence, mere reassembly of material parts will not get you a body. You would need a single soul to ground the nature of two bodies. But, that certainly isn't entailed by dualism, and I see no reason for adopting such a view.

Furthermore, suppose that a body at t_1 can remain numerically identical with a body at t_3 even though at t_2 that body did not exist. The dualist might hold that the body is a mode of the soul, where a mode is a concrete particular, a specific way something is.[91] A red vase has both a shape and a color, each of which are modes, ways the vase is. Likewise, for a human body to be a mode just is for the human body to be a way the person our soul is. Accordingly, I needn't be bodily. I can go from being embodied to being disembodied.

On this view, my being embodied is a modification that I undergo. Given this, one could adopt a type of immanent causal view that Corcoran defends, but with one important qualification. According to Corcoran,

> A human body B that exists in the future is the same as a human body A that exists now if the temporal stages leading up to B are immanent causally connected to the temporal stage of A now.[92]

I am not convinced that immanent causal connectivity is plausible, as it is not sufficient for numerical identity. However, here is an interesting possibility: a dualist could hold that BODILY IDENTITY is maintained in virtue of the immanent causal connectivity between the soul and body through time. One advantage of this view over Corcoran's view is that one of the relata, the soul, never goes out of existence. Alternatively, the dualist might hold that bodily continuity is maintained as follows. At death, my physical body is modified into or perhaps replaced with a nonphysical body. What I leave behind is a corpse, while I gain a nonphysical body, which will, at the resurrection, once again be modified to a physical body.[93] So long as either of these accounts is logically possible, and that is all I am suggesting, the dualist avoids Corcoran's objection.

CONCLUSION

Given my argument in this chapter, we have some important lessons. First, the doctrine of bodily resurrection is not reducible to problems of personal identity. Moreover, contrary to claims from Christian physicalists, dualism has substantive reasons for why a soul needs a resurrected body. These arguments have been repeatedly ignored by Christian physicalists. Lastly, there are currently no good bodily resurrection objections to dualism. We may continue to confess, in spirit and truth, that "my only comfort in life and death is that I belong, body and soul, to my faithful Savior Jesus Christ."[94]

NOTES

1. *Heidelberg Catechism, A New Translation* (Grand Rapids, MI: Christian Reformed Board of Publications, 1975), Qa1.

2. Regarding resurrection, the *Heidelberg Catechism* makes this more explicit, stating that "not only will my soul be taken immediately after this life to Christ its head, but even my flesh, raised by the power of Christ, will be reunited with my soul and made like Christ's glorious body" (*Heidelberg Catechism*, question and answer 57). Similar confessions are made in other catechisms, including *The Westminster Shorter Catechism* (question and answer 37), and *The Longer Catechism of The Orthodox, Catholic, Eastern Church* (question and answer 366). See also, *Luther's Large Catechism*, Q. 34; *Westminster Confession of Faith*, chapter 32; and *To Be a Christian: An Anglican Catechism*, Q. 142.

3. Dean Zimmerman, "Three Introductory Questions," in *Persons: Human and Divine*, eds. Dean Zimmerman and Peter van Inwagen (Oxford: Oxford University Press, 2007), 19.

4. Alvin Plantinga, "Materialism and Christian Belief," *Persons: Human and Divine*, 99.

5. Dallas Willard, *Renovation of the Heart: Putting on the Character of Christ* (Colorado Springs, CO: NavPress, 2002), 159. For an excellent treatment of Willard's view, see, J. P. Moreland, "Tweaking Dallas Willard's Ontology of the Human Person," *Journal of Spiritual Formation & Soul Care* 8, no. 2 (2015): 187–202.

6. Willard, *Renovation of the Heart*, 162.

7. Kevin J. Corcoran, *Rethinking Human Nature: A Christian Materialist Alternative to the Soul* (Grand Rapids, MI: Baker Academic, 2006), 121.

8. See, for example, N. D. Kelly, *Early Christian Doctrines*, 4th ed. (London: Adam and Charles Black, 1968), chapters 6 and 11.

9. See for example, Frances Young, *From Nicaea to Chalcedon: A Guide to the Literature and Background*, 2nd ed. (Grand Rapids, MI: Baker Academic, 2000), chapter 6; Christopher Beeley, *Unity of Christ: Continuity and Conflict in Patristic Tradition* (New Haven, CT: Yale University Press, 2012), chapters 1 and 6; and Edward R. Hardy, ed., *Christology of the Later Fathers* (Louisville, KY: Westminster John Knox Press, 1954), 124–125.

10. The council, for example, has some anathemas against it, as it affirmed that the soul is given a physical body. See Richard Price, *The Acts of the Council of Constantinople of 553* (Liverpool: Liverpool University Press, 2009), 278–279.

11. Fourth Council of Constantinople, canon 11.

12. I am grateful to Ryan Mullins who first drew my attention to these historical points.

13. Charles Taliaferro, "The Virtues of Embodiment," *Philosophy* 76/295 (2001): 111–125.

14. Howard Robinson, "A Dualist Account of Embodiment," in *The Case for Dualism*, eds. John R. Smythies and John Beloff (Charlottesville: University of Virginia Press, 1989), 43–58.

15. Richard Swinburne, "What is So Good about Having a Body?" in *Comparative Theology: Essays for Keith Ward*, ed. Timothy Walter Bartel (London: SPC Publishing, 2003), 134–142.

16. This has been noted by dualists themselves. See, for example, Keith E. Yandell, "Materialism and Post-Mortem Survival," in *Knowledge and Reality: Essays in Honor of Alvin Plantinga*, eds. Thomas M. Crisp and Matthew Davidson (Leiden, Netherlands: Springer, 2006), 262.

17. The most significant contributions are from John W. Cooper. Beyond his chapter in this volume, see his *Body, Soul, and Life Everlasting: Biblical Anthropology and the Monism-Dualism Debate* (Grand Rapids, MI: Eerdmans, 1989); "The Bible and Dualism Once Again: A Reply to Joel B. Green and Nancey Murphy," *Philosophia Christi* 9 (2007): 459–469; "The Current Body-Soul Debate: A Case for Holistic Dualism," *Southern Baptist Journal of Theology* 13 (2009): 32–50; "Exaggerated Rumors of Dualism's Demise: A Review Essay on Body, Soul and Human Life," *Philosophia Christi* 11 (2009): 453–464. See also, Brandon Rickabaugh, "Responding to N. T. Wright's Rejection of the Soul," *Heythrop Journal* (forthcoming).

18. Charles Taliaferro and Stewart Goetz, "The Prospect of Christian Materialism," *Christian Scholar's Review* 37, no. 3 (2008): 303–321, William Hasker, "Materialism and the Resurrection: Are the Prospects Improving?" *European Journal for the Philosophy of Religion* 3 (2011): 83–103. Dean Zimmerman has argued against Baker's constitution view. See Dean Zimmerman, "Rejoinder to Lynne Rudder Baker," in *Contemporary Debates in the Philosophy of Religion*, eds. Michael L. Peterson and Raymond J. Vanarragon (Oxford: Blackwell, 2003), 338–341.

19. For example, Baker argues against animalism. See her "Death and the Afterlife," in *The Oxford Handbook of Philosophy of Religion*, ed. William J. Wainwright (Oxford: Oxford University Press, 2005), 387–89. Corcoran thinks there is a problem for both Christian materialists and dualists. See his *Rethinking Human Nature*, 123, and "Dualism, Materialism, and the Problem of Postmortem Survival," *Philosophia Christi* 4, no. 2 (2002): 415–416.

20. See Trenton Merricks, "The Resurrection of the Body," in *The Oxford Handbook of Philosophical Theology*, eds. Thomas Flint and Michael Rea (New York: Oxford University Press, 2008), 476–490; Peter van Inwagen, "The Possibility of Resurrection," *International Journal for Philosophy of Religion* 9, no. 2 (1978): 114–121; and Lynne Rudder Baker, "Materialism with a Human Face," in *Soul, Body, and Survival: Essays on the Metaphysics of Human Persons*, ed. Kevin Corcoran (Ithaca, NY: Cornell University Press): 159–180.

21. However, some dualists argue that the resurrection makes dualism more attractive than Christian materialism. See William Hasker, "Emergentism," *Religious Studies* 18 (1982): 473–488; and Taliaferro and Goetz, "The Prospect of Christian Materialism."

22. Lynne Rudder Baker, "Death and the Afterlife," in *The Oxford Handbook for the Philosophy of Religion*, ed. William J. Wainwright (New York: Oxford University Press 2007), 368; and Lynne Rudder Baker, "Persons and the Metaphysics of Resurrection," *Religious Studies* 43 (2007): 339–340.

23. Stephen T. Davis lists these doctrines in his own words as assumptions of resurrection theology. See his *After We Die: Theology, Philosophy, and the Question of Life After Death* (Waco, TX: Baylor University Press, 2015), 49–50.

24. Trenton Merricks, "The Resurrection of the Body and the Life Everlasting," in *Reason for the Hope Within*, ed. Michael J. Murray (Grand Rapids, MI: Eerdmans, 1999), 267–271.

25. Lynne Rudder Baker, "Persons and the Metaphysics of Resurrection," *Religious Studies* 43 (2007): 340.

26. Merricks, "The Resurrection of the Body," 280.

27. Merricks, "The Resurrection of the Body," 280–281.

28. Merricks, "The Resurrection of the Body," 281.

29. N. T. Wright, *Resurrection and the Son of God* (Minneapolis, MN: Fortress Press, 2003), 31.

30. Luke 10:25–37; John 13:34–35; 15:12–17; Rom. 13:9; Gal. 5:14; 1 Pet. 1:22–23; 1 Jn. 3:10–24; 4:7–12.

31. Rom. 6:1–11, 8:5–13; 2 Cor. 3:18, 4:10–12, 16–18; Eph. 2:1–6; Col. 3:1–3.

32. John 3:36; 4:14; 5:24; 6:54; Matt. 12:28.

33. Matt. 22:32; John 5:24–25; 6:54.

34. Rom. 8:16–39; 2 Cor. 4:16–18; 5:16–17; Eph. 2:4–7; Phil. 1:6; 3:10–11.

35. Wright, *Resurrection and the Son of God*, 440.

36. Wright, *Resurrection and the Son of God*, 441.

37. Lynne Rudder Baker, "Christians Should Reject Mind-Body Dualism," in *Contemporary Debates in the Philosophy of Religion*, eds. M. Peterson and R. Van Arragon (Malden, MA: Blackwell, 2004), 337.

38. Further excellent treatments include Marilyn McCord Adams, "The Resurrection of the Body According to Three Medieval Aristotelians: Thomas Aquinas, John Duns Scotus, William Ockham," *Philosophical Topics* 20, no. 2 (1992): 1–33, and "Why Bodies as Well as Souls in the Life to Come," in *The Science of Being as Being: Metaphysical Investigations*, ed. Gregory T. Doolan (Washington, DC: The Catholic University of America Press, 2012), 264–297.

39. Robert Pasnau, *Metaphysical Themes: 1274–1671* (Oxford: Oxford University Press, 2011), 324.

40. Charles Taliaferro, "The Virtues of Embodiment," *Philosophy* 76 (2001): 111–125.

41. Richard Swinburne, "What's So Good about Having a Body?" in *Comparative Theology: Essays for Keith Ward*, 137.

42. Gordon Barnes, "Is Dualism Religiously and Morally Pernicious?" *American Catholic Philosophical Quarterly* 78, no. 1 (2004): 103.

43. Maximus the Confessor, *Epistulae* 6; and Maximus the Confessor, *Migne, Patroligia, Gracea* 91, 429 B–432 A. For helpful commentary, see Lars Thunberg, *Microcosm and Mediator: The Theological Anthropology of Maximus the Confessor*, 2nd ed. (Chicago: Open Court, 1995), 95–143.

44. Baker, "Persons and the Metaphysics of Resurrection," 341.

45. Baker, "Death and the Afterlife," 375.

46. Baker, "Death and the Afterlife," 375.

47. Oddly enough, Baker recognizes that her own account of personal identity is also circular. Although, she argues that her view offers a noncircular account of *human personal identity*. See, Lynne Rudder Baker, *Persons and Bodies: A Constitution View* (Cambridge, UK: Cambridge University Press, 2000), 132–141.

48. Trenton Merricks, "There are no Criteria for Identity Over Time," *Noûs* 32 (1998): 106–124.

49. E. J. Lowe, "Substance, Identity, and Time," *Proceedings of the Aristotelian Society* 62 (1988): 77–78. For his more recent defense of this view, see E. J. Lowe, "The Probable Simplicity of Personal Identity," in *Personal Identity: Complex or Simple*, eds. Georg Gasser and Matthias Stefan (Cambridge, UK: Cambridge University Press, 2012), 137–55.

50. Baker, "Death and the Afterlife," 375.

51. Baker, "Persons and the Metaphysics of Resurrection," 9.

52. There were Christians in the past that denied MIRACLE when arguing for the immortality of the soul on purely philosophical grounds. See, for example, George Berkeley, *A Treatise Concerning the Principles of Human Knowledge*, sec. 141.

53. Richard Swinburne, *The Evolution of the Soul*, rev. ed. (Oxford: Oxford University Press, 1986), 305–306.

54. Swinburne, *The Evolution of the Soul,* 311.

55. Swinburne, *The Evolution of the Soul*, 311–312.

56. See Karl R. Popper and John C. Eccles, *The Self and Its Brain: An Argument for Interactionism* (New York: Springer-Verlag, 1977), 556.

57. Robert Audi, "Personhood, Embodiment, and Survival Speculations on Life after (Biological) Death," in *Paradise Understood: New Philosophical Essays about Heaven*, eds. T. Ryan Byerly and Eric J. Silverman (Oxford: Oxford University Press, 2017), 204–206, 209.

58. Bruce Reichenbach, *Is Man a Phoenix? A Study of Immortality* (Grand Rapids, MI: Eerdmans, 1978), 49.

59. Richard Swinburne, "Soul, Nature, and Immortality of the," in *Routledge Encyclopedia of Philosophy*, ed. E. Craig (London: Routledge, 1998). Retrieved July 11, 2005 from http://www.rep.routledge.co m/article/K096

60. See, for example, Steven T. Davis, *Risen Indeed: Making Sense of the Resurrection* (Grand Rapids, MI: Eerdmans, 1993), 94–102.

61. See, for example, Baker, *Persons and Bodies*, 119–124.

62. For example, Paul Fiddes says of BODILY IDENTITY, "Together with virtually all modern theologians, I do not want to take this over-materialistic view of resurrection" (*The Promised End: Eschatology in Theology and Literature* [Oxford, UK: Blackwell, 2000], 80).

63. For an alternative dualist defense of BODILY IDENTITY or something very close, see, Davis *After We Die*, 52–59.

64. Trenton Merricks, "The Resurrection of the Body," in *The Oxford Handbook of Philosophical Theology*, 479.

65. Most dualists agree that the soul is not a part of the body. Swinburne, for example, holds that the body is a contingent part of the person while the soul is the essential part (Swinburne, "The True Theory of Personal Identity," 120).

66. C. Stephen Evans and Brandon Rickabaugh, "What Does it Mean to Be a Bodily Soul?" *Philosophia Christi* 17, no. 2 (2005): 315–330.

67. Edmund Husserl, *Ideas Pertaining to a Pure Phenomenology and to a Phenomenological Philosophy, Second Book: Studies in the Phenomenological Constitution*, trans. Richard Rojcewicz and André Schuwer (Dordrecht: Kluwer Academic Publishers, 1993), 185.

68. Husserl, *Ideas Pertaining to a Pure Phenomenology and to a Phenomenological Philosophy*, 193.

69. See Pasnau, *Metaphysical Themes: 1274–1671*, 549, 558, 560–565; and Dennis Des Chene, *Life's Form: Late Aristotelian Conceptions of the Soul* (Ithaca, NY: Cornell University Press, 2000), 67–113; 191–199.

70. E. J. Lowe, *Subjects of Experience* (New York: Cambridge University Press, 1996), 14–51; and "Non-Cartesian Mind-Body Dualism," in *After Physicalism*, ed. Benedikt Paul Göcke (Notre Dame, IN: University of Notre Dame Press, 2012), 48–71.

71. See J. P. Moreland, "Tweaking Dallas Willard's Ontology of the Human Person," 187–202; and "In Defense of a Thomistic-Like Dualism," in *The Blackwell Companion to Mind-body Dualism*, eds. Jonathan Loose, Angus Menuge, and J. P. Moreland (Oxford: Blackwell, forthcoming).

72. See, for example, A. G. A. Balz, *Cartesian Studies* (New York: Columbia University Press, 1951), 279–323; Stephen Voss, "Understanding Eternal Life," *Faith and Philosophy* 9, no. 1 (1992): 3–22; Gordon Barnes, "Is Dualism Religiously and Morally Pernicious?" *American Catholic Philosophical Quarterly* 78, no. 1 (2004): 99–106; and Gordon Barnes, "Should Property-Dualists Be Substance-Hylomorphists?" *Proceedings of the American Catholic Philosophical Association* 75 (2002): 285–299. This account also shares similarities to the view of Bolzano. See Bernard Bolzano, *Athanasia; oder Gründe für die Unsterblichkeit der Seele* (Sulzbach: J. G. v. Seidleschen Buchhandlung, 1838), 55–56, 101, 283–84.

73. Pasnau, *Metaphysical Themes* 1274–167, 99–114.

74. See J. P. Moreland, "Theories of Individuation: A Reconsideration of Bare Particulars," *Pacific Philosophical Quarterly* 79 (1998): 251–263.

75. Augustine, *On Freedom of the Will*, II, XVI.41; *The Trinity*, IV, I.3; *Confessions*, II, 6.

76. See, for example, Thomas Aquinas, *Summa Theologiæ*, I, Q75; and Thomas Aquinas, *Disputed Questions on Spiritual Creatures*, trans. Mary C. Fitzpatrick and John J. Wellmuth (Milwaukee, WI: Marquette University Press, 1949), IV.ad 9.

77. Francisco Suárez, *On the Formal Cause of Substance: Metaphysical Disputations XV*, trans. John Kronen and Jeremiah Reedy (Milwaukee, WI: Marquette University Press, 2000), 140–142.

78. Aristotle, *De Anima*, 415b.9 (cf. 412a22, 412a27–28).

79. See Pasnau, 549, 558 560–565.

80. For more on this as it relates to contemporary biology, see, Thomas J. Kaiser, "Is DNA the Soul?" *The Aquinas Review* 20 (2015): 90–92.

81. See Lowe, *Subjects of Experience*, chapter 2; and J. P. Moreland, "In Defense of a Thomistic-Like Dualism."

82. This does not preclude, of course, that facts about the body ground certain facts about the soul. For example, neurological facts likely ground certain developmental facts about the soul.

83. Merricks, "The Resurrection of the Body," 484.

84. This objection is also briefly raised by Patrick Lee and Robert P. George in, *Body-Self Dualism in Contemporary Ethics and Politics* (Cambridge, UK: Cambridge University Press, 2008), 75.

85. Lee and George offer a similar argument in their *Body Self Dualism in Contemporary Ethics and Politics*, 74–75.

86. Lee and George, *Body Self Dualism in Contemporary Ethics and Politics*, 75.

87. Gordon Barnes, "Is Dualism Religiously and Morally Pernicious?" *American Catholic Philosophical Quarterly* 78, no. 1 (2004): 103.

88. Corcoran, *Rethinking Human Nature*, 123–133; and "Dualism, Materialism, and the Problem of Postmortem Survival," *Philosophia Christi* 4, no. 2 (2002): 415–416.

89. Corcoran, *Rethinking Human Nature*, 123.

90. Peter van Inwagen, "Dualism and Materialism: Jerusalem and Athens?" *Faith & Philosophy* 12, no. 4 (1995): 485–486.

91. See, E. J. Lowe, *The Possibility of Metaphysics: Substance, Identity, and Time* (Oxford: Oxford University Press, 1998), 78–79.

92. Corcoran, *Rethinking Human Nature*, 128.

93. For a recent defense of the possibility of a nonphysical body, see, Robert Audi, "Personhood, Embodiment, and Survival Speculations on Life after (Biological) Death"; and Peter Drum, "On the Resurrection of the Body: Discussions with Merricks," *European Journal for Philosophy of Religion* 3, no. 2 (2011): 451–454.

94. I am grateful to the following individuals for helpful comments on an earlier draft of this chapter: Richard Swinburne, Alexander Pruss, J. P. Moreland, Trent Dougherty, Philip Swenson, Lori Morrow, and Ryan Mullins.

SUGGESTED FURTHER READING

Adams, Marilyn McCord. "Why Bodies as Well as Souls in the Life to Come." In *The Science of Being as Being: Metaphysical Investigations*, edited by Gregory T. Doolan. Washington, DC: The Catholic University of America Press, 2012.

Davis, Stephen T. "Physicalism and Resurrection." In *Soul, Body and Survival: Essays on the Metaphysics of Human Persons*, edited by Kevin Corcoran. Ithaca, NY: Cornell University Press, 2001.

———. "Resurrection." In *The Cambridge Companion to Christian Philosophical Theology*, edited by Charles Taliaferro and Chad Meister. Cambridge, UK: Cambridge University Press, 2010.

Evans, Stephen C., and Brandon Rickabaugh. "What Does It Mean to Be a Bodily Soul? *Philosophia Christi* 17, no. 2 (2015): 315–330.

Gundry, Robert H. "Addendum: A Biblical & Philosophical-Scientific Conversation with Christian Nonreductive Physicalists." In *The Old is Better: New Testament Essays in Support of Traditional Interpretations*. Tübingen, Germany: Mohr Siebeck, 2005.

Hasker, William. "Materialism and the Resurrection: Are there Prospects for Improving?" *European Journal for Philosophy of Religion* 1, no. 3 (2012): 83–103.

Moreland, J. P. *The Soul: How We Know It's Real and Why It Matter*s. Chicago: Moody Press, 2014.

———. "Tweaking Dallas Willard's Ontology of the Human Person." *Journal of Spiritual Formation & Soul Care* 8, no. 2 (2015): 187–202.

Robinson, Howard. "A Dualist Account of Embodiment." In *The Case for Dualism*, edited by John R. Smythies and John Beloff. Charlottesville: University of Virginia Press, 1989.

Swinburne, Richard. *Mind, Brain, and Free Will*. Oxford, UK: Oxford University Press, 2013.

———. "What's So Good About Having a Body?" In *Comparative Theology: Essays for Keith Ward*, edited by Timothy Walter Bartel. London: SPCK, 2003.

Taliaferro, Charles. "The Virtues of Embodiment." *Philosophy* 76, no. 1 (2001): 111–125.

Chapter 16

"Absent from the Body . . . Present with the Lord"

Is the Intermediate State Fatal to Physicalism?

John W. Cooper

This chapter considers the biblical doctrine of the intermediate state—the existence of human persons (also referred to as souls, spirits, egos, minds, selves, subjects, agents) between death and bodily resurrection—as the most compelling reason for Christians to affirm substance dualism of some sort as the true metaphysics of the human constitution. There are good philosophical reasons for inferring the metaphysical distinctness of soul and body or spirit and matter, many of them articulated in this volume. But the intermediate state entails that persons and their bodies are not merely distinct but separable. Disembodied existence undercuts not only physicalism and materialism but all philosophical anthropologies which regard the body as metaphysically necessary for personal existence, including emergent monism, psychophysical (property dualist, dual aspect) monism, multidimensional monism, and ontological holism. I refer to those anthropologies which essentialize embodiment collectively as *bodily monism.*[1]

The eschatological narrative of disembodied existence between death and final resurrection is not derived from experience or philosophical reflection. It is revealed in the Bible—implied by the Hebrew Scriptures and explicitly affirmed in the New Testament (argued later). Whatever our views on the right relation between reason and revelation, most Christian scholars seek to hold philosophical positions which at least are consistent with Scripture, if not shaped by it. Thus, if the Bible teaches the intermediate state in addition to the unity of body and soul in this life and the life to come, then we should embrace some sort of holistic or integral dualism and eschew bodily monism as Christian anthropology. Virtually all Christian churches and traditions since New Testament times have understood Scripture this way. Many Christian academics continue to affirm historic anthropology, as is apparent from this volume.[2]

Since the mid-twentieth century, however, bodily monism of various kinds has gained a significant following among Christian academics. Christian monists appeal to biblical-theological, practical, philosophical, and scientific reasons. They claim that the Christian tradition has misunderstood the anthropology of Scripture and interprets it according to Platonic dualism rather than Hebraic monism. As a consequence, they charge, human life and Christian practice have been distorted by various dualisms: spirit against body, intellect over feelings, inward-focused ego isolated from other people and the world, and spiritualized religion disengaged from the rest of life. Bodily monists also embrace trends in philosophy and science which regard our personal, mental, and spiritual capacities as products of our brains and organisms, and hold that human nature has evolved entirely by (theistic) evolution from the physical energy of the Big Bang. They see no need for a spiritual ingredient or supernatural agency to explain the soul or the image of God. Christian monists believe that Scripture, philosophy, and science converge to corroborate their perspective.[3]

The intermediate state is a serious challenge to bodily monists precisely because it involves disembodied existence. Monists typically ignore it, marginalize it as an incidental doctrine, or reject it as a misunderstanding of Scripture.[4] They especially are critical of the notion of an immortal soul. Instead of an intermediate state and final resurrection, most monists believe that human individuals cease to exist (except in the mind of God) at death and are reactualized by bodily resurrection. Some propose an immediate bodily resurrection, thereby avoiding a period of disembodiment. A few defend a bodily monistic account of personal existence in the intermediate state.

The main part of this chapter elaborates and defends the intermediate state as a clear and important teaching of Scripture and historic Christianity. The second part challenges monistic eschatologies as incompatible with Scripture and/or sound philosophy.[5]

APOLOGETICS FOR THE INTERMEDIATE STATE AND DISEMBODIED EXISTENCE

Physicalists and other bodily monists challenge the intermediate state at two points: they deny that it is an enduring and important doctrine of Scripture and the church, and they charge that it posits the objectionable notion of an immortal soul. I address these topics in order.

The Intermediate State in Scripture

In spite of ongoing debates, leading biblical scholars still conclude that Scripture envisions personal existence—largely undescribed—between death and future bodily resurrection.[6] Monists are correct that scholarship since the nineteenth century turned against overly Platonic, Cartesian, or idealistic views of the soul. But scholars have not shown that biblical anthropology is monistic—that humans consist of just one substance or basic ingredient—certainly not material or physical. The general consensus is that the Hebrew Bible reflects ancient near-eastern animism. Philosophical reflection on animistic anthropology extrapolates integral dualism—the holistic, existential unity of distinct spiritual and earthly ingredients (cf. Gen. 2:7). After physical death, identifiable individuals, such as Samuel, David, Hezekiah, and Job, continue to subsist in Sheol—the realm of the dead—even if their personal capacities are greatly reduced. The Psalmist anticipates "dwelling in the house of the Lord forever" (Ps. 23), and latter prophets envision the shades in Sheol returning to bodily life in the world on the great day of the Lord (Is. 26:19; Ezek. 37). Continuing existence of individuals with kinship groups beyond death is part of the narrative of God's people in the Hebrew Bible.

Intertestamental Judaism includes diverse perspectives on eschatology and anthropology, including a mortal soul with no afterlife, which the Sadducees held. The Pharisees and rabbis embraced an integral or holistic dualism, believing that souls or spirits exist after death until bodily resurrection. Jewish believers influenced by Plato, such as Philo, affirmed the immortality of the soul and regarded resurrection as spiritual elevation to a heavenly mode of existence. All three of these perspectives appear in the background of the New Testament.

But the New Testament clearly and consistently presents an anthropology and eschatology similar to that of the Pharisees, reoriented by the proclamation that Jesus is the Messiah and that human life as created and spoiled by evil and death has been transformed into everlasting life in union with Christ by the Holy Spirit. Bodily resurrection is the central hope because it completes the salvation of human nature that God created. But continuing personal existence until the resurrection is an integral phase of salvation history even though it is less prominent in the New Testament. Jesus himself assures the penitent thief on the cross, "today you will be with me in Paradise" (Luke 23:43). Jesus was in Paradise between his death on Friday and his resurrection on Sunday morning. The thief was with Jesus instead of in the place of punishment that he anticipated. The unity of the human nature they share was sundered by death, but they continued to exist in fellowship. Christ's path through death to resurrection is paradigmatic for all Christians (1 Cor. 15:20, 23).

The apostle Paul, educated as a Pharisee, shared their doctrine of (human) spirits and bodily resurrection (Acts 23:6–8). It is clear in his letters: the resurrection of our earthly bodies at the return of Christ in 1 Thessalonians 4:13ff and 1 Corinthians 15, as well as personal fellowship with the Lord between death and resurrection in 2 Corinthians 5:6–9 ("away from the body and at home with the Lord," v. 8) and Philippians 1:21–24 ("live in the flesh . . . or depart and be with Christ"). Paul does not use the terms *soul* or *spirit* in contrast to *the body* and *flesh* but uses personal pronouns: I live in the body or am with Christ—a person-body/flesh distinction. Pauline and New Testament anthropology and eschatology are expressed in diverse ways in the original languages, but they are completely consistent: persons in their core identity continue to exist between death and bodily resurrection.

This same sequence is evident in the apocalyptic texts of Revelation, which envision both lamenting martyrs and the church triumphant praising God before the final resurrection. Hebrews 12:22–24 likewise refers to "the spirits of the righteous made perfect" presently dwelling with God, Christ, and the angels in the heavenly Jerusalem, anticipating its eschatological descent to the new earth.

In sum, the biblical canon as a whole consistently moves toward a doctrine of the last things which includes an intermediate state, bodily resurrection, and cosmic renewal.[7] Modern theologians and biblical scholars rightly have criticized Platonic, idealistic, and otherworldly elements in some strands of traditional anthropology and eschatology, but they have not shown that the basic narrative, including the intermediate state, is a misinterpretation of Scripture or an incidental teaching that can be disregarded. They certainly have not established that bodily monism and nonexistence until resurrection are the correct readings of Scripture.

The Biblical-Theological Rationale for the Intermediate State

Monist scholars are convinced that the intermediate state and disembodied existence are foreign to Scripture and incompatible with the biblical view of God, humanity, and our relationship. Their rhetoric sometimes sounds as though obviously God himself is a monist and that any dualism—even the most integral and holistic—is perversely antithetical to his nature, will, and design for humanity.

But in Scripture quite the opposite is true. Temporary disembodied existence is entirely consistent with, if not essential to, the biblical narrative of God's enacting his eternal intention through Christ to redeem and glorify his people and the whole creation (e.g., Eph. 1:3–10; Col. 1:15–20). Although penultimate to the resurrection, the intermediate state is a crucial, integral phase of the history of salvation. From before creation, God wills that we humans constantly remain in the loving, reciprocal, covenantal relationship with him in which and for which he created us, until it is fully realized eschatologically. The image of God in Genesis 1 is three-dimensional. It involves our relationship with God, with other people, and with nature—all three intrinsic to human nature. Sin is treason against God, which leads to physical and spiritual death—alienation from God, from others, and nature—and potentially to complete annihilation and nonexistence. But in spite of sin and death, God loves the world, remains faithful to his original intention, and sends Jesus Christ to assume our human nature and save the world. Accordingly, God providentially upholds humans in all three dimensions during this life and restores all three after the resurrection. When we die, he temporarily sustains his people in two of the three relationships in which we were created: our relationship with him and with other humans (the saints in heaven), but not our bodily relationship with the earth. Faithful to his original intention, he preserves us continuously between death and resurrection and into endless future, which begins with the resurrection of our bodies at the return of Christ. The New Testament indicates that God also perpetuates the existence of those who reject him and whose destiny is not in his kingdom. In anthropological terms, God preserves humans as conscious and responsive beings, even though our physical organisms cease functioning and disintegrate. In this way we do suffer death as the consequence of our sin and fallen nature. In the intermediate state God graciously prevents death from completely obliterating the permanent relationship he ordained for us before he created the world. God's grace is stronger than sin and his love is stronger than death, even in the realm of the dead. In biblical theology, therefore, the intermediate state is an essential chapter in the history of salvation—integral to the Gospel.

In support of this perspective, consider two biblical themes which often are not related to the intermediate state: everlasting life in John and the invincible love of God in Romans 8.

In John's Gospel, the human life which was created by God the Logos (John 1:4) can be supernaturally transformed into everlasting life by regeneration of the Holy Spirit (John 3:6, 15–16). Those who are "born from above" by the spirit possess everlasting life already now. The crucial point is that transitory earthly life becomes everlasting life during our lives, not after death. And if life in Christ is truly *everlasting*, then it endures forever. An existential gap is logically impossible. Life which begins, lasts a few years, ceases to exist for millennia, and then begins again at the end of the world is not everlasting. Understanding the relation of life and everlasting life in John's Gospel also illuminates Jesus's teaching at the death of Lazarus. Jesus assures Mary that Lazarus will rise again. Mary expresses her belief in the resurrection "on the last day." Then Jesus proclaims: "I am the resurrection and the life. Those who believe in me even though they die, will live, and everyone who lives and believes in me will never die" (John 11:25–26). The only way to make coherent sense of Jesus's words is to recognize that he is speaking both of earthly life and everlasting life. He is saying that even though we die physically, those who believe in him will not die spiritually. They already have everlasting life, which implies that they will not cease to exist between physical death and bodily resurrection.[8] We might wonder what Lazarus experienced while he was dead, but there is no doubt that he continued to exist. The same assurance is meant for all who have everlasting life in Christ.

Romans 8 proclaims the role of the Holy Spirit in implementing the cosmic power of the Gospel—the assurance of salvation, the power of prayer, the renewal of creation, and finally the certainty that God will bring his chosen people to ultimate salvation through all the tribulations of life in this world. The chapter climaxes with the assurance that nothing can separate us from God: "For I am convinced that neither death, nor life, nor angels, nor rulers, nor things present, nor things to come, nor powers, nor height, nor depth, nor anything else in all creation, will be able to separate us from the love of God in Christ Jesus our Lord" (Rom. 8:38). What does it mean that "death cannot separate us from the love of God in Christ"? It cannot mean that when we die, we cease to exist but God will resurrect us someday. True, it is possible for us humans to love a pet who has died and no longer exists. But in the context of Paul's theology and the rest of Scripture, Romans 8 cannot mean that—for two reasons. First, the love of God is interpersonal and reciprocal: he loves us, and we love him in return. In Scripture, love is covenantal—like marriage, the parent-child bond, and deep friendship (Buber calls it an I-Thou relationship). Loving the memory of someone nonexistent is a weak parody of the love that Paul has in mind. Second, the love of God is "in Christ," and union with Christ is an ontological status which, like everlasting life in John, begins already in our current lives to impart all the benefits of salvation, including everlasting life, the renewed image of God (2 Cor. 5:17), the new nature which already now participates in what is eternal (2 Cor. 4:16–18) and heavenly (Eph. 2:6), and eventually the immortal resurrection body. It is extremely difficult to reconcile Paul's doctrine of union with Christ with personal nonexistence between death and resurrection.

In sum, the intermediate state is an integral part of the biblical-theological narrative of God's salvation of creation, and it is an essential part of the personal history of each of God's children. Ignoring or denying it conflicts with the central message of Scripture.

The Intermediate State in the History of Christianity

Some monists suggest that the intermediate state was not an early, widely held doctrine but hovered in the background until Calvin made an issue of it.[9] But this notion is demonstrably false.[10] Wolfson's masterful response to Cullman's famous *Immortality or Resurrection* documents that virtually all the church fathers both affirmed the soul's survival of physical death and insisted on bodily resurrection.[11] Those who adapted Plato's view of the soul for this purpose followed Scripture by rejecting Plato's view that the soul is essentially immortal, perfect, and capable of knowing ultimate truth. Bynam's *The Resurrection of the Body in Western Christianity* confirms the endurance and ubiquity of the intermediate state and resurrection eschatology.[12] It is detailed in Aquinas's "Treatise on the Resurrection" in the *Summa Theologica Supplement* and in Dante's graphic depiction of purgatory, hell, and heaven in the *Divine Comedy*. It was a daily concern of ordinary people, constantly reminding them of their mortality. It obviously was not an obscure and marginal theme. It has remained alive to the present in the doctrinal standards and liturgies of almost all branches of Christianity in spite of differences on particular issues, such as purgatory and whether souls sleep or are active.[13] The first major deviation was the idealist eschatology of modern theology following Kant and Schleiermacher in the nineteenth century, which ironically veered back toward Platonism. Bodily monism first was asserted by Hobbes in the seventeenth century but did not gain traction until scientific-naturalistic monism became acceptable to mainline theology in the twentieth century. Bodily monism cannot seriously claim to represent the Christian tradition or the contemporary church—especially not in Latin America, Asia, and Africa, where belief in the souls of the dead is ubiquitous.

Are "Immortal" Souls an Unbiblical Notion? Mere mention of the intermediate state triggers a barrage of monist complaints that it posits an "immortal soul," which they take as proof that dualism is antithetical to biblical anthropology. The intermediate state is alleged to deny the reality of death because souls do not cease to exist.[14] The notion of a disembodied soul is condemned as Platonic idealism or Gnostic spirituality. It nurtures delusions of self-sufficiency and autonomy because immortal souls do not need God.[15] It might even usurp a divine attribute, because only God is immortal. In addition, monists claim that belief in separable souls generates all sorts of dichotomies and dualisms—spirit versus body, intellect versus emotions, inner self versus others and the world, and religion isolated from the rest of life—dualisms which fracture our humanity and distort the Christian life.[16] Fortunately, these caricatures and false charges can be dealt with in short order.

First, the monists' stipulation that death must result in total annihilation of body and soul is arbitrary and begs the question of the monism-dualism debate. A less questionable definition of mortality is cessation of organic life. By that standard, holistic dualists clearly affirm that humans die. We are unities of body and soul, not souls incidentally attached to bodies other than ourselves. As such we undergo biological death, even though as core persons we continue. In fact we *suffer* death because the constitutional unity of body and soul is dichotomized—torn apart. Death truncates our existence, is metaphysically destructive, and often traumatic. Thus, persons or souls do not avoid death because they continue to exist, as monists falsely allege.

If *immortality* simply means *perpetual existence*, then the intermediate state does imply that persons or souls are immortal. But this fact in no way attributes metaphysical self-sufficiency as an intrinsic or essential property of souls. Christians confess that only God is immortal in this sense—having the essential attribute of *aseity*, self-existence, intrinsic necessity, and utter independence of anything other than himself for existence. Plato attributed aseity to souls, holding that they are eternal and uncreated like the ideal forms in the intelligible world which they contemplate. However, since the earliest church fathers, Christian thinkers emphatically rejected this notion of immortality and asserted instead that the human soul receives its nature and existence from God and cannot exist without God's constant providence.[17] To my knowledge, no Christian theologian has ever held that the human soul possesses aseity.

If *immortal* means *having a nature which can endure forever*, then most classical and some current dualists affirm the soul's immortality. In classical metaphysics, souls are regarded as immaterial and simple—basic unitary wholes rather than composed of form and matter or immaterial stuff. On this view, souls do not naturally decompose because they are not composed, whereas bodies are composed of matter, and thus they can decompose. However, Christians who affirm this kind of immortality also hold that the soul receives its created nature from God, who could have made it otherwise, and also that immortal souls cannot exist an instant without God sustaining them. Thus, the Christian doctrine of the soul's immortality is essentially different than Plato's theory.

This concept of the God-given immortality of the soul fits well with other biblical doctrines. One is the image of God. If God is eternal and immortal spirit, then it is not surprising that the human spirit which images God is everlasting, immortal, and capable of perpetual bodily life. Another doctrine is human destiny. Because God from the beginning intends humans for everlasting life, it makes sense that he created us with intrinsic immortality. In fact, it is hard to see why an all-wise, omnipotent, and provident Creator would not design something to fulfill its purpose, especially foreknowing sin and death. These reasons for affirming immortality are revealed in Scripture, however, not philosophical inferences from the nature of the soul. Most Christian thinkers since Descartes realize that the soul's simplicity and distinctness from the body do not prove its immortality.

Intrinsic immortality is not a necessary implication of the intermediate state. In fact many current Christian dualists do not affirm it. Instead they hold that because God created humans as integral body-soul unities, soul and body are interdependent and therefore neither is naturally capable of existence without the other. Modern psychology and brain science corroborate this position. Disembodied persons in the intermediate state are sustained entirely by the supernatural power of God in spite of their natural dependence on their bodies. But these dualists do affirm the immortality of the soul in the sense that it does not cease to exist.

A final issue is whether an immortal soul and the intermediate state imply or promote the functional dichotomies and existential dualisms that monists allege: soul versus body, intellect versus emotions, inner self versus others and the natural world, and spiritualistic religion divorced from the rest of secular life. The answer is No.

Christian anthropology—traditional and current—is quite different than Platonic, Gnostic, or caricatured Cartesian dualism, and otherworldly spirituality in that it emphasizes the holistic integration of soul and body and their multiple functions and relations. Most dualists emphasize the correlation of thinking, willing, and feeling, and some offer intricate analyses of the reciprocal effects of the passions, the intellect, and the will during this life.[18] Of course, they also acknowledge that the disembodied soul has altered capacities and limitations. The point is that there is no connection between the soul's surviving death and a tendency toward intellectualism or emotional disengagement.

Does the intermediate state implicitly spiritualize the image of God and consequently secularize life? In classical Christian anthropology, the image of God is seated in the soul because God is spirit and does not have a body. But the image also permeates the body because the soul permeates the body, and thus the body also images God as humans engage each other and the natural world. Because the entire human being images God, a person's religious or spiritual orientation toward God fills and directs one's entire life and is not limited to its transcendent supernatural dimension.[19] Otherworldly spirituality and religion isolated from life are impossible for biblical holistic dualism. All of life is religious. Religious neutrality or secularism in any aspect of life is impossible.

Furthermore, it is a false dichotomy to pit a substantial view of the image of God against relational or functional views, as many monists do.[20] Both historic Christian anthropology and sound metaphysics affirm the correlativity of substance, functions, and relations: substances function and relate; (human) beings are not sets of functions or bundles of relations. Thus functional and relational ontologies are incomplete. Worse, they are deceptive because they actually are substantial ontologies: they assume bodily monism as the metaphysical nature of the human beings whose functions and relations image God.

In sum, integral Christian dualism neither entails nor inclines toward any of the pernicious dichotomies alleged by monists even though some Christians (dualists and monists alike) live by spiritualistic, reductionistic, or unintegrated versions of the faith. In fact dualists—Abraham Kuyper and

John Paul II among them—often have promoted comprehensive and integral visions of Christian faith and practice for this life and the life to come.

I conclude that none of the common criticisms and caricatures of the immortality of the soul raised by monists are valid, given how the doctrine as it is actually explained by thoughtful dualists. In fact it is difficult to take many of the charges seriously. Although not all the views of immortality held by traditional Christian anthropology can be conclusively derived from Scripture, all are consistent with the biblical view of human nature. If immortality merely means that God faithfully and sovereignly sustains people in existence, then it is a biblical doctrine.

Philosophical Speculations about the Nature of the Intermediate State

The supernatural and metaphysical processes by which God accomplishes this astounding feat are a mystery. Disembodied existence seems contrary to and impossible given our holistic, integral constitution as body-soul unities. Perhaps the intermediate state is a supernatural miracle beyond explanation by science and philosophy. But it is not impossible for a God whose power transcends as well as creates, sustains, and is immanent (omnipresent) in the natural order. God can sustain disembodied souls in existence just as easily as he can sustain detached heads.

When Christian believers follow Scripture in speaking about absence from the body and presence with the Lord, we are not using philosophical terms or offering metaphysical explanations. We are using ordinary language about our earthly existence as psychophysical unities to refer to a mode of existence that is difficult to understand or imagine in nonbodily terms. Like parables do, we use earthly language to express heavenly realities. Our language for the afterlife, similar to our language for God, is analogical—figurative yet referentially realistic. We can make true inferences about our general condition even without clear and distinct concepts. For example, if we ourselves actually fellowship with Christ and the blessed dead between death and the resurrection of our bodies, then we are not identical with our bodies or absolutely dependent on them for existence, experience, and action even though God created us as bodily beings. Body-soul or body-person dualism is logically and metaphysically necessary, even if we rely on ordinary language and do not have metaphysical definitions and explanations of bodies, souls, and persons.

Given these epistemic limitations and possibilities, it is possible to speculate about bodiliness during the intermediate state. In Scripture, we are bodily beings. Even the dead appear in bodily form. Samuel returns from Sheol wearing a robe (1 Sam. 28:13–14), as do the martyrs in Revelation 6, and the disciple's mistook the resurrected Jesus for a ghost (Lk. 24:36–37). When I think of my deceased mother, I imagine her in bodily form, especially her face. Although Scripture teaches that we are with the Lord "apart from" our earthly bodies, it is possible that interim embodiment is real and not merely an imaginary projection of earthly life.

One possibility is that God temporarily joins our souls to bodies which consist of the matter or elemental stuff of the place of the dead. This model accommodates both Augustinian-Cartesian substance dualism and Thomistic hylemorphism. The latter regards the soul not as a distinct substance but as the subsistent form of the body, which it actualizes from matter.[21] Another Thomistic hypothesis is that interim bodiliness is the real but temporarily inoperative power of the soul to form a body—the animal aspect of the essence *rational animal*—which is evident to other souls in the intermediate state and to living humans as a ghost. A final scenario for interim embodiment (considered later) is presented by material constitutionists, who hold that persons are constituted by their material bodies but not identical with them. It is possible that at death God causes one's body to divide into two bodies so that one body continues to constitute the person in the intermediate state and the other disintegrates. All these hypotheses are speculative because Scripture does not clearly affirm interim embodiment even though it depicts the dead in bodily form.

Speculation is also possible about the location and duration of interim existence. Paul was "caught up to the third heaven . . . to Paradise" (2 Cor. 15:2–4), which is the location of faithful spirits awaiting final resurrection in eschatology of the Pharisees. This heaven is not a distant place in the universe, but a transcosmic dimension sustained by God with the conditions necessary for interim existence—beyond our earthly understanding.

The period between death and resurrection is indexed to cosmic time but not an extension of it. Scripture consistently correlates the final resurrection with the second coming of Christ at the end of cosmic history (which rules out an immediate resurrection). But the dead in the intermediate state transcend earth time. Their duration and succession is not measured by cosmic motion. Neither are they in eternity. Only God is eternal. The intermediate state is a mode of duration distinct from cosmic time and eternity which many medieval theologians termed the *aevum*, *aeviternity*, or *sempiternity*, elaborating Augustine and Boethius.[22] It is the duration of nonphysical creatures who have a beginning but no end (that is, who are everlasting): angels and human souls.

What Difference Does It Make?

Monists who believe humans do not exist between death and resurrection find little of importance in the intermediate state and see no great loss in denying it. At best they regard it as a comforting fiction for individuals who cannot face temporary extinction. Monists try to brighten the picture by noting that it is not possible to experience one's own nonexistence. From the first-person point of view, they say, death is like falling into dreamless sleep and waking again in what seems like an instant, no matter how much time has passed. We need not fret about ourselves or our deceased loved ones, because we will all immediately experience resurrection together. Nothing we hope for is lost.

This scenario is subjectivistic, however, focused entirely on experience. In reality, it entails a number of significant losses from biblical eschatology—the culmination of God's plan for creating and redeeming his human family, outlined earlier.

First, if God were to let us pass out of existence, it would mean that sin and death temporarily defeat his precreational intention that humanity be a family in everlasting fellowship with one another and with him. Even if ontological annihilation were the just and natural consequence of sin, the God of John 3:16 has chosen to limit its effects to physical death and to sustain humanity in spite of it. The incarnation, death, and resurrection of Jesus Christ graciously assume, atone for, and transform death sufficiently to allow for an intermediate state consistent with God's justice. Annihilation is unnecessarily avoidable. Faithful Israelites were "gathered to their fathers" when they died in anticipation of national renewal when God restores Jerusalem (Isa. 65:17ff.). New Testament believers expect to join those who have died in the Lord and await resurrection at his return (1 Thess. 3:13–18). From Genesis 1 to Revelation 22, the people of God never cease to exist. Temporary extinction would diminish the achievement of God in creating, redeeming, and perfecting humanity.

Nonexistence would also ontologically separate individuals from the human community. Monists sometimes caricature dualism's substantial soul as an isolated "essence" or self-absorbed monad for which relationships are incidental to existence and identity. Instead they promote "relational ontology" and "narrative identity" as constitutive of personal existence and identity.[23] But if there are no people of God in the intermediate state, then it is monists who posit an existential-ontological gap in every individual's continuous participation in the human community, not to mention their relation to God and their very identity (explained later). Historic Christian anthropology affirms that self-identical individuals never cease to be in dynamic, growing relationships with other humans and God.

Furthermore, if there is no intermediate state, then there are no humans in heaven praising God, as envisioned in the book of Revelation. The church's ancient hymn, *Te Deum* ("We praise you, O God") envisions the apostles, prophets, saints, and martyrs along with the angels and heavenly hosts ceaselessly singing "holy, holy, holy" to the Lord God of Hosts. This hymn also reflects Hebrew 12:22–24, which lists "the spirits of righteous men made perfect" with God, Christ, and the angels in the heavenly Jerusalem. But all of this is pious fiction if there is no intermediate state. Eliminating the intermediate state would drastically reduce the praise that God receives from his people, the church.

Finally, denying the intermediate state is incompatible with creedal Christology and the doctrine of the incarnation. The Nicene and Athanasian Creeds assert that Jesus Christ is truly and fully God and truly and fully human. The Athanasian Creed concurs and adds that human nature subsists of "a reasonable soul and human flesh."[24] The Council of Chalcedon (451) further declares that Jesus's divine and human natures are "without confusion, without change, without division, and without separation."[25] According to bodily monism, however, humans entirely cease to exist at physical death unless there is immediate resurrection.[26] But if Jesus's human nature was nonexistent between his death and resurrection, then it was separated from his divine nature. (He also did not actually "descend into hell," the realm of death.) His resurrection reactualized his human nature and reunited it with his divinity. Easter was a reincarnation of God the Son. According to classical orthodoxy, Jesus's human nature was dichotomized, but his human soul was not separated from his divinity. Creedal Christology is not an insignificant issue.

In sum, denying the intermediate state makes a huge doctrinal and practical difference to the Christian faith. In addition to disturbing the vast majority of Christians who believe it, denying it implicitly diminishes the effects of salvation, breaches God's love and faithfulness to his people, reduces their praise of God, separates individuals from the human community, and conflicts with creedal Christology. The cost of bodily monism is high.

EVALUATION OF BODILY MONIST ESCHATOLOGIES

Bodily monists have proposed three alternative eschatologies. The most widely held is nonexistence between death and resurrection. A second is immediate resurrection—a new body in another dimension the instant that the earthly body dies. Finally, there is a monist account of the intermediate state and final resurrection narrative involving an intermediate body. Here follow brief summaries and evaluations.[27]

Nonexistence until Resurrection

The most widely held monist eschatology is temporary nonbeing between death and resurrection. Persons do not actually exist, but they will be reactualized when their bodies are resurrected. According to Polkinghorne, the soul is "the form, or immensely complex information-bearing pattern, of the body. That form is dissolved at death. . . . God will remember and reconstitute the pattern that is a human being, in an act of resurrection taking place beyond present history."[28]

One fatal flaw of this scenario is its denial of the biblical doctrine of the intermediate state. Although Scripture does not reveal much about it, and our attempts to imagine it are speculative, there is little room for doubt that the biblical authors affirm it and intend their readers to believe it. Christian anthropology at least ought not to contradict it.

A second flaw which undermines this account is a tenuous view of personal identity. It seems axiomatic that persons are necessarily self-identical, not accidentally or contingently so. It is absolutely impossible that there be two of me or that I could become someone else even if my personality and body changed radically. An adequate anthropology ought to account for the logical necessity of personal self-identity.

For dualism, the soul is the locus of self-identity. It endures continuously as the self-same substantial or subsistent entity throughout life, during the intermediate state, and after the resurrection of the body. It remains numerically identical even if one's personality or body changes radically, or one loses awareness of self-identity, or is chronically misidentified by other people, or is completely unconscious. It cannot fail to be self-identical.

But the nonexistence until resurrection eschatology leaves personal identity contingent and dubious. Physicalism and all other kinds of bodily monism must ground personal identity in the body, which does remain the self-same living entity during life. To account for the identity of earthly and resurrected persons, however, monists must explain the numerical identity of the earthly body with the resurrection body in spite of the gap in their existence. Perhaps the resurrection body is reconstructed from a sufficient amount of matter from the earthly body.

Identical matter is not a secure basis for the identity of human bodies because the material composition of organisms changes over time, and the same matter can sequentially constitute different organisms. Thus, most bodily monists avoid locating the identity of the earthly and resurrection bodies (and persons) in shared matter. Instead most hold that identity consists in the uniqueness of one's bodily and personal characteristics—a personal essence or Polkinghorne's "information-bearing pattern."[29] At the resurrection, God will recreate each bodily person with his or her unique spiritual, personal, and physical characteristics, sanctified and perfected. Each person will have a sense of self-identity and be recognized by others as the very same person who lived on earth.

But this explanation involves an inadequate notion of personal identity. Logically and metaphysically, multiple replication is possible. Even though

God would not create multiple instances of John Cooper on resurrection day, hypothetically an evil genius could. Any number of persons instantiating the unique Cooper essence are metaphysically possible, and each would have an equally legitimate claim to being Cooper. The issue is not whether God would do such a thing but the nature of personal identity, which is contingent on there being only one claimant. Even then, it is indeterminable whether the earthly and resurrected Coopers are numerically identical or different but exactly similar persons. This is a far weaker view of personal identity than the absolute identity which common sense and dualism affirm. Theologically, it seems odd for God to give such flimsy identity to humans destined for everlasting life, while granting organisms substantial self-identity as long as they endure.[30]

In sum, bodily monism's extinction-recreation eschatology contradicts biblical eschatology and has a philosophically inferior account of personal identity. It does not qualify as a Christian philosophical anthropology.

Immediate Resurrection

Some bodily monists postulate immediate resurrection to avoid disembodiment and the dualism it entails.[31] At the instant of death, God resurrects us and thus our personal existence continues without a gap, like switching an operating system from one computer to another without an interruption in the program's functioning.

One major problem with immediate resurrection is inconsistency with Scripture, which teaches that the resurrection is a general event correlated with the return of Christ and not an individual event for each person at death. In addition, immediate resurrection implicitly posits two different bodies with no continuity, which contradicts the biblical doctrine that the resurrection body is the earthly body transformed (1 Cor. 15:42ff). Immediate resurrection further implies that the new heaven and earth, the dwelling of resurrected people, already exists. These are serious inconsistencies with the narrative of biblical eschatology.

Philosophically, immediate resurrection founders on personal identity in its own way. Bodily monism posits that persons are generated by and dependent on their organisms. If identity is substantial, that is, an intrinsic property of an entity, then immediate resurrection implies that there are two different persons because there are two different bodies. One could solve this problem by claiming that the very same person instantaneously switches bodies. But that solution implies dualism—one self-identical person separates from one body and unites with another. Alternatively, if identity merely consists in the earthly and resurrected person having the identical set of properties, then immediate resurrection commits to the same dubious account of identity as the extinction-recreation eschatology.

For these biblical and philosophical reasons, immediate resurrection does not enable bodily monism to qualify as a sound Christian anthropology.

A Bodily Monist Account of the Intermediate State and Resurrection

Although almost all bodily monists endorse either temporary nonexistence until resurrection or immediate resurrection, a monist account of the intermediate state is available.[32] At death, God causes a person's body to divide by a process analogous to organic budding or nuclear fission. One body continues to constitute the person and the other becomes his corpse. The continuity of the bodily person could explain an immediate final resurrection. But it also allows for intermediate existence after which God transforms the intermediate bodily person into the resurrected bodily person. Personal continuity and identity seem assured by the continuity and identity of the body. This is a monist account of the intermediate state.

Thus it conforms to the biblical narrative of the afterlife. Although Scripture states that we are "absent from the body" and refers to the dead as souls or spirits, it also depicts the dead in bodily form, as noted above. The idea of postmortem bodily persons can explain absence from the earthly body and continuing bodily form. So this account is consistent with Scripture.

It has been challenged philosophically, however. If one motive of bodily monism is to articulate and defend Christian belief in terms of current science and philosophy, there is nothing in this eschatology that science or philosophy would take seriously. Bodies that bud or fission at death are imaginative metaphors for a supernatural miracle.

More directly, this eschatology has its own problems with personal identity. If God causes the body to bud or divide, are both bodies identical with the original? Is what grows from a bud identical with the organism that budded? Is a clone identical with its parent? Are both halves of a cell that divides identical with the parent cell? If not, then by analogy any person generated by a postmortem body is a different person than the one who lived and died. If the answer is unclear or doubtful, then this scenario is no better than the others regarding personal identity. It is likewise vulnerable to multiple replication. An evil genius could keep both bodies alive and generating persons, or produce multiple bodily persons from the original, each with identically strong claims to being the predeceased person, which is impossible. At best, personal identity is contingent on there being only one postmortem bodily person.

Perhaps personal identity is not an essential property of persons; perhaps it is contingent on other things, or evidence, or wholly on the will of God. If so, then this account of the intermediate state and final resurrection allows bodily monism to qualify as a Christian philosophical anthropology. But conceding a strong view of personal identity is a high cost for Christians, both for doctrinal and pastoral issues about the life to come, and also for a philosophy of personal identity that is as clear and sound as for the identity of physical objects and living things.

CONCLUSION

The Bible teaches that persons exist without their bodies during the intermediate state, which is an integral phase of God's plan of redemption. Because no anthropology that is inconsistent with Scripture can be regarded as Christian (or true), any anthropology endorsed by Christians must allow for the separation of existing persons or souls from their bodies between death and the general resurrection. Bodily monism, which includes materialism and physicalism, either precludes this possibility or cannot provide an adequate philosophical account of it, in particular the identity of earthly persons with resurrected persons. Thus bodily monism is either defeated or seriously undermined as a Christian philosophical anthropology. And theologically, it is hard to understand why the God who creates, redeems, and perfects his human image-bearers for everlasting fellowship in spite of sin and death would choose bodily monism instead of holistic dualism as the metaphysics most conducive to his project.

NOTES

1. Prominent representatives include the following. For nonreductive physicalism, see Nancey Murphy, *Bodies and Souls, or Spirited Bodies?* (Cambridge, UK: Cambridge University Press, 2006); for material constitution, see Kevin Corcoran, *Rethinking Human Nature: A Christian Materialist Alternative to the Soul* (Grand Rapids, MI: Baker Academic, 2006); for emergent monism, see Philip Clayton, *Mind and Emergence: From Quantum to Consciousness* (Oxford: Oxford University Press, 2004); for psychophysical monism, see John Polkinghorne, *Science and Theology: An Introduction* (London: SPCK, and Minneapolis, MN: Fortress Press, 1998), chapter 3; for multidimensional monism, see Velli-Matti Karkainen, "'Multidimensional Monism': A Constructive Theological Proposal for the Nature of Human Nature," in *Neuroscience and the Soul: The Human Person in Philosophy, Science, and Theology*, eds. Thomas Crisp, Steven Porter, and Gregg Ten Elshof (Grand Rapids, MI: Eerdmans Publishing, 2016); for an endorsement of ontological monism in spite of affirming the intermediate state in Scripture, see N. T. Wright, "Mind, Spirit, Soul and Body: All for One and One for All. Reflections on Paul's Anthropology in his Complex Contexts," presented at the Society of Christian Philosophers Regional Meeting, Fordham University (March 18, 2011). Available online at (www.ntwrightpage.com/

Wright_SCP_MindSpritiSoulBody.htm). Joel Green, *Body, Soul, and Human Life: The Nature of Humanity in the Bible* (Grand Rapids, MI: Baker Academic, 2008), 46, is open to any view in which embodiment is essential for but nonreductive of persons.

2. Stephen Yates, *Between Death and Resurrection: A Critical Response to Recent Catholic Debate Concerning the Intermediate State* (New York: Bloomsbury Academic, 2017) is a recent example. I thank Joshua Farris for bringing it to my attention.

3. Joel Green summarizes these arguments in "The Bible, the Natural Sciences, and the Human Person," in *Body, Soul, and Human Life*, 1–34.

4. For example, Trenton Merricks ignores it in "How to Live Forever without Saving Your Soul: Physicalism and Immortality," in *Soul, Body, and Survival: Essays on the Metaphysics of Human Persons,* ed. Kevin Corcoran (Ithaca, NY: Cornell University Press, 2001); Murphy marginalizes it as "not . . . central to Christian teaching" in "Human Nature: Historical, Scientific, and Religious Issues," in *Whatever Happened to the Soul? Scientific and Theological Portraits of Human Nature*, eds. Warren Brown, Nancey Murphy, and H. Newton Malony (Minneapolis, MN: Fortress Press, 1998), 19; Green argues that it is a flawed reading of Scripture in *Body, Soul, and Human Life* and "Eschatology and the Nature of Humans: A Reconsideration of Pertinent Biblical Evidence," *Science and Christian Belief,* 14, no. 1 (April 2002): 33–50.

5. An important anthology of various positions is *Soul, Body, and Survival*, and *Personal Identity and Resurrection: How Do We Survive our Death?*, ed. Georg Gasser (Farnham, UK: Ashgate Publishing, 2010).

6. Most recent and comprehensive is N. T. Wright, *The Resurrection of the Son of God* (Minneapolis, MN: Fortress Press, 2003).

7. This is Wright's conclusion, in *The Resurrection of the Son of God*, after examining ancient near-eastern views, the Old Testament, Second Temple Judaism, and the New Testament.

8. Wright comments on John 11 in *Resurrection of the Son of God*: "the believer now possesses, already, a divinely given immortal life which will survive death and be re-embodied in the final resurrection" (444).

9. Murphy, *Bodies and Souls, or Spirited Bodies?*, 16. Calvin wrote against Anabaptist belief in soul-sleep, but the intermediate state was integral to Catholic doctrine before the Reformation and to Catholics and Protestants alike thereafter.

10. Stewart Goetz and Charles Taliaferro, *A Brief History of the Soul* (Oxford, UK: Wiley-Blackwell, 2011) is a clear summary of the philosophical debates and most of the theological issues addressed in this chapter. I have benefited from it.

11. Harry Wolfson, "Immortality and Resurrection in the Philosophy of the Church Fathers," in *Immortality and Resurrection*, ed. Krister Stendahl (New York: Macmillan, 1965), 54–96. Wolfson challenged Oscar Cullman's famous lecture, "Immortality of the Soul or Resurrection of the Dead," also in *Immortality and Resurrection.* Cullman is the source of much confusion because his title casts immortality and resurrection as exclusive alternatives although he actually affirms an unconscious soul between death and resurrection.

12. Carolyn Walker Bynam, *The Resurrection of the Body in Western Christianity, 200–1336* (New York: Columbia University Press, 1995).

13. *Catechism of the Catholic Church* (1994), exposition of Articles 11 and 12 of the Apostles' Creed; website of The Greek Orthodox Archdiocese of America, Our Faith: "Death: The Threshold to Everlasting Life" (http://www.goarch.org/ourfaith/ourfaith7076), accessed July 2017; *The Heidelberg Catechism*, Question/Answers 1 and 57 on the unity of body and soul in life, their separation at death, and their reunion at the resurrection.

14. Bruce Reichenbach, *Is Man the Phoenix?* (Grand Rapids, MI: Eerdmans, 1983), 52–54.

15. Wright, "Mind, Spirit, Soul and Body," charges that dualism's immortal soul is residual Platonism, conducive to otherworldly spiritualization, and "the ontological equivalent of works-righteousness in its old-fashioned sense: something we possess which enables us to establish a claim on God, in this case a claim to 'survive'." These charges caricature most historic Christian theology.

16. For example, Warren Brown and Brad Strawn, *The Physical Nature of Christian Life: Neuroscience, Psychology and the Church* (Cambridge, UK: Cambridge University Press), chapters 1 and 2.

17. Wolfson, "Immortality and Resurrection in the Church Fathers."

18. Aquinas is an astute observer of the multiple mutual influences of the intellect, the will, and the various affections, passions, and appetites. He is not a crude intellectualist. Peter King, "Aquinas on the Passions," *Aquinas' Moral Theory* (Ithaca, NY: Cornell University Press, 1999), 101–32.

19. A Reformed elaboration of this view is Herman Bavinck, "The Whole Person as the Image of God," *Reformed Dogmatics*, vol. 2, ed. John Bolt, trans. John Vriend (Grand Rapids, MI: Baker Academic, 2004), 554–62.

20. For example, Joel Green, "Why the *Imago Dei* Should Not Be Identified with the Soul," in *The Ashgate Research Companion to Theological Anthropology*, ed. Joshua Farris and Charles Taliaferro (Farnham, UK: Ashgate Publishing, 2015), 179–90.

21. Brian Leftow, "Souls Dipped in Dust," in *Soul, Body, and Survival*, 120–38, is a clear account of Aquinas' anthropology.

22. Aquinas, *Summa Theologica*, I.Q.10, a.5.

23. Joel Green makes this argument repeatedly in *Body, Soul, and Human Life*, chapters 2, 4, and 5.

24. Jesus Christ is "*perfectus homo: ex anima rationali et humana carne subsistens*."

25. Henry Bettenson, *Documents of the Christian Church* (Oxford: Oxford University Press, 1947), 73.

26. Glenn Andrew Peoples, "The Mortal God: Materialism and Christology," *Ashgate Research Companion to Theological Anthropology*, 331–43, affirms this position.

27. I have benefited from J. P. Moreland and William Lane Craig, "Personal Identity and Life after Death," *Philosophical Foundations for a Christian Worldview* (Downers Grove, IL: InterVarsity Press, 2003), chapter 14; and Stewart Goetz and Charles Taliaferro, *A Brief History of the Soul*, chapter 4, which summarizes theories of personal identity in modern philosophy.

28. John Polkinghorne, "Human Destiny," in *Science and Theology*, 115–16. Others who endorse this view are Nancey Murphy, *Bodies and Souls, or Spirited Bodies*,

132–42; Joel Green, *Body, Soul, and Human Life*, 178–80; Trenton Merricks, "How to Live Forever without Saving Your Soul," in *Soul, Body, and Survival*, 183–200, and Velli-Matti Karkainen, "Multidimensional Monism," in *Neuroscience and the Soul*, 221–22.

29. Notice that monists, not dualists, are guilty of reducing souls or persons to "essences."

30. Organisms are self-identical because they are enduring structural wholes, not because their material components remain the same.

31. John Hick, *Death and Eternal Life* (Louisville, KY: Westminster John Knox Press, 1994), chapter 15, and Wolfhart Pannenberg, *Systematic Theology*, vol. 3 (Grand Rapids, MI: Eerdmans, 1998), chapter 15, sec. 3. Hick locates resurrected persons in another dimension of creation; Pannenberg locates them in eternity.

32. Kevin Corcoran presents this possibility even though he does not endorse it. See "Physical Persons and Postmortem Survival without Temporal Gaps," in *Soul, Body, and Survival*, 201–17; and "The Constitution View of Persons," in *In Search of the Soul*, ed. Joel Green (Downers Grove, IL: Intervarsity Press, 2005), 153–76, esp. 167–68. I have benefited from the summary and critique of this view by Jonathan Loose, "Constitution and the Falling Elevator: The Continuing Incompatibility of Materialism and Resurrection Belief," *Philosophia Christi,* 14, no. 2 (2012): 439–49.

SUGGESTED FURTHER READING

Bavinck, Herman. "The Whole Person as the Image of God." In *Reformed Dogmatics*, vol. 2, edited by John Bolt, translated by John Vriend, 554–62. Grand Rapids, MI: Baker Academic, 2004.

Brown, Warren, and Brad Strawn. *The Physical Nature of Christian Life: Neuroscience, Psychology and the Church.* Cambridge, UK: Cambridge University Press, 2012.

Bynam, Carolyn Walker. *The Resurrection of the Body in Western Christianity, 200–1336.* New York: Columbia University Press, 1995.

Clayton, Philip. *Mind and Emergence: From Quantum to Consciousness.* Oxford, UK: Oxford University Press, 2004.

Corcoran, Kevin. *Rethinking Human Nature: A Christian Materialist Alternative to the Soul.* Grand Rapids, UK: Baker Academic, 2006.

———, ed. *Soul, Body, and Survival: Essays on the Metaphysics of Human Persons.* Ithaca, NY: Cornell University Press, 2001.

Farris, Joshua, and Charles Taliaferro, eds. *The Ashgate Research Companion to Theological Anthropology.* Farnham, UK: Ashgate, 2015.

Gasser, Georg, ed. *Personal Identity and Resurrection: How Do We Survive our Death?* Farnham, UK: Ashgate, 2010.

Green, Joel. *Body, Soul, and Human Life: The Nature of Humanity in the Bible.* Grand Rapids, MI: Baker Academic, 2008.

———. "Eschatology and the Nature of Humans: A Reconsideration of Pertinent Biblical Evidence." *Science and Christian Belief* 14, no. 1 (April 2002): 33–50.

Hick, John. *Death and Eternal Life*. Louisville, KY: Westminster John Knox Press, 1994.

Loose, Jonathan. "Constitution and the Falling Elevator: The Continuing Incompatibility of Materialism and Resurrection Belief." *Philosophia Christi* 14, no. 2 (2012): 439–49.

Murphy, Nancey. *Bodies and Souls, or Spirited Bodies?* Cambridge, UK: Cambridge University Press, 2006.

———. "Human Nature: Historical, Scientific, and Religious Issues." In *Whatever Happened to the Soul? Scientific and Theological Portraits of Human Nature*, edited by Warren Brown, Nancey Murphy, and H. Newton Malony, 1–30. Minneapolis, MN: Fortress Press, 1998.

Pannenberg, Wolfhart. *Systematic Theology*, vol. 3. Grand Rapids, MI: Eerdmans, 1998.

Reichenbach, Bruce. *Is Man the Phoenix?* Grand Rapids, MI: Eerdmans, 1983.

Wright, N. T. *The Resurrection of the Son of God*. Minneapolis, MN: Fortress Press, 2003.

Yates, Stephen. *Between Death and Resurrection: A Critical Response to Recent Catholic Debate Concerning the Intermediate State*. New York: Bloomsbury Academic, 2017.

Chapter 17

Physicalism and Sin

Charles Taliaferro

I have argued against physicalism in print elsewhere, employing a modal and a knowledge argument for a rather different philosophy of mind: dualism. In this volume, I propose that substance dualism is better placed than physicalism to make sense of a central Christian teaching about the death of Jesus Christ (see chapter 9 of this volume).[1] In this chapter, I am more accommodating to physicalism. I concede that, at first glance, there is reason to think that a certain form of physicalism has a more elegant, holistic approach to a Christian theology of sin, including original sin, than substance dualism. The aim of the chapter will be to take note of the ostensible advantage of physicalism, but then to defend a form of dualism, *integrative dualism*, that provides a framework that is as good, and in some respects, a better framework for thinking about sin.

The kind of physicalism that can best accommodate a theology of sin is one that is nonreductive. Forms of physicalism that either eliminate or do not appreciate the reality of desire, love, and hate qua emotions (in addition to, for example, brain states) are too austere to entertain the reality of sin. So, in this chapter I assume (if only for the sake of argument) that there is a plausible, nonreductive form of physicalism that allows for persons to have beliefs, desires, sensations, reason, powers to act for reasons, moral awareness, the capacity for self-control, the ability to be tempted, and the power to resist temptation. I propose that insofar as physicalism can be stretched to include these elements, plus perhaps also accommodating the awareness of God's commands and the possibility of life after life, then the topic of sin need not be a threat to Christian physicalists. There are reasons for thinking physicalism faces serious challenges with making that theological venture, but these are not part of the concerns of this chapter. An example of a version of physicalism that would, in my view, foot the bill theologically is the hylemorphic animalism articulated and defended by Patrick Toner.[2]

To appreciate the importance of Christian physicalists being able to take a nonreductive route, consider first this overview of the state of play of philosophy of mind by Jaegwon Kim, which I go on to amend to take on the subject of sin:

> For most of us, there is no need to belabor the centrality of consciousness to our conception of ourselves as creatures with minds. But I want to point to the ambivalent, almost paradoxical, attitude that philosophers have displayed toward consciousness . . . Consciousness had been virtually banished from the philosophical and scientific scene for much of the last century and consciousness bashing still goes on in some quarters, with some reputable philosophers arguing that phenomenal consciousness, or "qualia" is a fiction of bad philosophy. And there are philosophers . . . who, while they recognize phenomenal consciousness as something real do not believe that a complete science of human behavior, including cognitive psychology and neuroscience, has a place for consciousness in an explanatory/predictive theory of cognition and behavior.[3]

Kim (rightly) laments this derogatory, sterile approach to consciousness, for he then goes on to note the odd juxtaposition of those who denigrate consciousness with those philosophers who work on ethics and value theory:

> Contrast this lowly status of consciousness in science and metaphysics with its lofty standing in moral philosophy and value theory. When philosophers discuss the nature of the intrinsic good, or what is worthy of our debate and volition for its own sake, the most profoundly mentioned candidates are things like pleasure, absence of pain, enjoyment, and happiness. . . . To most of us, a fulfilling life, a life worth living, is one that is rich and full in qualitative consciousness. We would regard life as impoverished and not fully satisfying if it never included experiences of things like the smell of the sea in a cold morning breeze, the lambent play of sunlight on brilliant autumn foliage, the fragrance of a field of lavender in bloom, and the vibrant, layered soundscape projected by a string quartet.[4]

Given the topic of this chapter, sin, allow me to adjust the previous observation:

> When philosophers or theologians discuss sin, the most profoundly mentioned candidates are things like free will, responsibility, temptation, weakness of will, failures of self-mastery, disordered desire and pleasure, vanity, (unwarranted) anger, lust, envy, sloth, avarice, gluttony, greed, self-deception, self-destruction, malice, malignant hatred and (harmful, dangerous) fears along a myriad of fronts involving race, gender, age, economics, class, sexual orientation . . . . To

> most of us, our concept of a sinful life, a life that is unworthy of image-bearers of God to live, is one that is rife in qualitative consciousness. Such a life would include experiences like feeling deep resentment about the happiness of others, a keen desire to destroy the innocent, a desire to smell the fear in others whom we intimidate to fulfill our narrow desires for personal satiation, the fragrance of a building burning that we have lit on fire, the screams of our victims.

Provided that physicalists can recognize a full range of such depraved, sinful experiences, I suggest that they are in good shape in terms of Christian theological anthropology. The one exception to note (for the record) is whether physicalists can accommodate belief that there are evil, immaterial person-like beings, Satan and devils, but I do not wish to defend Satan (even philosophically as a possible being) in this chapter. Satan has too many advocates these days, in my view.

In this chapter, I focus on two matters: whether physicalism or substance dualism is better able to take seriously the embodiment of sin and grace, and which of the two philosophies of mind is better equipped to address original sin. I first address sin, and then original sin.

SIN FROM A PHYSICALIST AND DUALIST PERSPECTIVE

The concept of sin is not, primarily, a secular concept. It is, rather, a principle element in the Abrahamic tradition. A sin is either an act or omission or state of character (such as vanity) that is in violation of the will and nature of God. Sins involve individual persons, but Judaism, Christianity, and Islam each recognize that groups of peoples (families, tribes, cities, nations, kingdoms) can and do sin. A paradigm case of a sin would be an act or state of character in which the subject knowingly violates a divine command, as one finds in the biblical narratives in which Satan tempts Jesus to sin by violating the will of God. But one may also sin by failing to know or by neglecting God's laws.

A reason for thinking that nonreductive physicalism may be well placed to address sin theologically is that sins in the Abrahamic tradition usually involve what might be called concrete embodiment, bodily acts or omissions themselves, or the intention or lack of intention to commit or omit bodily action. Matters of intention and the heart are widely recognized as a domain of either sin or sanctity. So the mere going through the motions of worshipping God is sacrilege because the action is not backed up by the right motives and thoughts. Famously, Jesus claims that even a person who lusts after someone commits adultery (Matt. 5:8), but lust and debasing worship both refer to wrongful matters of embodiment. The sinful acts of so-called "worship" are not actually acts of worship, but idolatry (perhaps a degrading form of trying to appear pious in order to manipulate others) and looking at someone lustfully is a failure to embody chastity. Because physicalists do not distinguish between the person (soul or mind) and the body, there is no danger of treating sin as a matter of pitting the person against the body or vice versa. The question of which is more important, the soul or the body, does not arise. With substance dualism, however, there can be such a question.

Such a question need not be ruinous, as I shall argue. Here is a slightly amusing depiction of the plight of human embodiment from a dualist point of view, in which Augustine laments the way in which, after the fall, the body is unresponsive to the mind:

> When the first man transgressed the law of God, he began to have another law in his members which was repugnant to the law of his mind, and he felt the evil of his own disobedience when he experienced in the disobedience of his flesh a most righteous retribution recoiling on himself. . . . When it must come to man's great function of the procreation of children the members which were expressly created for this purpose will not obey the direction of the will, but lust has to be waited for to set these members in motion, as if it had legal right over them, and sometimes it refuses to act when the mind wills, while often it acts against its will! Must not this bring the blush of shame over the freedom of the human will, that by its contempt of God, its own Commander, it has lost all proper command for itself over its own members? Now, wherein could be found a more fitting demonstration of the just depravation of human nature by reason of its disobedience, than in the disobedience of those parts?[5]

This complaint about incontinence, impotence, or lack of sexual self-mastery may be harmless (even quaint) from the mind-body perspective, but matters prove to be darker when taking into account extreme asceticism.

The Abrahamic traditions have included practices of asceticism in which the soul and body have been seen in profound bifurcation with the body taking most of the punishment. In its extreme forms, the body is subject not just to fasting and various exercises of self-mastery, but to outright self-laceration, flagellation, and even castration and female genital mutilation. To be sure, physicalism is neither necessary nor sufficient to avoid these dangers. A penitent might well engage in self-mutilation while believing that he and the body he is mutilating are one and the same thing. Still, physicalism will avoid the at least naïve or errant view that, for example, the body as flesh is a natural enemy of the spirit or soul, weighing us down toward mundane, rather than heavenly, goods. On this front, I suggest that nonreductive physicalism has some advantage over substance dualism. Such a form of physicalism provides an elegant, holistic picture impeding the idea that mind and body are natural enemies.

Having made this concession, however, I believe that integrative dualism has equal merit and some advantage over nonreductive physicalism. According to integrative dualism, while the person and body are metaphysically distinct, in a healthy form of embodiment the person and body function as one. So, while it is possible for a person to treat his body as a mere container or a machine or a home that he inhabits, in a healthy form of embodiment, for you to see, touch, and hear the persons you are seeing, touching, and hearing is to interact with a unified, embodied subject. Because of its being dualistic, integrative dualism does not attribute to the body a separate set of (potentially) sinful desires, base appetitive urges, lusts, and so on. When a person is subject to base desires, it is the integrated whole, embodied person who has been or is sinful. While integrative dualism does not guarantee the prevention of extreme asceticism, neither does nonreductive physicalism.

While integrative dualism and nonreductive physicalism can be evenly matched in addressing the topic of sin, I suggest that integrative dualism offers a more elegant understanding of the Christian antidote to sin which is for penitent sinners to confess, forgive and be forgiven, repent, make whatever restitution is possible for past sins, and through commitment and renewed desires to become part of the church as understood as the Body of Christ. The concept of the Body of Christ is the idea that baptized Christians make up a community that is to so follow the teaching and model the life of Jesus (the imitation of Christ), inspired by the Holy Spirit, that Christians in the world function as Christ's body. So, when the followers of Christ feed the poor, visit those in prison, care for the sick and vulnerable, and so on, this is to be understood theologically as Christ himself feeding the poor, visiting those in prison, and so on, in which the followers of Christ and Christ himself form or function as an integrated whole. Integrative dualism provides an analogous model of the relationship between Christ and the body of believers. In 2 Corinthians 2:16, the invocation to follow "the mind of Christ" hints at this integration of mind and body which is very different from a nonreductive physicality account that would literally (or metaphysically) identify Christ with (or as) the body of believers. The latter would involve the conceptual absurdity of Christ qua Second Person of the Trinity, necessarily existing, omnipresent, and so on, becoming metaphysically identical with contingent, finite persons, and so on. True, the Christian physicalists can still claim that their model of Christ's relationship with the church is an analogy or metaphor (and not literal), but I suggest that integrative dualism is the more fitting model. On integrative dualism, just as persons are not metaphysically identical with their bodies, but can function as a unified being, so Christ and the church are not metaphysically identical, but can function as a unified being. Let us now consider the teaching of original sin.

ORIGINAL SIN FROM A PHYSICALIST AND DUALIST PERSPECTIVE

The belief in original sin is subject to great philosophical and theological attention. The basic idea is that all human beings are, in some fashion, implicated in the first sin of Adam (or Adam and Eve). The Western and Eastern Christian traditions have shown some tension on what this implication amounts to. The Eastern Orthodox have tended to construe original sin as the original sin of Adam and Eve and the consequence of that sin is death, but not guilt. On this view, you and I do not inherit any of the guilt of the Adamic sin, only its consequence, whereas the West has claimed that we, in some sense, share in the guilt or sin of Adam and Eve. *The Catechism of the Catholic Church* has this entry:

> How did the sin of Adam become the sin of all his descendents? The whole human race is in Adam "as one body of one man" [Thomas Aquinas]. By this "unity of the human race" all men are implicated in Adam's sin, as in all are implicated in Christ's justice. Still, the transmission of original sin is a mystery that we cannot fully understand. But we do know by Revelation that Adam had received original holiness and justice not for himself alone, but for all human nature. By yielding to the tempter, Adam and Eve committed a personal sin, but this sin affected the human nature that they would then transmit in a fallen state. It is a sin which will be transmitted by propagation to all mankind, that is, by the transmission of a human nature deprived of original holiness and justice. And that is why original sin is called "sin" only in an analogical sense: it is a sin "contracted" and not "committed"—a state and not an act.[6]

It seems, initially, that nonreductive physicalists and substance dualists can equally accommodate either the Eastern or Western views of original sin. Or, more modestly, it seems that neither (at first glance) adds an impediment to making a philosophical case for the coherence of believing in original sin. So, it seems (prima facie) that either philosophy of mind could claim with Augustine that all of humanity was (mysteriously) present in Adam.[7] Either might make use of a view adopted by some Protestants that involved a form of federalism, according to which Adam had the authority to act on behalf of all human beings. Kant's appeal to a fall that takes place with a noumenal self is perhaps more favorable to dualism, but his distinction between a noumenal and phenomenal self is so obscure that it has few adherents. Kierkegaard's notion that we all recreate the Adamic sin could be worked out on either philosophy of mind, and Richard Swinburne's notion that original sin involves our inheriting a natural disposition to sin seems neither helped nor aided by either philosophy of mind.

There is, however, a plausible, but controversial interpretation of the notion of original sin that appears to give some advantage to nonreductive physicalism. This model may be called the *originalist model*, according to which the consequences of sin are transmitted as well as wrongfulness or sinfulness, but without transmitting personal guilt. I articulate this model with a thought experiment.

Imagine your conception involved a grave sin such as rape. As a grave sin, let us fully acknowledge that the rape should not have occurred. If *it should not have occurred*, does it follow that *your conception should not have occurred*? Assuming that the rape was integral to the intercourse that led to the pregnancy, I think it is plausible to think that, *ceteris paribus*, your conception should not have occurred. There are many causes of your birth, but given the enormity of the causal role of conception, I think it plausible to claim that if your conception should not have occurred, your birth should not have occurred. Given that your birth should not occur, does it follow that, once conception has taken place, your birth should have been prevented by abortion or infanticide? No. One might adopt a "pro-life" stance according to which the fetus is a person and, thus, terminating fetal life would be homicide. Or one may prohibit termination on the grounds of the fetus being a potential person or sacred.

Even apart from these considerations, other reasons may come into play as to why your birth should occur. Imagine that your mother was the one raped and yet she was resolved to insure your birth and life occur. However, allowing *that the pregnancy even ought to be brought to term, the child reared and to have a full life,* there would still be a blight or tarnish to the life lived; *it would still forever be the case that the conception should not have taken place*. This is not a matter of personal guilt borne by the individual who came into being due to the sinful act of the assailant. But it would involve recognizing the idea that there would remain a sense in which a person's life is in some sense marked by (a) sin. The originalist model can then be extended by proposing that it is highly likely (and not merely logically possible) that all or most human beings have such a tarnished origin. While I have used rape as the example of a primeval sin in terms of one's origin (given the irreplaceable biological role of conception), a host of other types of sin may be in play in terms of one's birth; perhaps sins of deception, manipulation, exploitation, undue social pressures, and so on, may have played a role in one's ancestry. Such a hypothesis about the extent that all or most human beings have come about (somewhere in their ancestry) through sin would be virtually impossible to assess, but given the vast evidence of human sin, it is not (I suggest) unreasonable to believe vast numbers of us are now alive due to sin.

On behalf of the originalist model, it should be noted that it works with values that come into play in the case of restoration justice. If your ancestors stole some property from a people and gave it to you, you might be totally innocent personally in terms of your being made a recipient of the land (you played no role in the original theft), but (arguably) you would not have a right to the land. In a clear case of theft, and when the rightful owners are still alive at the time of your receiving your bequest of the land, we might well conclude that you share in the wrongfulness (or sinfulness) of your ancestors insofar as you refuse to restore the land to those who had been unjustly deprived of what is properly theirs. I raise this case not to suggest that someone's life might be analogous to stolen land, but to highlight a case of when a person may be innocent personally in receiving some good, and yet that good is tarnished.

If the above model has even remote plausibility as an interpretation of original sin, I suggest that nonreductive physicalism may have some at least apparent advantage insofar as physicalism, rather than dualism, is often committed to the essentiality of origin. That is, physical objects, such as this desk, seem to be such that it could not have originated by some other means or be constituted by other bits of matter. If we ourselves are the very same as our physical bodies, then our bodies are plausibly regarded as having the origin that they have essentially. Given physicalism, you could not have been conceived of by different parents or even by your current parents but with a different sperm and egg. Dualists (typically) treat the person-body relationship as contingent. As Richard Swinburne articulates his version of person-body dualism in which the person and body are in a contingent relationship, the whole history of the world could be the same with one exception: he could have had a different body, yours perhaps, and vice versa. That is, it is a metaphysical possibility that persons have lives in different embodiments.[8]

Nonreductive physicalism, on the other hand, would lead us to understand our very being or essence as tied to our sinful origin; we could not have come into being through different ancestors. This later claim needs to be qualified slightly: given our (presumed) actual origin through sinful ancestry, our lives are (in some sense) tarnished by sin, but it might still be a contingent matter. One might be well aware that one's life was partly caused and conditioned by sin, and yet one might seriously and rightly wish that matters had been otherwise. One might well wish that one's parents or, indeed, wish that the parents of us all, Adam and Eve and their progeny, had not sinned in procreating generations of humans.[9]

I present the above scenario as an apparent advantage for nonreductive physicalism, though substance dualism can use the originalist model. All that is required is that the dualist maintains that if a key causal role in one's origin is sin, then one has a life that is tarnished by sin. So, even if the dualist accepts creationism (the thesis that each person or soul is a direct creation by God), insofar at the causal account of a person's birth involves sin, the subsequent life would be sinful, albeit not a matter of the person herself being personally guilty or blameworthy.

In this chapter, I have articulated some of the advantages of nonreductive materialism in addressing sin in general and original sin in particular. I have also highlighted some of the merits of integrative dualism. In the end, I have proposed that nonreductive physicalism only appears to have an advantage over dualism, especially integrative dualism.

NOTES

1. See *Consciousness and the Mind of God* (Cambridge, UK: Cambridge University Press, 1994).
2. See Patrick Toner's "Hylemorphic Animalism," *Philosophical Studies* 155, no. 1 (2011): 65–81.
3. Jaegwon Kim, *Physicalism or Something Near Enough* (Princeton, NJ: Princeton University Press, 2005), 10–12.
4. Kim, *Physicalism or Something Near Enough*, 10–12.
5. Augustine, *Anti-Pelagian Writings*, in *Nicene and Post-Nicene Fathers*, series 1, vol. 5 ed. Philip Schaff (Grand Rapids, MI: Christian Classics Ethereal Library, 2009), see "On Marriage and Concupiscence," 773–774.
6. *Catechism of the Catholic Church*, 404. http://www.vatican.va/archive/ENG0015/_INDEX.HTM [accessed on August 30, 2017]
7. For an overview of the major positions on original sin, see Philip Quinn, "Sin and Original Sin," in *A Companion to Philosophy of Religion*, 2nd ed., eds. Charles Taliaferro, Paul Draper and Philip L. Quinn (Oxford: Wiley-Blackwell, 2010).
8. Richard Swinburne, *The Evolution of the Soul*, rev. ed. (Oxford: Oxford University Press, 1997). *Mind, Brain, and Free Will* (Oxford: Oxford University Press, 2013).
9. Whatever model of original sin is adopted, some attention would have to be given to the traditional Roman Catholic teaching that neither Jesus nor Mary were bearers of original sin.

SUGGESTED FURTHER READING

Augustine. *Anti-Pelagian Writings*. In *Nicene and Post-Nicene Fathers*, series 1, vol. 5, edited by by Philip Schaff. New York: Charles Scribner's Sons, 1902.

Kim, Jaegwon. *Physicalism or Something Near Enough*. Princeton, NJ: Princeton University Press, 2005.

Taliaferro, Charles. *Consciousness and the Mind of God*. Cambridge, UK: Cambridge University Press, 1994.

Taliaferro, Charles, Paul Draper, and Philip L. Quinn, eds. *A Companion to Philosophy of Religion*, 2nd ed. Hoboken, NJ: Wiley-Blackwell, 2010.

Toner, Patrick. "Hylemorphic Animalism." *Philosophical Studies* 155, no. 1 (2011): 65–81.

Swinburne, Richard. *The Evolution of the Soul*, rev. ed. Oxford, UK: Oxford University Press, 1997.

———. *Mind, Brain, and Free Will*. Oxford, UK: Oxford University Press, 2013.

Chapter 18

Christian Materialism and Christian Ethics

Moral Debt and an Ethic of Life

Jonathan J. Loose

Is the decision to adopt a contemporary version of Christian physicalism or dualism neutral with respect to important Christian ethical concerns? In this chapter, I argue it is not. The common dismissal (rather than refutation) of dualist anthropology and the attractiveness of various versions of Christian materialism (due to their quasinaturalistic nature and perceived harmony with a contemporary scientific worldview) can draw attention away from the relevance of nonphysicalist accounts to ethical concerns and their superior ability to ground moral claims that flow from Christian theology. In this chapter, I first challenge David Shoemaker's claim that a "soul" criterion of personal identity is irrelevant to practical concerns.[1] Following this, I consider two moral claims arising from Christian theology and Scripture: moral accountability as the settling of a debt or record, and a robust ethic of life as a response to the divine image. In each case, I hope to show that one or more versions of Christian physicalism is inconsistent with (or unable sufficiently to ground) these moral claims and to indicate that substantive, nonphysicalist views of human persons that entail both a simple view of personal identity and the personhood of all human beings provide a contrastingly robust grounding for them. Thus dualist anthropology should not be dismissed, and the claim that physicalism is an attractive alternative that is consistent with science and neutral with respect to Christian ethics should be treated with extreme caution.

PRACTICAL IRRELEVANCE: A PROBLEM FOR DUALISM?

In surveying the area of personal identity and ethics, David Shoemaker argues that the Platonic-Cartesian "soul" criterion of personal identity entails an epistemological problem which, while not fatal for it as a *metaphysical* criterion of identity, renders it irrelevant to our practical concerns. Since souls cannot be directly detected we can never know if the criterion has been met. Shoemaker writes,

> Holding people responsible, compensating them, determining the moral relation between fetuses and the adult humans into which they develop, determining the moral relation between early- and late-stage Alzheimer's patients, and . . . rationally anticipating some future experience(s)—all of these practical concerns and commitments presuppose our ability to identify and track whatever criterion of identity turns out to ground them; they presuppose a tight connection, that is, between the metaphysical and epistemological senses of "criterion of personal identity." Consequently, any theory of personal identity to which we lack this kind of epistemological access is just going to be *practically* irrelevant.[2]

Shoemaker describes this issue as "crippling" since the strong physical and psychological continuities that we typically rely on when making judgments of personal identity have no necessary connection to the presence or absence of a particular soul.[3] While our practical concerns "presuppose that we *can* make correct judgments about when the identity relation obtains,"[4] the soul criterion entails that our reidentification practices are "likely ungrounded and potentially wildly mistaken."[5] The soul criterion should be rejected because it is irrelevant to our practical concerns.

Insofar as this problem is real and crippling, it also applies to physicalist views. For example, Peter van Inwagen is a Christian animalist who has built his account of the metaphysical possibility of resurrection on the coherence of the idea that God could replace a human body instantaneously and undetectably.[6] If this idea is coherent, then physicalists also lack the epistemological access that Shoemaker requires, since they could be deceived by the instantaneous, undetectable replacement of living bodies with physical and psychological duplicates at any time.

However, this problem of epistemological access need not arise given Kwan's "Principle of Critical Trust."[7] Application of this principle suggests that even though strong physical and psychological continuities do not *entail* the continued existence of the same person, they nevertheless offer sufficiently good *evidence* of it that the dualist's reidentification practices can be considered robust. The Principle of Critical Trust includes both the claim that certain beliefs are prima facie justified and that the best explanation of a phenomenon is probably the simplest. Swinburne applies these points to the question of what we might justifiably believe about personal identity given these strong continuities. He writes,

> It is a fundamental epistemological principle that (apparent) memory beliefs are probably true (in the absence of counter-evidence), and my personal memories

(that is memories "from the inside" about what I did and experienced) concern the actions and experiences of the person who had a brain strongly continuous with my present brain.[8]

The reason to hold apparent memory beliefs prima facie justified is our deep and widespread reliance on such beliefs in themselves, without further evidence, for a vast proportion of what we think we know. We rely on memories for everything that we have been taught and more besides—perhaps for everything that is not presently available to our senses. A skeptical approach to our memory beliefs would be epistemically devastating. Of course we may have reasons to doubt the probable truth of certain beliefs. For example, we may take our memory beliefs about an event to be unreliable because we have been the subject of intense suggestive questioning about them. However, in the absence of special reasons for doubt we are justified in holding that memory beliefs are probably true, and this applies to what Swinburne describes as "personal memories" as much as to any others. Therefore I am prima facie justified in believing that I did and experienced those things that I remember doing and experiencing, and which are the actions and experiences of a person who had a brain strongly continuous with my present brain.

The Principle of Critical Trust also includes a commitment to *simplicity* when explaining phenomena, including explaining the phenomenon of strongly continuous memory and character sustained by the same brain. Swinburne writes,

the *simplest, and so most probable*, hypothesis supported by the strong continuity of memory and character sustained by the same brain is that these are mental properties belonging to the same person. It would be less simple, and so less probably true, to suppose that the memory and character strongly continuous with the previous memory and character sustained by a brain having strong continuity with the previous brain are the memory and character of a different person.[9]

Given a critical trust approach, then, the presence of these various strong continuities makes it enormously probable that I continue to exist under normal circumstances. If a person with a brain that was strongly continuous with my present brain did and experienced things that I remember doing and experiencing then—in the absence of evidence to the contrary—I am justified in believing that that person was me. Furthermore, it is *simpler* to hold that these continuities are explained by the ongoing existence of a single individual rather than to hold that there are two persons involved.

So Shoemaker's objection to the "soul" criterion does not justify its dismissal. It is also at odds with a contemporary emphasis on souls as functionally dependent on the bodies with which they function as deeply integrated unities such that body and soul must be considered holistically. While the departure of the soul from the body is a logical possibility, the soul's continued existence is not entailed by its natural properties and so would rely on a miraculous divine gift. Dualisms are considered minimal, integrative, holistic, or emergent and all this provides further reason to believe that physical and psychological continuity offer good evidence of the persistence of a particular ensouled human being.[10]

The "soul criterion" cannot be dismissed on the grounds that it entails an understanding of personal identity that is irrelevant to practical concerns. Nor is this criterion less able to reveal the facts about identity in a given situation as compared with a complex criterion based on empirical (typically physical or psychological) continuity. In either case there are conceivable situations in which we could be deceived or uncertain about the identities of persons. The reasonableness of the claim that empirical continuities provide evidence of identity establishes that the soul criterion is not at a disadvantage in this respect.[11]

MORAL DEBT: A PROBLEM FOR MATERIALISM?

We can turn, then, to the practical question of moral responsibility, which is an important consideration for theological anthropology. As Birch notes, Scripture assumes that humans created by God are moral agents, capable of responsibility, and able not only to be shaped by relationships with God and others, but also to make moral decisions that affect those relationships. As such, humans are *accountable* both for their identities (in the psychological sense) and for their actions as individuals and in the community.[12] The New Testament includes a notion of *moral debt* that can be removed only by being repaid or else through the answerability of the debtor to an auditor or judge.[13] For example, Paul writes of a *record* that is created and erased:

> God made you alive together with him, when he forgave us all our trespasses, erasing the record that stood against us with its legal demands. He set this aside, nailing it to the cross. (Col. 2:13b–14)

The Fruitless Search for Complex Criteria

How should we understand human beings as morally responsible creatures; creatures accountable for their actions?

David Shoemaker addresses the question of accountability and its relationship to personal identity.[14] He holds that it is a necessary condition of my being accountable for an action that that action be *attributable* to me and notes that, typically, attributability is understood in terms of personal identity so that, "whatever makes a past person's actions mine is whatever made him me."[15] So one approach to the attribution of moral responsibility is to recognize with Locke that personal identity is a *forensic* term. If the Christian physicalist can offer a plausible criterion of personal identity, then she has offered a plausible criterion of attributability and hence a way to ground moral responsibility. However, while neglecting the soul criterion, Shoemaker evaluates each of the candidate criteria for attributability and concludes that "all proposed criteria of attributability are implausible."[16]

Shoemaker works through popular complex accounts of identity, offering counterexamples in which its predictions about attributability are out of step with intuition.[17] The *Lockean* consciousness (memory) criterion is not necessary for attributability since we might take an action to be properly attributable to a person who cannot remember doing it, and it is not sufficient given the possibility of implanted memories that would cause an individual to recall performing acts that we have no reason to believe belong to her. The *biological* criterion is not necessary since we assume that the owner of an action would leave the organism with the cerebrum if it were transplanted; neither is it sufficient, since we intuit that a radical and instant change in psychology would produce a different person despite biological continuity. The *psychological* criterion is more attractive given that it involves connections directly related to attributability such as the fulfillment of intentions in action. Nevertheless, a case of violent physical damage that left only the weakest of psychological links intact would seem to entail persistence despite the criterion's prediction that it does not, and a *complete* psychological transformation occurring sufficiently gradually would lead us to believe that there were different persons at either end of the process despite the criterion's contrary verdict. Thus the psychological criterion, like the others, is neither necessary nor sufficient for attributability. The *narrative* criterion[18] takes identity to be a matter of *characterization* rather than *reidentification* and so is immune from the foregoing objections. In assuming, however, that lives are characterized by self-told narratives, it is excessively subjective. For reasons of narrative coherence an individual may have acts attributed to her that we intuit she does not own, while acts with which she has become identified passively in response to experience may be excluded or not even recognized.

This picture of the fruitless search for an adequate complex criterion of attributability is instructive. In offering it Shoemaker both summarizes some of the objections that have plagued the wider quest for such criteria and expresses a strong sense of the problematic nature of the discussion as a whole.

Attributability: From Identity to Similarity

However, if these issues were not enough, Shoemaker goes on to raise a further problem that he takes to be serious and which applies to all of the criteria that seek to establish numerical identity. He cites Wiggins's famous example of *physical fission*.[19] In *physical fission* a person (call him Alfred) splits down the middle, with each fission product subsequently regrowing the missing half of its body. Since identity is a one-to-one relation, Alfred cannot be numerically identical with the two products of fission, and so he has fissioned out of existence and two new individuals—call them Boris and Charles—each physically and psychologically indistinguishable from Alfred have begun to exist instead. If the attributability of actions has anything to do with numerical identity, then any actions performed by Alfred cannot be attributed to Boris or Charles, since they are numerically distinct people. However, the intuition that is expected to arise in the face of this example is that we believe that this *cannot* be right. Intuitively, we are supposed to believe that Alfred's actions should be attributable to Boris and Charles. They are indistinguishable from Alfred in every way; they even recall performing actions attributable to Alfred and expect to receive any reward or punishment due for the things he did.

Shoemaker's intuition accords with this expectation, and he holds that the actions of the prefission individual are attributable to the two fission products despite the fact that neither of them existed when those actions were planned and performed.[20] He also justifies his turn from the claim that a criterion of identity is at the core of attributability by noting that numerical identity and attributability are different kinds of relation. While identity is a transitive, one-to-one relation obtaining between a single individual at different times, attributability is an intransitive, one-to-many relation obtaining between one act and one or more agents. Just as there are objects, such as houses or companies, that may be owned by more than one person, so there are acts such as singing a duet or winning a tug-of-war that may be, indeed must be, attributed to multiple agents. So identity and attributability cannot be the same thing.

Having removed numerical identity from the core of attributability, Shoemaker seeks to sketch a way to fill the resulting gap and offer some conditions governing the attributability of actions. If an action is to be attributable to an agent then, when the action is performed, it must depend in some way on the will of that agent, since the will is the central component of agency. In addition, the act of will establishes the attributability of the action to a particular agent because it is an act in which that agent's true self is implicated; it is an act arising from the particular volitional "web of beliefs, judgments, attitudes, desires and cares" that constitute the agent's "real self."[21] If the agent's core self is implicated in the act of will, then that act can reasonably be taken to be *his*.

The question remains of how to establish that an act is attributable to an agent later on. Having removed numerical identity entirely from his scheme, Shoemaker must establish a diachronic connection of a different sort. He suggests that the relation between the agent at a later time and the agent who performed the act is one of *similarity* between relevant elements of the real selves of each of the two agents, and this relation exists in virtue of the causal dependence of the later real self on the former. The similarity relation, like the attributability relation, is neither one-to-one nor transitive and given this connection Shoemaker believes he has sketched a plausible account of attributability, and hence accountability without requiring personal identity.

Similarity and the Fading of Accountability

This account of attributability presents a dilemma for various versions of Christian physicalism. On the one hand, a physicalist view typically requires a satisfactory complex criterion of identity, but Shoemaker's instructive critique suggests the long search for such a criterion, involving numerous proposals and counterexamples, has proved fruitless. This is further evidenced by the subsequent willingness to turn away from the overwhelmingly intuitive idea of numerical identity in order to avoid the need for a complex criterion.[22] This seems indicative of a crisis, and this is the first horn of the materialist's dilemma: a satisfactory complex criterion of identity is required, yet there is little reason to hope that such a criterion will be forthcoming.[23]

On the other hand, the Christian physicalist is not at liberty to reject numerical identity. First, it seems highly implausible to do so. In turning from numerical identity, Shoemaker ensures that the constraints on the attributability of actions are much reduced. On his view, it is possible that an action be attributed to individuals who *did not even exist* at the time it was performed (individuals such as Boris and Charles). For those of us who cannot bring ourselves to hold that an action should be attributed to someone simply in virtue of his being highly similar to the person who performed it, it may seem that it just *has* to be better to defend the importance of numerical identity, however difficult that may be, rather than to be forced to such a conclusion; yet the crisis for complex criteria is such that Shoemaker takes the rejection of numerical identity to be the better path.

Second, the *Christian* physicalist has an important theological reason not to turn against numerical identity in explaining the attributability of actions. To see this, consider again Shoemaker's account. While his understanding of attribution at the time of the performance of the action seems unobjectionable (i.e., attribution to the extent that the agent's act of will is involved in the action and the extent to which that act of will expresses the agent's real self), the attribution of the act at a later time is problematic. Given the rejection of numerical identity, attributability of an action to an agent at some later time is explained in terms of a *similarity* relation. (Thus Alfred's actions are attributable to Boris and Charles, since Boris and Charles are maximally similar to Alfred without being numerically identical to him.) However, if the attributability of an act to an agent depends on that agent's similarity to a previous agent then, inevitably (and as Shoemaker recognizes), accountability will wane over time as similarity to the individual identified with the action reduces.[24] However, the waning of moral responsibility due to the passage of time is inconsistent with the New Testament notion of a record of moral debt and credit that accrues and then remains until it is repaid or dealt with by an auditor or judge. The mere passage of time has no power to reduce such a debt or to change the record. This notion of a moral record also has a depth of ingression in Christian theology that would suggest that its absence would be profoundly destabilizing in a range of areas.[25]

Thus another way out of this dilemma is required. A solution is available, but it comes at a price. First, note that the fact that attributability is a one-to-many relation does not entail that it must have *nothing* to do with the one-to-one relation of identity. Consider an individual *X* who has a share in the ownership of an act at the time it was performed (t_1) because she is identified with it in the way that Shoemaker describes. We are at liberty to say that *X*'s identification with the act can be attributed to a later person *Y* at t_2 if and only if that person is *identical* to *X*. The act is attributable to *Y* because *Y* is identical to *X* who is the person to whom a share in the act was attributed at t_1. While the attribution of the act to *X* at t_1 involves the one-to-many relation of attributability, the identification of *X* at t_1 with *Y* at t_2 involves the one-to-one relation of identity. Attribution to *Y* is a two-step process of determining that the act is attributable to *X* (one-to-many) and that *X* is identical with *Y* (one-to-one); we need not choose between these relations since they can coexist.

Given that the nature of attributability does not necessitate the rejection of numerical identity, we must ask how numerical identity is to be retained given the pressure exerted by the crisis for complex criteria. There is an alternative to the search for a complex criterion of numerical identity that is not considered by Shoemaker, perhaps because it implicates the "soul" criterion previously dismissed as irrelevant to practical concerns. Rather than denying numerical identity, we simply deny that the correct criterion is a complex one. A simple view of personal identity may be taken in which identity is held to consist not in the empirical continuities that we make use of when recognizing it, but in a "further fact."[26] That further fact may well consist in the persistence of a simple soul that, as we have seen, we have good reason to hold is evidenced by physical and psychological continuities associated with a particular brain and body and is thus relevant to practical concerns. While there is much to do to explicate this alternative, it avoids the need for a complex criterion of identity and maintains the importance of numerical identity, thus promising to ground a notion of moral responsibility that does not fade over time, which is thereby consistent with the notion of a persistent moral record and thus a New Testament notion of moral debt. Although there are ostensibly materialist views that seek to recruit this approach,[27] the simple view is most closely and plausibly associated with immaterialism, and this raises the prospect that in order to hold on to a theologically robust account of moral responsibility, the Christian should leave behind a commitment to materialism.

AN ETHIC OF LIFE: A PROBLEM FOR CONSTITUTIONALISM?

It flows from a theological commitment to creation in the *imago Dei* that all human beings possess an intrinsic value and there is significant scriptural support for the claim that vulnerable humans including the unborn, those with disabilities, and the elderly possess this value equally. Summarizing the biblical material, Paul Copan writes: from "the womb to the end of life, human beings possess dignity and worth."[28] The notion of dignity is tied very deeply to the reflection of the divine image and the special (exceptional) nature of human persons.[29]

While this affirmation of dignity and worth is not accompanied in Scripture by explicit prohibitions against abortion, it is notable that such opposition is found in both early Judaism (Philo, Josephus) and early Christianity. As Gorman notes: "opposition to abortion, exposure, and infanticide became an ethical boundary marker for both groups in their pagan cultures."[30] For example, the command to love one's neighbor leads to the claim in both the *Didache* and the *Gospel of Barnabas* that, "Thou shalt not murder a child by abortion" (Did. 2.2; Barn. 19:5), a point reiterated by later Church Fathers such as Athenagoras who took the unborn to be "the object of God's care."[31]

Thus, it seems Christians should affirm an obligation to protect vulnerable human lives including the lives of the unborn. This is an *ethic of life* that may well be expressed in an opposition to abortion. For example, Corcoran writes: "I don't want the following fact to go unnoticed: . . . *abortion, from the moment of conception, is* prima facie *morally wrong.*"[32] How is such an ethic to be sustained?

AN ETHIC OF LIFE AND MATERIALISM

Metaphysical theories of the human person are often assumed to have significant moral implications, especially in relation to an ethic of life. However, Corcoran argues that metaphysical theories—whether dualist or physicalist—do not offer any advantage in establishing it. He writes,

> I acknowledge that materialist views of human persons like CV [the Constitution View] do fail to provide *metaphysical* resources necessary or sufficient for generating moral obligations or moral expectations to protect and care for the life of a human fetus or a PVS patient. However, *any* metaphysical view of persons, be they metaphysic dualist or materialist in nature, is impotent to provide such resources. Other resources, metaphysically neutral with respect to dualism and materialism, must be added to such views to generate moral obligations or moral expectations to care for and protect human life.[33]

These additional resources are ethical principles. According to Corcoran, a moral obligation such as an ethic of life will be generated by a combination of a metaphysical theory *and* an ethical principle. Whatever metaphysical theory of humans is chosen, the question is whether there is a justifiable ethical principle that can be combined with it to generate an ethic of life. For example, Corcoran's own preferred Constitution View entails that human beings lacking certain psychological capacities, including early stage fetuses and PVS patients, do not constitute persons. The question thus becomes whether a moral principle can be provided that justifies the protection of a human nonperson. Corcoran offers just such a principle, suggesting that the result is an obligation that would support a rational protest against the abortion of an early term fetus:[34]

> (1) it is *prima facie* morally wrong to destroy a person *in potentia*, and a normal human fetus is just such a being, or (2) even if the fetus is defective and does not qualify as a potential person, it is still a member of the *human* community, and to terminate the existence of a member of the community would diminish the kind of bond essential to the preservation and health of the community.[35]

Whether or not these principles turn out to be robust, Corcoran's point is correct in principle. A metaphysical view of human persons need not be the determining factor in one's position on one side or the other of an ethical dispute, if that view can be combined with an appropriate moral principle or principles to establish the desired ethical conclusion. Ganssle labels the view that ethical conclusions are determined not only by metaphysical principles but also by empirical facts and metaphysical theories the *combination thesis*.[36] The combination thesis holds just so long as there are suitable ethical principles available to supplement a metaphysical view, and so long as we can reject any unsuitable ethical principles that entail a conclusion other than the one we desire.

What, then, is the contribution of a metaphysical theory to a moral conclusion? At least part of the answer is that a metaphysical theory establishes the *scope* of application of a moral principle; it establishes to what entities that principle will apply. It therefore has an important bearing on whether or not a particular ethical principle will generate an obligation in a particular situation. David Shoemaker writes that there are

> important ways for metaphysical facts to be relevant to moral *arguments*. Here is how: in making the case for some moral conclusion, among one's premises will have to be both normative *and* descriptive assertions, and the truth of these descriptive assertions may well depend on metaphysical considerations.[37]

Thus, if an ethical principle is accepted that establishes particular obligations to persons, and if a metaphysical theory of persons is accepted that establishes that there is a person present in a given situation, then that ethical principle applies in that situation. For example, consider Singer's metaphysical account that holds that it is the members of the class of *sentient* beings that are worthy of moral consideration because they can be bearers of morally significant properties, namely being able to experience pain and pleasure. In this way, Singer ensures that the scope of his utilitarian moral theory will extend to both human and nonhuman animals.[38] Metaphysical and ethical theories together generate moral conclusions that are obligations to maximize pleasure and minimize pain within the group as a whole, even if this entails the destruction or significant harm of a minority of individuals. Thus metaphysics plays its role in establishing to what the ethical theory will apply. In order to maintain human moral exceptionalism, the class of beings worthy of moral consideration would have to be understood in a different way (e.g., consider Baker's efforts in this regard, discussed later).[39]

So Corcoran is correct to assert that metaphysics alone is typically insufficient to determine a moral conclusion, but this does not mean that metaphysics has *no role* to play in determining moral conclusions, since—at the very least—a metaphysical theory will establish whether a given situation contains those things to which a particular ethical principle applies.

As the foregoing indicates, metaphysical theories entail or relate closely to judgments of value. Ganssle makes the important point that moral obligations are not all or nothing affairs, but are *degreed*, being more or less weighty. For example, I may accept that as a driver I have an obligation to avoid causing harm to a deer that wanders onto the forest road, and so all things being equal I will steer my car to avoid it. However, if doing so would put the life of my passenger at risk, then the obligation to avoid causing harm to a human person outweighs my obligation to the deer. This distinction in weight is grounded in the relative value of the different things which I am obligated to protect. The question, then, is not simply whether Christian physicalism can ground an ethic of life that applies all things being equal, but whether it can ground an ethic of life that establishes obligations to human nonpersons that are sufficiently weighty in the face of conflicting obligations to other things.

Here we find the problem for the Christian physicalist who holds that abortion of an early stage fetus is prima facie morally wrong. It seems that the outcome of conflicting obligations to different beings will depend—probably to a great degree—on which has the higher metaphysical status. (My obligation to the passenger in my car overrides that to the deer in the road.) Where obligations to protect the life of a vulnerable human *nonperson* conflict with obligations to protect the life of a vulnerable human *person*, we find that the latter will typically be taken to carry greater weight than the former. As Ganssle says,

> We ought to note that the presumption in the abortion debate has often been that the obligation to protect the human fetus will give way to other obligations unless the fetus has the highest metaphysical status that is possible for it and, therefore, the right to the highest degree of protection . . . the protection of the fetus will go only so far if it is less than a person.[40]

Of course it might be objected that appeals to the metaphysical status of personhood in the abortion debate are simply mistaken. Perhaps the same intrinsic worth and dignity is properly ascribed to *humans* irrespective of whether or not they are *persons* and so an obligation to a human person will not override an obligation to a human nonperson. This objection seems unavailable to the theistic moral realist (such as the Christian physicalist) who holds that it is in God's nature to be perfectly good and thus God is the most valuable being. In this case, there is an unavoidable hierarchy of value dependent upon the extent to which a being reflects intrinsically the likeness of a nonhuman personal God. Given a view in which there can be human nonpersons, a being that possesses the property of personhood will be further up that hierarchy than a being that lacks it. It does not seem reasonable to object that personhood is irrelevant to God's perfectly good nature in virtue of which human beings are possessors of the divine image and inherent value, in part because some of the common desiderata for any account of the image of God would seem to presuppose it.[41]

So it seems that Corcoran is right to object to the claim that, given a materialist metaphysics of human persons, "we have *no* moral obligations or responsibilities with respect to nonperson-constituting human organisms."[42] If there are such organisms then it is certainly possible to claim that we have obligations toward them, since it is clear that there are viable ethical principles that will generate such obligations. However, this is beside the point in the face of an attempt to provide an ethic of life entailing an obligation to vulnerable human nonpersons (fetuses or PVS patients) that will be weighty enough sometimes to avoid being overridden by conflicting obligations to human persons. While value can be ascribed to all human beings irrespective of personhood, those who are persons will have a higher metaphysical status in virtue of the degree to which they reflect the divine image and in that way obligations toward them will be strengthened. It may seem that this comparatively greater value applies only *ceteris paribus*, but just as it is difficult to think of circumstances in which I would risk the life of the human passenger in my car to save the life of a deer in the road, so it may not be possible to think of a situation in which obligations to a human nonperson would override those to a human person.

Seemingly the only sure way to provide a robust ethic of life is to ensure that human beings are necessarily human persons. Here, a substance view of the *imago Dei* underpinned by a traditional metaphysical picture of human persons as simple souls has a great deal to offer. For example, Farris offers a view flowing from the Western/Augustinian tradition in which the image is taken to refer to a natural, individual substance that is a "stable thing that truly images God."[43] On this view, we can say that an individual human being *is* an image and persons, as soul-substances, are able to actualize their powers teleologically within compound dynamic structures. On this view of persons as substances rather than properties there are no human nonpersons, and thus there is no image-relevant categorical distinction between humans at different developmental stages that would ground a distinct metaphysical status. In this way, substance dualism offers a stronger ground for an ethic of life than does Christian physicalism.

CAN A CONSTITUTIONALIST HOLD AN ETHIC OF LIFE?

Once it is granted that, *ceteris paribus*, obligations to persons will trump obligations to nonpersons in cases of conflict, the problems for an ethic of life on the Constitution View (CV) appear deeper still. First, it may be that on CV one is unable to hold that a prepsychological fetus is *not* a person after all. This may seem encouraging, but in accepting that consequence CV itself is undermined because it becomes clear that a first-person perspective—that property which is held to be essential to persons—is in fact neither necessary nor sufficient for personhood.

Baker recently has provided a detailed account of the property of a *first-person perspective*. She takes it to be a complex dispositional property consisting of two stages: first is the *rudimentary* stage, which amounts to the capacity for conscious and intentional interaction with the environment. The second *robust* stage involves the capacity for self-consciousness, that is, the ability to conceive of oneself simply as oneself. For example, the second use of *I* in the sentence, "I wonder if I will miss my train this evening" requires the ability to think of myself as myself, simply as an individual at a future time seeking to catch a train. One important distinction between these two stages of the first-person perspective is that while the rudimentary stage is possessed by both human infants and higher-order nonhuman animals, the robust stage is possessed uniquely in the animal kingdom by humans postinfancy. Thus, in order to maintain human exceptionalism, Baker considers that a person must be of a *kind* for which exemplification of the rudimentary stage is normally developmentally preliminary to exemplification of the robust stage. (The qualifier "normally" is included to address the further problem of ensuring that mentally impaired humans are ruled *in* to the category of persons.) The result is a conception of the property of a first-person perspective that maintains human exceptionalism by ruling out higher animals and ruling in infants and mentally impaired humans.

However, this is not the end of the story. As Ross Inman has observed, this development of the idea of personhood seems to have a further problematic consequence.[44] Human personhood is found in the possession of a capacity for a first-person perspective in its *robust* stage, but in order to recognize human infants as persons we hold that a being is also a person if it is of a kind that will normally *progress* to this robust stage even though it is presently in the *rudimentary* stage which is also possessed by higher nonhuman animals. In Baker's terms, while the adult has an *in-hand* (that is first-order) capacity for a robust first-person perspective, the infant has a *remote* (that is second-order) capacity for a robust first-person perspective. But this has the further problematic consequence that if human infants are persons in virtue of a *remote* capacity, then implanted embryos also have a remote capacity for a rudimentary and then robust first-person perspective, and so it is hard to resist the claim that they too are persons.

Inman's observation highlights an instance of a more general problem that arises for any view that takes personhood to reside in the potential to become something later on: just how far back do we go in assigning personhood to the progenitors of the entity bearing the property we consider essential? In short, if personhood today is reliant on being the kind of thing that will normally possess a robust first-person perspective tomorrow, then why extend personhood only to infancy? This development might seem welcome to a constitution theorist like Corcoran who holds that abortion from the moment of conception is prima facie morally wrong. However, Baker resists this further extension of personhood, and she must do so because if it were accepted then it would be clear that exemplification of a first-person perspective is neither a sufficient condition of personhood (since there are nonhuman animal nonpersons that possess rudimentary first-person perspectives) nor a necessary condition of personhood (there are human persons without first-person perspectives) and so the premise of the constitution view would be undermined:

first-person perspectives (at whatever stage) would no longer be essential for personhood.

Christian materialists do not seem able to avoid holding that in an ethical conflict involving obligations to a person and a nonperson, the person is at an advantage that is likely to be decisive. For this reason, the Christian physicalist's ethic of life is less weighty than that of the dualist and may be too flimsy to survive any serious conflict with an obligation to a human person. On Baker's Constitution View, the need to mark out human persons as distinctive and exceptional as compared with nonhuman animals possessing similar capacities leads to an understanding of personhood in terms of a disposition rather than a first-order capacity. However, this suggests that all human beings should be considered to be persons, including those lacking a first-person perspective. While this may be a welcome development for those Christians seeking a robust ethic of life, it comes at the expense of undermining the essential nature of the first-person perspective and thus undermining the constitution view itself.

CONCLUSION

Dualism is not irrelevant to practical concerns, and once this is accepted it becomes clear that it can ground important concerns in Christian ethics in a more robust way than materialist theories are able to do. If human persons are wholly material, then an account of moral debt is likely to be caught between the seemingly fruitless quest for an acceptable complex criterion of identity (and thus attributability of actions) and the theologically objectionable notion of grounding accountability in a similarity criterion entailing that moral debts fade with the passage of time. The simple view offers a way out of this dilemma, but probably at the cost of rejecting materialism. When it comes to the scripturally and historically important Christian commitment to an ethic of life, materialist views entailing the existence of human *nonpersons* as well as human *persons* seem unable to avoid a distinction in metaphysical status between them that places obligations to the human nonperson at a decisive disadvantage when in conflict with obligations to persons. A substance view of the *imago Dei* underpinned by a dualist ontology of persons entails that there are no human nonpersons and so this categorical distinction in value does not arise.

I hope to have shown that there are good reasons for Christians to affirm traditional and widespread nonmaterialist views of human persons, and among these are concerns to ensure a robust grounding for Christian ethics. We cannot claim that a turn to Christian materialism is neutral with respect to important Christian ethical concerns.[45]

NOTES

1. David Shoemaker, *Personal Identity and Ethics: A Brief Introduction* (Ontario: Broadview Press, 2009).

2. Shoemaker, *Personal Identity and Ethics*, 32–33.

3. By "strong continuity" of brain, memory, or character, I refer to an overlapping chain in which there is a high level of similarity between successive links. Continuity of memory and character are assumed to be causally sustained by strong continuity of the brain. See Richard Swinburne, "How to Determine Which is the True Theory of Personal Identity," in *Personal Identity: Complex or Simple?*, ed. Georg Gasser and Matthias Stefan (Cambridge, UK: Cambridge University Press, 2012), 107.

4. Shoemaker, *Personal Identity and Ethics*, 32.

5. Shoemaker, *Personal Identity and Ethics*, 33.

6. Peter van Inwagen, "The Possibility of Resurrection," *International Journal for the Philosophy of Religion* 9, no. 2 (1978): 114–21.

7. Kai-Man Kwan, *The Rainbow of Experiences, Critical Trust, and God: A Defense of Holistic Empiricism* (New York: Continuum, 2011).

8. Swinburne, "How to Determine Which is the True Theory of Personal Identity," 107.

9. Swinburne, 107, italics added.

10. See C. Stephen Evans, "Separable Souls: Dualism, Selfhood, and the Possibility of Life after Death," *Christian Scholar's Review* 34, no. 3 (2005): 327–40; Charles Taliaferro, *Consciousness and the Mind of God* (New York: Cambridge University Press, 1994); John W Cooper, *Body, Soul and the Life Everlasting: Biblical Anthropology and the Monism-Dualism Debate* (Grand Rapids, MI: Eerdmans, 2000); William Hasker, *The Emergent Self* (Ithaca, NY: Cornell University Press, 1999).

11. Note that in separating the metaphysical criterion of identity from the evidence of its presence, the soul criterion has the advantage that epistemic uncertainty is guaranteed not to be a reflection of metaphysical vagueness. There are determinate facts about identity whether or not we know them.

12. Bruce E. Birch, "Scripture in Ethics: Methodological Issues," in *Dictionary of Scripture and Ethics*, ed. Joel B. Green (Grand Rapids, MI: Baker Academic, 2011), 30.

13. Bonnie Howe, "Accountability," in *Dictionary of Scripture and Ethics*, 39.

14. David Shoemaker, "Responsibility without Identity," *The Harvard Review of Philosophy* XVIII (September 2012): 109–32.

15. Shoemaker, "Responsibility without Identity," 111.

16. Shoemaker, "Responsibility without Identity," 110.

17. Parfit coined the distinction between complex and simple views of personal identity, writing that "the fact of personal identity over time just consists in the holding of certain other facts. It consists in various kinds of psychological continuity, of memory, character, intention, and the like, which in turn rest upon bodily continuity" (Derek Parfit, "Personal Identity and Rationality," *Synthese* 53, no. 2 [1982]: 227).

18. Marya Schechtman, *The Constitution of Selves* (Ithaca, NY: Cornell University Press, 2007).

19. David Wiggins, *Identity and Spatio-Temporal Continuity* (Oxford: Blackwell, 1967), 50.

20. Should one respond with a four-dimensionalist metaphysic and thus hold that numerical identity can be preserved through fission because there were two people there all along, Shoemaker raises a similar "branch-line" case, derived from Parfit, that is intended to produce the same intuitive response as the physical fission case while involving the destruction of an individual rather than fission. The branch-line case is then considered immune to the four-dimensionalist response (Shoemaker, "Responsibility without Identity," 119–20).

21. Wiggins, *Identity and Spatio-Temporal Continuity*, 124.

22. Of course, Shoemaker follows Parfit in this, and his discussion of physical fission prompting his rejection of numerical identity (Derek Parfit, "Personal Identity," *The Philosophical Review* 80, no. 1 [1971]; Derek Parfit, *Reasons and Persons* [Oxford: Oxford University Press, 1984]). While Parfit holds that "identity is not what matters in survival," so Shoemaker holds that identity is not what matters in attributability.

23. The futility of the search for complex criteria of identity and the seeming ease with which proposals are counterexampled is most likely due to the inability to render personal identity *determinate* (Joseph Butler, "Of Personal Identity," in *The Works of Joseph Butler* [Oxford: Clarendon Press, 1897], 317–25; Parfit, "Personal Identity and Rationality"; Swinburne, "How to Determine Which is the True Theory of Personal Identity").

24. See and compare Lewis's "Survival and Identity," in *The Identities of Persons* (Berkeley: University of California Press, 1976). Methuselah case for an instructive example of the way in which personal identity might wane over time if identity is grounded in a similarity relation.

25. Shoemaker's rejection of numerical identity leaves us with a theologically unacceptable notion of moral attributability and hence responsibility. In the same way, Parfit's rejection of numerical identity (after which Shoemakers's account is shaped) leaves us with an unacceptable notion of survival; what matters at the resurrection is that the future individual is not just indistinguishable from me, but that he is numerically identical to me.

26. Parfit, "Personal identity and rationality"; Swinburne, "How to determine which is the true theory of personal identity."

27. See, for example, Lynne Rudder Baker, "Personal Identity: A Not-so-Simple Simple View," in *Personal Identity: Complex of Simple?* 179–91.

28. Paul Copan, "A Protestant Perspective on Human Dignity," in *Human Dignity in Bioethics: From Worldviews to the Public Square*, eds. Stephen Dilley and Nathan J. Palpant (New York: Routledge, 2013), 74. Copan notes the valuation of persons in the womb and as gifts from the Lord (Ps. 139:14; Isa. 49:5; Jer. 1:5; Luke 1:15, 44; cf. 1:42; 2:21; Ps. 127:3), obligations to the elderly and frail (Ps. 71:9, 18; Prov. 16:31; 20:29; 23:22)., concluding that, "With God, there are no 'potential human beings' (as opposed to 'actual human beings'—no more so than a woman being 'somewhat pregnant')" (Copan, "A Protestant Perspective," 74).

29. David H. Calhoun, "Human Exceptionalism and the *Imago Dei*: The Tradition of Human Dignity and the Imago Dei," in *Human Dignity in Bioethics*, 19–45.

Calhoun emphasizes the importance of understanding human dignity as a tradition exploring human value and existence within the constraints of human exceptionalism and the *imago Dei*. Human exceptionalism is the view that "humans are different from and superior to all other living things, primarily due to the power of reasoning" (20), while the *imago Dei* is the view that "humans reflect the divine image in a way unique in the created order" (20). The tradition emerged from classical thought, was synthesized in early Christianity, developed through the Middle Ages and Renaissance, was modified and then attacked in the modern period and then restored in the twentieth century.

30. Michael J Gorman, "Abortion," in *Dictionary of Scripture and Ethics*, 36.

31. See Michael J. Gorman, *Abortion and the Early Church: Christian, Jewish & Pagan Attitudes in the Greco-Roman World* (Eugene, OR: Wipf & Stock, 1998). As a consequence of the obligation to love one's neighbor, Gorman notes that Rabbinic literature permits abortion to save the woman's life.

32. Kevin J. Corcoran, "A Critical Appraisal of Francis Beckwith's Defending Life," *Philosophia Christi* 12, no. 2 (2010): 452.

33. Kevin J Corcoran, *Rethinking Human Nature* (Grand Rapids, MI: Baker, 2006), 84–85.

34. Elsewhere, Corcoran writes: "I don't want the following fact to go unnoticed: I am in wholehearted agreement with Beckwith—*abortion, from the moment of conception, is* prima facie *morally wrong*" (Corcoran, "A Critical Appraisal of Francis Beckwith's Defending Life," 452).

35. Corcoran, *Rethinking Human Nature*, 88.

36. Gregory E. Ganssle, "Metaphysics, Ethics and Personhood: A Response to Kevin Corcoran," *Faith and Philosophy* 22, no. 23 (2005): 373.

37. Shoemaker, *Personal Identity and Ethics*, 121.

38. Peter Singer, "All Animals Are Equal," *Philosophical Exchange* 1, no. 5 (Summer 1974): 243–57.

39. Additionally, a different ethical theory might avoid obligations to cause significant harm to individuals (e.g., consider Regan's *The Case for Animal Rights* [Oakland: University of California Press, 1983], "rights" view).

40. Ganssle, "Metaphysics, Ethics and Personhood," 374.

41. See Marc Cortez, *Theological Anthropology: A Guide for the Perplexed* (London: T & T Clark, 2010), 16–17. Cortez notes wide agreement that an adequate account of the image (or, likeness) of God will include reflecting God in creation, possession by all humans, the possibility of being affected by sin, and a christological and teleological nature.

42. Corcoran, "A Critical Appraisal of Francis Beckwith's Defending Life," 86.

43. Joshua R. Farris, "A Sustantive (Soul) Model of the Imago Dei: A Rich Property View," in *The Ashgate Companion to Theological Anthropology* (Farnham, UK: Ashgate, 2015), 168; see also J. P. Moreland, *The Recalcitrant Imago Dei: Human Persons and the Failure of Naturalism* (London: SCM Press, 2009).

44. Ross Inman, "Against Constitutionalism," in *The Blackwell Companion to Substance Dualism*, eds. Jonathan J. Loose, Angus J. L. Menuge, and J. P. Moreland (Oxford: Wiley-Blackwell, *forthcoming*, 2018).

45. I am grateful to Angus Menuge for comments on an earlier draft of this chapter.

SUGGESTED FURTHER READING

Birch, Bruce E. "Scripture in Ethics: Methodological Issues." In *Dictionary of Scripture and Ethics*, edited by Joel B. Green. Grand Rapids, MI: Baker Academic, 2011.

Calhoun, David H. "Human Exceptionalism and the *Imago Dei*: The Tradition of Human Dignity and the Imago Dei." In *Human Dignity in Bioethics: From Worldviews to the Public Square*, edited by Stephen Dilley and Nathan J. Palpant. New York: Routledge, 2013.

Copan, Paul. "A Protestant Perspective on Human Dignity." In *Human Dignity in Bioethics: From Worldviews to the Public Square*, edited by Stephen Dilley and Nathan J. Palpant. New York: Routledge, 2013.

Corcoran, Kevin J. *Rethinking Human Nature*. Grand Rapids, MI: Baker Publishing Group, 2006.

Cortez, Marc. *Theological Anthropology: A Guide for the Perplexed*. London: T & T Clark, 2010.

Ganssle, Gregory E. "Metaphysics, Ethics and Personhood: A Response to Kevin Corcoran." *Faith and Philosophy* 22, no. 23 (2005): 370–376.

Gorman, Michael J. *Abortion and the Early Church: Christian, Jewish & Pagan Attitudes in the Greco-Roman World*. Eugene, OR: Wipf & Stock, 1998.

Inman, Ross. "Against Constitutionalism." In *The Blackwell Companion to Substance Dualism*, edited by Jonathan J. Loose, Angus J. L. Menuge, and J. P. Moreland. Hoboken, NJ: Wiley-Blackwell, forthcoming, 2018.

Moreland, J.P. *The Recalcitrant Imago Dei: Human Persons and the Failure of Naturalism*. London: SCM Press, 2009.

Shoemaker, David. *Personal Identity and Ethics: A Brief Introduction*. Ontario: Broadview Press, 2009.

———. "Responsibility without Identity." *The Harvard Review of Philosophy* XVIII (September 2012): 109–32.

Swinburne, Richard. "How to Determine Which is the True Theory of Personal Identity." In *Personal Identity: Complex of Simple?*, edited by Georg Gasser and Matthias Stefan. Cambridge, UK: Cambridge University Press, 2012.

Chapter 19

The Incompatibility of Physicalism with Physics

Bruce L. Gordon

The physicalist thesis, in broadest terms, is an ontological claim: whatever exists or occurs is *ultimately* constituted out of physical entities, where physical entities are anything wholly composed of material simples, namely, the most fundamental constituents of material reality as determined by elementary particle physics.[1] This ontological physicalism (materialism) comes in reductionist, eliminativist, and nonreductionist varieties. Reductive physicalists contend that psychological and biological properties are ultimately explainable in the language of particle physics; eliminativists claim that psychological and biological properties are illusory and the only real properties are physical; and nonreductive physicalists maintain that, while there isn't anything that isn't physical, efforts to explain psychological and biological properties in *purely* physical terms are misguided and will inevitably miss laws, generalizations, and explanations that can only be formulated using psychological or biological concepts. Regardless of whether the physicalist understands his explanatory project in terms of ultimate reducibility to the physical, complete elimination of the nonphysical, or irreducible explanatory levels ultimately dependent only on what is physical, it is clear that the tenability of the physicalist thesis has, as a necessary condition, the existence and causal closure of the material realm under the aegis of modern physical theory. At a minimum, this entails that macroscopic material objecthood and behavior either be reducible to, supervenient upon, or emergent from the existence and behavior of material reality at the microscopic level, and for microphysics itself to be consistent with the existence and causality requirements of physicalism. In short, it requires a fundamental physical theory—that is, quantum physics—to parse matter in terms of material simples and to cast all events in the purely material realm as dependent upon them and emergent from them by efficient material causation.

And therein lies the rub, for as we shall see, the physicalist thesis is rendered untenable by the phenomena of quantum physics. Neither quantum entities nor the macroscopic objects that depend upon them have metaphysical identities compliant with physicalist requirements, and nonlocal quantum correlations violate sufficient material causality in ways showing *either* that physical reality is explanatorily incomplete *or* that the principle of sufficient reason (that every contingent event has an explanation) is false. If the latter horn of the dilemma is grasped, human knowing in general and scientific knowledge in particular are undermined to such an extent that physicalist doctrine becomes unsustainable. If the former horn of the dilemma is grasped, physicalist strictures on explanation are violated, and physicalism itself is seen to be false: in light of the principle of sufficient reason, the explanatory insufficiency of material causation leads directly to the need for transcendent causation in the form of divine action.[2] Specifically, to be consistent with quantum physics, the divine action that grounds physical regularity must be conceived in occasionalist rather than secondary causational terms. Furthermore, the quantum-theoretic dissolution of material substances mandates that this occasionalism be realized in an idealist (immaterialist) metaphysical context, rendering material identity a phenomenological rather than a substantial construct. Needless to say, such conclusions have profoundly antiphysicalist implications both for the nature of human persons and the nature of reality itself.

These are strong claims and an extended argument is needed. We will begin our inversion of the physicalist narrative by showing how the physics on which physicalism depends renders it untenable. After a brief discussion of the significance of the irreducibly probabilistic nature of quantum descriptions, nonlocal quantum phenomena, and the measurement problem, we will work our way through supervenience and emergence accounts of ways that the macroproperties of material objects might depend on their microproperties, showing that no notion of material substance or nomological necessity survives the quantum-theoretic dissolution of material identity and causality. What remains of the world in fundamental physics are phenomena that conform to certain structural constraints but do not have their genesis in a substantial material reality. This ontic structural realism is incompatible with physicalism and requires a radically different metaphysical orientation that flips the physicalist dependency relation on its head: mental properties do not supervene on physical properties; rather, physical properties depend on mental properties. The argument will culminate with an explanation of this dependency, making the case for an occasionalist idealism in which material objects, qua material, are mere phenomenological entities that we are caused to perceive by God and which have no nonmental reality. Ironically, physicalism turns out to be not only incompatible with physics, but physics mandates a kind of *immaterialist* quantum idealism (neo-Berkeleyan occasionalist idealism) in which *esse est percipi aut percipere*: to be is to be perceived or to be a perceiver.

A PRÉCIS OF QUANTUM PHYSICS

Quantum theory—which is a pillar of modern physics that includes quantum mechanics and various quantum field theories—is the mathematical theory describing the behavior of reality at the atomic and subatomic level.[3] At dimensions this small, the world behaves very differently than the world of our ordinary experience. This peculiarity is a consequence of the basic quantum hypothesis: energy does not have a continuous range of values but is absorbed and radiated *discontinuously* in units (*quanta*) that are multiples of Planck's constant. While this quantum hypothesis was put forward by Max Planck in 1900 to explain black body radiation (energy emitted by a nonreflecting body due to its own heat), the work of Albert Einstein, Niels Bohr, and others soon showed it was foundational to the whole of physics.[4]

The peculiarity of the quantum realm is evident in the classic *double-slit experiment* demonstrating the *wave-particle duality* of light.[5] To visualize the situation, consider two waves of the same size (amplitude) traveling through water in opposite directions. Each wave has a crest (its highest point) and a trough (its lowest point). When they meet, they move through each other in various phases of *superposition*. Since they have the same size, when a crest meets a crest or a trough meets a trough, it will amplify respectively to twice its height or depth, and when a crest meets a trough, each cancels the other and the water is level. The former behavior is called *constructive interference* and the latter *destructive interference*. Light exhibits these kinds of interference—manifested as closely spaced light and dark bands on a projection screen—when passed through two narrow parallel slits. So light has a wave nature. But light also knocks electrons out of a variety of metals and therefore, as Einstein's 1905 explanation of this "photoelectric effect" demonstrated, exists as packets of energy called *photons* that behave like particles. This strange quantum-mechanical wave-particle duality is displayed in the double-slit experiment. When *very* low-intensity light is directed through narrow parallel slits, an interference pattern builds up on a photographic plate one spot at a time, manifesting the wave nature of light in the emerging interference pattern and the particle nature of light in its spotty accumulation. The pattern emerges if only one photon is in the apparatus at a given time and it disappears if one of the slits is covered. So, *each* photon behaves as though it passes through *both* slits and *interferes with itself*, something that, from the standpoint of classical (nonquantum) physics and our ordinary experience of the world, is impossible. What is more, matter particles display this same wave-particle duality under similar experimental conditions, as the Davisson-Germer experiment demonstrated for electrons.[6]

The way that quantum mechanics deals with such things is to set aside classical conceptions of motion and the interaction of bodies and to introduce acts of measurement and probabilities for observational outcomes in an *irreducible* way, that is, in a way that *cannot* be resolved by an appeal to our inability to observe what is actually happening (in fact, quantum theory shows this peculiarity is *intrinsic* to reality rather than an artifact of our limited knowledge). In classical mechanics, the state of a physical system at a particular time is completely specified by giving the precise position and momentum of all its constituent particles, after which the equations of motion determine the state of the system at all later times. In this sense, classical mechanics is *deterministic*. But quantum mechanics does *not* describe systems by states in which particle position and momentum, for example, have simultaneously defined values. Instead, the state of the system is described by an abstract mathematical object called a *wavefunction*.[7] As long as the system is *not* being measured, the wavefunction develops deterministically through time, but it only specifies the *probability* that various *observables* (like position or momentum) will, when measured, have a particular value. Furthermore, not *all* such probabilities can equal zero or one (be absolutely determinate). This fact is expressed in *Heisenberg's indeterminacy/uncertainty principle*: no mathematical description of the state of a quantum system assigns probability 1 (determinateness) to the simultaneous existence of exact values for certain "complementary" pairs of observables. The particular value resulting from the measurement of a quantum observable is therefore *irreducibly probabilistic* in the sense that no sufficient condition is provided for *this* value being observed rather than another that is permitted by the wavefunction. This is *one* sense in which quantum theory is *indeterministic*. Also, since *all* the information about a quantum system is contained in its wavefunction, no measurement of the *current* state of a system suffices to determine the value that a *later* measurement of an observable will reveal. This is *another* (related) sense in which quantum theory is *indeterministic*. Applied to the double-slit experiment, the quantum wavefunction gives a probability distribution for measurement outcomes associated with a photon being observed to hit the photographic plate in a certain region when a measurement is made. This probability distribution describes the interference pattern on the plate that results when both slits are open, even if just one photon is sent through at a time.

This way of describing physical systems has further paradoxical consequences that conform to experimental observations. Albert Einstein, Boris Podolsky, and Nathan Rosen pointed out one of these paradoxes in 1935, arguing that the quantum description of physical systems must be incomplete because there are elements of reality that quantum theory does not recognize. To make this case, they considered a situation in which two quantum particles interact so as to "entangle" their spatial coordinates with each other and their linear momenta with each other.[8] As a result of this wavefunction entanglement, measuring either the position or the momentum for one particle instantaneously fixes the value for that same observable for the other particle, no matter how far apart they are. If one assumes, as the 1935 paper did, that what counts as an element of reality for the second particle is independent of which measurement is performed on the first particle, then reality can be attributed to both the position *and* the momentum of the second particle since measuring the position or the momentum of the first fixes the position or the momentum of the second without disturbing it and without any signal (subject to the limiting velocity of light) having passed between them. As Einstein, Podolsky, and Rosen (EPR) put it, "[i]f, without in any way disturbing a system, we can predict with certainty (i.e., with probability equal to unity) the value of a physical quantity, then there exists an element of physical reality corresponding to this physical quantity."[9] Since quantum theory does not allow the second particle to have both position and momentum simultaneously, it is incomplete.

By way of response, Bohr argued that EPR missed the point of quantum-mechanical descriptions by ignoring the *different contexts of measurement*.[10] He agreed that measuring *either* the position *or* the momentum of one particle would render *either* the position *or* the momentum of the other particle an element of reality, but *denied* that the results from these different experimental contexts could be *combined*. In other words, if we try to make context-independent claims about what is real in a distant system, we will violate quantum-mechanical predictions and run afoul of experiment. This amounts to the claim that *measurement* of the first particle *can constitute what is real* about the second particle, even when they are separated by a distance that would prohibit any signal (subject to the limiting velocity of light) from passing between them.

While Bohr's attempt to justify these claims generated much confusion,[11] John Bell's work on the EPR argument and missing elements of reality,[12] along with subsequent experimental tests,[13] have shown that Bohr was essentially correct and Einstein wrong about the completeness of quantum mechanics. As we have noted, the wavefunctions of interacting quantum systems can become entangled in such a way that what happens to one of them instantaneously affects the other, no matter how far apart they have separated. Since local effects obey the constraints of special relativity and propagate at speeds less than or equal to that of light, such instantaneous correlations are called *nonlocal*, and the quantum systems manifesting them are said to exhibit *nonlocality*. What John Bell showed is that, if quantum theory is correct, no hidden variables (empirically undetectable elements of reality) can be added to the description of quantum systems exhibiting *nonlocal* behavior that would explain these instantaneous correlations on the basis of local considerations. As indicated, subsequent experiment showed that quantum theory is correct and complete as it stands. But since all *physical* cause-and-effect relations are local, the completeness of quantum theory implies the *physical incompleteness* of reality: the universe is shot through with mathematically predictable nonlocal correlations that, on pain of experimental contradiction, have *no* physical cause.[14]

The radicalness of nonlocality is actually deeper than this because it extends to isolated quanta as well. Stated roughly, it has been shown that if one makes the reasonable assumptions that an individual quantum can neither serve as an infinite source of energy nor be in two places at once, then that particle has *zero* probability of being found in any bounded spatial region, no matter *how* large.[15] In short, unobserved quanta do not exist *anywhere* in space, and so, to be honest, have no existence at all apart from measurement![16] Hans Halvorson and Robert Clifton closed some minor loopholes and extended this argument by demonstrating that the Hegerfeldt-Malament result holds under even more general conditions—including when the standard relativistic assumption that there is *no* privileged reference frame is dropped.[17] The proper conclusion seems to be that there is no intelligible notion of microscopic material objects: particle talk has pragmatic utility in relation to measurement results and macroscopic appearances, but *no* basis in an unobserved and independent microphysical reality.

So how should we understand the relationship and transition between the microscopic and the macroscopic world? This question leads to the second famous paradox of quantum theory, *the measurement problem*, which was first described in Erwin Schrödinger's famous "cat paradox" paper.[18] In Schrödinger's iconic example, a radioactive atom with an even chance of decaying in the next hour is enclosed in a chamber containing a cat and a glass vial of poison. If a Geiger counter detects the radioactive decay of the atom in that hour, it triggers a relay that causes a hammer to smash the vial and release the poison, thus killing the cat; otherwise, the cat survives. After an hour, the quantum wavefunction for the whole system (atom + counter + relay + hammer + vial + cat) is in an unresolved superposition that involves the cat being neither dead nor alive. The question of *where* and *how* the superpositions in the wavefunction "collapse" into a determinate result is the essence of the measurement problem. Is such a determinate result the consequence of some special random process? Is it due to the quantum system's interaction with a macroscopic measurement device? Is it somehow connected to the *act* of observation itself? Is determinateness perhaps not manifested until the result is recognized by a *conscious* observer? This issue arises because every quantum wavefunction is expressible as a superposition of different states in which the thing it describes, say an alpha particle that could be ejected from an atomic nucleus, fails to *possess* the properties specified by those states. At any given time, then, some features of a quantum object occupy an ethereal realm between existence and nonexistence. Nothing subject to a quantum description *ever* has simultaneously determinate values for *all* its associated properties. And these ethereal superpositions percolate upward into the macroscopic realm because anything composed of quanta is always also intrinsically in a superposition of states, even though destructive interference (what physicists call *environmental decoherence*) may give the appearance that the wavefunction has "collapsed" into the single reality we observe.[19] What is more, under special conditions in the laboratory, we can create *macroscopic* superpositions. A clear example is provided by Superconducting Quantum Interference Devices (SQUIDs). SQUID states have been combined in which over a billion electrons move in a clockwise direction around a small superconducting ring, while another billion or more electrons simultaneously move around the ring in an anticlockwise direction, meaning that *the two incompatible currents are in superposition.*[20] With respect to this macroscopic quantum realization superposing classically incompatible states, the pressing question is: in what direction are the electrons *supposed* to be moving? Which of these classically incompatible macroscopic states is supposed to be the *real* one?

So it is that quantum theory raises fundamental questions about the existence of material identity, individuality, and causality that pose a prima facie problem for physicalist metaphysics: if material reality is sufficient unto itself, as physicalists insist, then, provided that quantum theory is correct, in *what* does the *intrinsic* substantial nature of material reality consist? What is more, given the irreducibly probabilistic nature of quantum outcomes and their demonstrable nonlocality, and given relativistic constraints on material causality, in *what* does the *causal integrity* and *sufficiency* of material reality consist? Why, in naturalistic metaphysics, if quantum outcomes lack any material explanation, does the physical universe cohere at all, let alone in a way that makes science possible?

ON REDUCIBILITY, SUPERVENIENCE, AND EMERGENCE

The basic dependency relation for the ontological physicalist must be that of the world of medium-sized dry goods upon the microphysical realm, and this dependency relation will have to be spelled out in terms of either a reducibility thesis, or a more relaxed supervenience and/or emergence thesis.

We begin by examining the explanatory prospects associated with various ways in which macroscopic properties might supervene on microscopic properties, demonstrating their inadequacy for physicalist purposes, then turn to a concept of emergence to evaluate its (in)ability to elucidate the transition between the microscopic and the macroscopic realm in ways useful to the physicalist.

Not So Super Supervenience

The physicalist account of the dependency relation of the macroscopic on the microscopic realm, understood as a supervenience thesis, must therefore be a kind of mereological supervenience in which the macroproperties of material things supervene on their microproperties. We may think of this mereological supervenience relation in a variety of ways. If it is to be distinct from a nomological or even broadly logical (metaphysical) reducibility, the relationship between the macro and micro realms will need to be anomalous in much the same way that Davidson's anomalous monism[21] tries to account for the supervenience of the mental on the physical. Using a distinction due to Jaegwon Kim, this requires (as we will presently explain) *weak* as opposed to *strong* supervenience, but as we shall see, weak mereological supervenience is insufficient to ground the *de re* objecthood of macroscopic objects in relevant microproperties.

Strong mereological supervenience is what the physicalist requires, and this commits him to the *reducibility* of the macroworld to the microworld through causal bridge laws of some sort. Nonreductive physicalism is not an option in this context. What the physicalist needs is a nomological specification of how macroscopic material objecthood emerges necessarily from intrinsic facts concerning the parts which compose it, together with the spatial relationships among those parts. But this is problematic in light of quantal nonlocalizability, which denies spatiotemporal location to the fundamental quantum "parts" of macroscopic objects, thereby preventing their possession of any of the other determinate properties requisite to mereological contributions to macroscopic objecthood. In short, the quantum "components" of a macroscopic system cannot participate in the kind of causal analysis that the physicalist requires since they have no intrinsic identities, they are not individuals, and arguably not material objects at all. If quantum physics is correct, it would seem that the identities of material objects, if there be such, do not reside in a *de re* physical substantiality that grounds an intrinsic identification.

Let's examine this conclusion more carefully by setting out the definitions of weak, strong, global, and mereological supervenience, noting that strong supervenience implies global supervenience. These distinctions originate with the work of Jaegwon Kim[22] and have become standard. Mereological (whole-part) supervenience may be defined in this way:

> The macroproperties of material things are supervenient on their microproperties.

If *A* and *B* are families of properties, weak supervenience may be defined as follows:

> *A weakly supervenes* on *B* just in case, necessarily, for any *x* and *y*, if *x* and *y* share all properties in *B* then *x* and *y* share all properties in *A*, that is, indiscernibility with respect to *B* entails indiscernibility with respect to *A*.

If we let *A* and *B* be families of properties closed under Boolean operations, this definition has the following equivalent formulation:

> *A weakly supervenes* on *B* just in case, necessarily, for any property F in *A*, if an object *x* has F, then there exists a property G in *B* such that *x* has G, and if any *y* has G it has F.

So we can say that:

> If *A* weakly supervenes on *B*, then necessarily, for every F in *A* there is a property G in *B* such that $\forall x(Gx \rightarrow Fx)$.

If infinite conjunction and disjunction are assumed, then this last statement can be strengthened to a biconditional so that every *A*-property has a coextension in *B*.

Strong supervenience can be constructed from weak supervenience when it is realized that weak supervenience does not carry the right kind of modal force—fixing the base properties (those in *B*) does not fix the supervenient properties (those in *A*). Putting it in terms of possible world semantics, weak supervenience requires that *within* each world the G–F generalization must hold. But this generalization does not have to be stable *across* worlds. To ensure the stability of the relationship between the base properties and the supervenient properties, we need to introduce a suitable modal operator. If *A* and *B* are families of properties closed under Boolean operations, then strong supervenience can be defined as:

> *A strongly supervenes* on *B* just in case, *necessarily*, for each *x* and each property F in *A*, if *x* has F, then there is a property G in *B* such that *x* has G, and *necessarily*, if any *y* has G, it has F.

Note that the modal term "necessarily" occurs twice in this definition and different understandings will result depending on how it is read in each instance (logically, metaphysically, nomologically). In the case of strong mereological supervenience, Kim suggests that the most plausible construal is to take the first occurrence to signify metaphysical necessity, and the second to represent nomological/physical necessity (assuming there is a coherent notion of physical necessity, an assumption we will find presently to be philosophically problematic and quantum-theoretically untenable). In analogy with weak supervenience we may write:

> If *A* strongly supervenes on *B*, then for every F in *A* there is a property G in *B* such that $\Box\forall x(Gx \rightarrow Fx)$.

With infinite conjunction and disjunction assumed, this can be strengthened as before to a biconditional so that for every *A*-property there is a necessarily coextensive *B*-property. Strong supervenience entails weak supervenience, but not vice versa. It is also worthwhile to note that the concept of global supervenience can be shown to be implied by that of strong supervenience. Global supervenience may be defined as:

> *A* globally supervenes on *B* just in case worlds that are indiscernible with respect to *B* are also indiscernible with respect to *A*.

This gives us all the tools we need to make our argument. The question we want to address is whether a *de re* property of macroscopic objecthood can supervene on a microphysical realm where there is no objecthood. Setting aside the absence of objecthood in the quantum realm for a moment, it is clear that weak supervenience is not the right kind of supervenience to provide for the possibility of modality *de re* in the macroscopic physical realm. Essential properties are possessed by an object in every world in which it exists, and material objecthood is an essential property of every material object (it could not be a material object and lack this property), but if we suppose that the macroscopic material objecthood of an object O weakly mereologically supervenes on a family Q of properties constituting its quantum properties in the *actual* world, then weak supervenience still allows for these possibilities:

1. In the actual world, any O possessing the properties in Q is a material object, but in another possible world, every such O is not a material object;
2. In the actual world, any O possessing the properties in Q is a material object, but in another world exactly like it in respect of the distribution of the properties in Q, *nothing* is a material object; and
3. In another possible world just like the actual world in respect of what has *or lacks* the properties in Q, *everything* is a material object.

Lest we doubt that weak mereological supervenience allows for these possibilities, remember that all it requires[23] is that *within* any possible world there not be two things agreeing with respect to Q but diverging with respect to macroscopic material objecthood, and this condition is satisfied in all three of these cases. In particular, weak mereological supervenience does not require that if in another world an object has the same Q-properties it has in the actual world, it must have the *same* property of macroscopic material objecthood. Any specific associations between Q-properties and macroscopic objecthood in the actual world—if there were such—would not necessarily carry over into other worlds. So we can see that weak mereological supervenience will not do as an account of *de re* macroscopic objecthood supervening on the quantum realm.

What the physicalist needs is for the base properties to *determine* the supervenient properties in the sense that once the former are fixed for an object, there is no freedom to vary the latter. This leaves strong mereological supervenience, or by entailment, global mereological supervenience, as the remaining option. But, as we shall see, the fact that there are no quantum objects becomes critical at this point. Strong mereological supervenience can be expressed as:

> For every property M in the macroscopic realm, there is a (possibly conjunctive or disjunctive) property Q in the quantum realm such that, *necessarily*, for every material object O, if O has Q, then O has M.

This condition must hold for the macroscopic property of material objecthood when it is predicated of O, but what will the base property Q be that provides a sufficient condition for O possessing macroscopic material objecthood *necessarily*? The type of necessity involved here seems to be broad logical (metaphysical) necessity, so what the physicalist needs here is an explicit account of how macroscopic material objecthood for the whole is necessitated by the intrinsic facts concerning the parts that compose it, together with the spatial relationships among these parts. This, in effect, is the criterion put forth by Terence Horgan,[24] though he frames it in terms of global supervenience rather than strong supervenience when the latter might have served his purposes better. In any case, Horgan takes things like electrons, protons, neutrons, and quarks to be natural kinds and characterizes the relationship of physical accessibility between the actual world and other possible worlds in his supervenience thesis this way:

> Let a *P-world* be a possible world that is *physically accessible* from the actual world. All P-worlds are worlds in which (1) all the entities are *either* entities whose specific natural kinds are actual-world natural kinds *or* are fully decomposable into such; and (2) all fundamental microphysical properties are properties explicitly cited in the laws of actual-world microphysics.[25]

Horgan is not just concerned with whole P-worlds, however, but more specifically, with particular spatiotemporal *regions* within P-worlds. He thus arrives at a broadly logical localized supervenience principle he states as follows:

> There do not exist any two P-regions that are exactly alike in all qualitative intrinsic microphysical features but differ in respect of some other qualitative intrinsic feature.[26]

Stated positively, this implies that

> The macroscopic properties of P-region *A* globally supervene on the intrinsic microphysical features of P-region *A* just in case worlds which are indiscernible with respect to the intrinsic microphysical features of P-region *A* are also indiscernible with respect to the macroscopic properties of P-region *A*.

Since global supervenience is implied by strong supervenience, this characterization is entailed by the criterion that for every macroscopic property M of P-region *A* there is a quantum property Q of P-region *A* such that, in a broadly logical sense, *necessarily*, everything exemplifying Q exemplifies M. If P-region *A* is taken to be coterminous with macroscopic object O, we are back to the expression of strong mereological supervenience that I gave above. Defending the supervenience of the macroscopic realm on the quantum realm therefore seems to commit the physicalist, at least in principle, to the broadly logical and nomological reducibility of the macroworld to the microworld.

Since the physicalist must maintain the supervenience of medium-sized dry goods on the quantum realm, he needs to give a metaphysical specification of how the macroscopic material objecthood of the whole is necessitated by intrinsic facts concerning the parts which compose it, together with the spatial relationships among those parts. But now our earlier observations about the nature of quantum reality come to bear. The fundamental quantum "parts" of macroscopic objects don't have well-defined spatiotemporal location and are not subject to this kind of metaphysical analysis. They are not autonomous material objects, they do not possess a complete set of determinate properties, they have no intrinsic identities, they are not individuals, and have no substantial material existence. As if this were not enough, where nonlocal phenomena are concerned, no supervenience of nonlocal quantum systems on the properties of various subsystems taken separately or in other combinations is possible (the relevant joint probabilities are not factorizable), nor, for this very reason, are there objective properties of the system immediately prior to measurement that can provide the nomologically necessary connection to measurement results required by any viable supervenience explanation of the macroscopic on the microscopic realm (postulation of such objective properties leads to empirically false consequences for both local deterministic and local stochastic models).[27] Whence, then, the supervenience explanation of macroscopic material objecthood? Nowhere at all, it would seem;

supervenience explanations are a complete nonstarter in the quantum context. In a delicious irony, physical theory has rendered impossible any physicalist account of the supervenience of macroscopic objects on microphysical entities and laws.

"Not so fast," the physicalist objects, "you haven't countenanced the possibility that macroscopic material objecthood is an irreducible *emergent* property of certain quantum systems!" True enough, so let's do that.

Emergence as Limit Behavior: Descriptively True but Metaphysically Unhelpful

There is a sense of "emergence" appropriate to quantum physics in which classical (Maxwell-Boltzmann) statistical behavior can be understood to emerge from quantum (bosonic and fermionic) statistics in both classical-mechanical and classical statistical limits. While these limits are useful in understanding how quantum descriptions can give rise to classical appearances, they are metaphysically unenlightening where relevant, and irrelevant in the case of nonlocal behavior. Let me briefly explain.[28]

With the standard definitions of the Poisson and commutator brackets, the *classical mechanical limit* (CM limit) of a quantum system is defined to be

$$\lim_{\hbar \to 0} \frac{1}{i\hbar}\left[\hat{A},\hat{B}\right] = \{A,B\}.$$

This limit is fictional, of course, because $\hbar$ is a physical constant. The limit represents the *transition* between the quantum and classical *descriptions* of a system; classical behavior "emerges" when quantum effects are dampened to the point of negligibility. It is important to note, however, that there are still residual effects (dependent on Planck's constant) even *after* the classical mechanical limit is taken and the underlying reality is still quantum-mechanical in character.

Statistical mechanics mathematically relates the thermodynamic properties of macroscopic objects to the motion of their microscopic constituents. Since the microscopic constituents obey quantum dynamics, the correct description must lie in principle within the domain of *quantum statistical mechanics*. Under thermodynamic conditions of high temperature (T) and low density (n), however, classical statistical mechanics serves as a useful approximation. With this in mind, the *classical statistical limit* (CS limit) may be defined as the situation represented by:

$$T \to \infty \text{ and } n \to 0.$$

These are the same conditions as those governing the applicability of the ideal gas law ($pV = nRT$), so the CS limit could equally well be called the ideal gas limit. Unlike the CM limit, the conditions governing the CS limit are subject to experimental control. In respect of quantum statistical behavior, both the CM and the CS limits are continuous, so the quantum indistinguishability arising from permutation symmetry is not removed, even though it is dampened in the limit. Quantum "particles" retain their indistinguishability even when their aggregate behavior can be approximated by a Maxwell-Boltzmann distribution.

These reflections lay the ground for understanding why any emergentist account of the dependence of the macroscopic realm on the microscopic realm, while perhaps descriptively interesting, will be unenlightening as a metaphysical explanation. It is environmental decoherence (essentially, statistical damping through wavefunction orthogonalization) that gives quantum-mechanical ephemera a cloak of macroscopic stability, but decoherence is not a real solution to the measurement problem. The apparent solidity of the world of our experience is a mere epiphenomenon of quantum statistics; the underlying phenomena retain their quantum-theoretic essence while sustaining classical appearances.

Emergence and Irreducibility

The essence of emergentism is a layered view of nature. The world is divided into ontological strata beginning with fundamental physics and ascending through chemistry, biology, neuropsychology, and sociology. The levels correspond to successive organizational complexities of matter, and at each successive level there is a special science dealing with the complex structures possessing the distinguishing causal characteristics of that level. Higher-level causal patterns necessarily supervene on (are dependent upon) lower-level causal interactions, but are not reducible to them. The picture, then, is of emergent nomological structures irreducible to lower-level laws, with emergent features that not only affect the level at which they appear, but also exercise "downward causation" on lower-level phenomena.

Moving beyond hand-waving declarations of the "lawful" character of emergence requires giving an account of the relationship between basal physical conditions and emergent properties. Brian McLaughlin[29] and Jaegwon Kim[30] have both attempted articulations of emergence in terms of what O'Connor and Wong[31] term "synchronic strong supervenience": given basal conditions *C* at time *t*, an emergent property *P* strongly supervening on conditions *C* will appear at time *t*. McLaughlin[32] defines such emergent properties in terms of strong supervenience as follows:

> If P is a property of w, then P is emergent if and only if (1) P supervenes with nomological necessity, but not with logical necessity, on properties the parts of w have taken separately or in other combinations; and (2) some of the supervenience principles linking properties of the parts of w with w's having P are fundamental laws.

McLaughlin defines a fundamental law as one that is *not* metaphysically necessitated by any other laws, even together with initial conditions. While Kim[33] also understands emergence as a form of strong synchronic supervenience, it is important to note that he also regards emergent properties as epiphenomenal and challenges the tenability of nonreductive physicalism on this basis (he is a physical reductionist). These arguments need not concern us here but have received responses from Barry Loewer[34] and Sydney Shoemaker.[35] The property-fusion account of emergence developed by Paul Humphreys[36] circumvents Kim's objections because it is not synchronic and because emergent properties are fusions of the basal properties, which then cease to exist.

The supervenience account of emergence will not suffice in the quantum context for reasons we have already discussed, most notably because nonlocal phenomena do *not* supervene on the properties of the various subsystems taken separately or in other combinations, and any viable account of nomological necessity in the quantum realm would have to connect objective properties of the system immediately *prior* to measurement with the measurement results obtained. Such restrictions lead to empirically false consequences for both local deterministic and local stochastic models. As we shall see momentarily, however, a nonsupervenient description of quantum emergence suffers from a sort of explanatory vacuity, and also founders on ontological contradictions arising from the postulation of nonlocal wholes or—if a privileged reference frame is postulated—on the nonlocalizability of individual quanta.

Property Fusion as an Account of Emergent Ontological Hierarchies

Paul Humphreys has developed a concept of emergence in terms of "property fusion" that he suggests can be used to describe entangled states in quantum theory.[37] His account *assumes* the existence of a hierarchy of distinct ontological levels, which he expresses in the form of a "level-assumption":

> *(L)* There is a hierarchy of levels of properties, L_0, L_1..., L_n... of which at least one distinct level is associated with the subject matter of each special science, and L_j cannot be reduced to L_i for any $i < j$.

A property P^i is then defined to be an "*i*-level property" just in case *i* is the *lowest* level at which instances of the property occur. A set of properties $\{P_1^i, P_2^i, \ldots, P_m^i, \ldots\}$ is associated with each level *i*, where P_m^i denotes the mth property at the *i*-level. A parallel hierarchy of entities is postulated: x^i is an *i*-level entity just in case *i* is the *lowest* level at which it exists and x_m^i denotes the mth entity at the *i*-level.

To characterize the property-fusion operation, Humphreys uses the notation $P_m^i(x_r^i)(t)$, which denotes an instantiation of property P_m^i by entity x_r^i at time *t*, because he regards property instances as being more fundamental than properties. We will suppress references to specific individuals and times to simplify the notation. The fusion operation [.*.] is defined by Humphreys as a process that combines two *i*-level properties P_m^i and P_n^i to form an (*i*+1)-level property $[P_m^i * P_n^i]$; this fusion could equally well be represented by the notation $P_{m,n}^{i+0}$. Once the basal properties have fused in this manner, they cease to exist and the new emergent property is all that remains.

Humphreys argues that entangled (or nonseparable) states in quantum mechanics lend themselves to description in terms of property fusion, maintaining that the emergent entangled state will remain intact so long as nonseparability persists. He thinks this can be the case even after the interaction ceases, whereas Fred Kronz and Justin Tiehen[38] adopt Humphreys's conception of property fusion but argue that persistence of the interaction is necessary for continued emergence. The arguments for this difference need not concern us. The more pressing concern is whether this technical account of emergence is explanatorily useful and metaphysically tenable in relation to nonlocal phenomena.

The Kronz-Tiehen Taxonomy for Quantum Mereology

On the basis of their discussion of fusion in the context of quantum chemistry, Kronz and Tiehen[39] suggest that there are at least three ways that philosophers could develop a metaphysical account of emergence in mereological terms; they advocate the last of the three. Since it is instructive to do so, we will briefly consider all three options.

Before examining these accounts, however, we need definitions of two background ideas employed by Kronz and Tiehen: *independent characterizations* of entities and *contemporaneous parts*. A *characterization* of an entity is an exhaustive list of the properties that are instantiated by that entity and this characterization is said to be *independent* just in case the elements on the list of its properties make no essential reference to some other entity. Secondly, an entity is said to be a *contemporaneous part* of some whole just in case that part exists while the whole does. (In relativistic contexts, Kronz and Tiehen make this relation reference-frame dependent in order to preserve standard interpretations of Lorentz invariance in terms of the relativity of simultaneity.) So armed, they define three conceptions of emergence:

Prototypical Emergence

The idea here is that every whole consists of contemporaneous parts that have independent characterizations, but there is some criterion for distinguishing between part-whole relationships that are emergent and those that are merely resultant. The British emergentists take this line and use additivity as the relevant characterization of a resultant as opposed to an emergent property.[40] The difficulty with this view is that it seems to trivialize the notion of emergence when quantum mechanics is brought into view, either rendering every part of the universe emergent because it is entangled through past interactions with everything else in the universe, or nothing emergent, because the universe is an undivided whole that has no parts with independent characterizations. A proper interpretation of quantum theory would seem to require grasping the second horn of this dilemma.

Radical Emergence

The idea behind radical emergence is that only resultant wholes have contemporaneous parts, emergent wholes do not. Kronz and Tiehen interpret this as Humphreys's view. Emergent wholes are produced by fusion of entities that can be likened to parts, but these parts cease to exist upon fusion, only existing when the whole does not, and vice versa. An example of this sort of thing presumably would be a nonseparable quantum state. Prior to interaction, quantum "particles" might be taken to have independent existence, but after they interact and their wavefunctions become entangled, they cease to exist as "parts" and a new entity at the next level in the ontological hierarchy comes into being. Again, it is hard to see on this view why there is not only one quantum entity: the universe itself.

Dynamic Emergence (Relational Holism)

Kronz and Tiehen proclaim themselves advocates of what they call "dynamic emergence," which, although unacknowledged, seems clearly to be a reinvention of Paul Teller's idea of relational holism.[41] Teller defines a relationally holistic property as one in which the relevant property of the whole does *not* supervene on the nonrelational properties of the relata, as for example, the tallness of Wilt Chamberlain relative to Mickey Rooney supervenes on the nonrelational height of each. In the Kronz-Tiehen reformulation, emergent wholes have contemporaneous parts, but these parts cannot be characterized independently of their respective wholes. These wholes are produced by an essential, ongoing interaction of their parts. Ultimately, of course, quantum theory is going to imply that every contemporaneous part of the universe, at least in its "material" respects, cannot in the final analysis be characterized independently of the whole universe, though for all practical purposes we can often treat subsystems of the universe as proximately independent.

Relational Holism and Quantum Nonlocality: A Very Holey Story

Granted that relational holism (to use Teller's term) seems the most reasonable description of quantum ontology, what more can be said? As Kronz and Tiehen have noted, speaking of contemporaneous parts for nonlocal wholes requires, in view of the relativity of simultaneity, a relativization of contemporaneousness to reference frames. Though they do not discuss how this is to be done, the most plausible candidate is Gordon Fleming's theory of hyperplane dependence,[42] in which judgments of simultaneity are relativized to hyperplanes constituted by three-dimensional temporal slices of space-time; this is the solution appropriated by Teller (1995 and elsewhere). The difficulty here is that the properties of a nonlocal quantum system can be different depending on which hyperplane is in view. In some hyperplanes, for example, the wavefunction of the system may have collapsed, while in others this will not yet have happened. But there are an infinite number of such hyperplanes, some of which intersect, and it will be the case at the point of intersection that ontologically inconsistent properties are attributed to the quantum system, for example, that it has *both* collapsed and not collapsed. I take this situation to indicate that particle ontologies are incompatible with fundamental physical theory. If we take the ontological contradiction as a harbinger of the nonlocalizability of any so-called "particles," then particle ontologies are not ultimately tenable because the particles don't exist anywhere. On the other hand, if we take the contradiction to indicate the metaphysical necessity of an undetectable privileged reference frame, the "particles" remain nonlocalizable,[43] so the same conclusion follows. Nor is this situation ameliorated by switching to a quantum field ontology for the same ontological contradictions arise with respect to states of the field and the fields themselves exhibit states of superposition of contradictory numbers of quanta that not only render the intrinsic substantiality of the quanta impossible, but also that of the field itself. For this reason and some others as well, field ontologies are as inadequate as particle ontologies for interpreting our most basic physical theories (relativistic quantum field theories).[44]

As a characterization of quantum nonlocality, therefore, while relational holism (dynamic emergence) may be descriptively accurate and revelatory of the challenge to ontological interpretation that quantum theory poses, it is explanatorily vacuous. If the objection is raised that the "individual" described by quantum theory must ultimately be the quantum system itself, with its Hilbert Space of states, and the ontological difference between particle and field a mere matter of representation for a selected set of states, all of which are allowed and used by quantum field theory, then the appropriate reply is that the question of ontology is not obviated, nor is the fact of systematic, predictable correlation without causation. Instantaneous adjustment of nonlocal relational wholes to local systemic changes, whether called "emergence" or some other term of art, remains a flagrant violation of relativistic causality that lacks a physical explanation and is present, if anything, to an even greater extent in quantum field theory than quantum mechanics.[45] Invoking "emergence" in such contexts seems little more than a terminological gambit to obscure things for which no adequate physical explanation exists and for which no adequate physicalist explanation is possible. While it pays lip service to a variety of ontological levels, the explanatory vacuity of emergentist metaphysics reinforces the untenability of physicalism.

THE NULLIFICATION OF NOMOLOGICAL NECESSITY

What are the implications of all this for physical law and physicalist conceptions of the natural order? There are various conceptions of physical laws that try to give an account of them as *natural necessities* of one variety or another. For obvious reasons, these nomological theories are called *necessitarian.* Alvin Plantinga has provided a cogent philosophical critique of the role of necessity in accounts of physical law.[46] Though some philosophers have argued that natural laws are broad logical necessities similar to statements like *no equine mammals are mathematical propositions*,[47] there seems little to no basis for this claim. If we take Coulomb's Law of electric charges, for instance, the fact that two like (or different) charges repel (or attract) each other with a force proportional to the magnitude of the charges and inversely proportional to the square of the distance between them gives no hint of being metaphysically necessary. We can easily conceive of a different mathematical relationship holding between the charges. This has led other philosophers to assert that the laws of nature are *contingently* necessary and to develop an account of natural laws based on this assumption.[48] But quite apart from the oxymoronic appearance of such a claim, no coherent account of its substance has ever been put forward. One cannot just call natural laws "contingent necessities" and expect it to be true "any more than one can have mighty biceps just by being called 'Armstrong,'" as David Lewis famously quipped.[49] Finally, other advocates of natural laws as physical necessities have proposed an account of physical laws deriving from innate causal powers:[50] laws of nature are grounded in the essential natures of things inherent in their material substance and manifested through forces or fields that express necessary capacities or emanations from these natures and mediate or constrain physical interactions in a way that also is necessary. But again, it is difficult to see why *this* causal power must necessarily flow from the essential nature of *that* material substance. Calling it necessary or essential doesn't make it so; we could imagine it otherwise.

Even if these necessitarian accounts were not philosophically inadequate, they all, without exception, fail to work in the quantum context. The essential causal powers account and the relation among universals account both require that physical systems and material objects objectively possess properties that are capable of being connected together in a law-like fashion. At a minimum, necessitarian and/or counterfactual physical law theorists have to maintain that quantum systems, or their components, objectively possess properties prior to measurement, whether these properties are determinate or indeterminate (probabilified dispositions), and that it is the *objective possession* of these properties that necessitates (or renders probable) their specific behavior. Bell's theorem demonstrates that this assumption leads to empirically false consequences in the case of local deterministic and local stochastic models.[51] As we have also seen, this assumption either leads to an ontological contradiction in the nonlocal stochastic case embodied by relational holism (dynamic emergence), or if an undetectable privileged reference frame is invoked, succumbs to the nonlocalizability and insubstantiality of the intended possessors of the requisite properties.[52]

What we are left with, therefore, is a situation in which there are *no* objective physical properties in which to ground necessitarian/counterfactual relations. So necessitarian theories of natural law cannot gain a purchase point in fundamental physical theory and must be set aside. All that remains in such a case is the so-called *regularist* account of natural laws, which asserts that while there are regularities present in the phenomenology of the world on a universal scale, there are no real *laws* of nature, that is, there is *no necessity* that inheres in the natural relationships among things or in the natural processes involving them. In short, nature behaves in ways we can count on, but it does so for no discernible *physical* reason. This state of affairs requires an explanation.

THE NECESSITY OF SUFFICIENCY

Before we proceed to discuss an explanation adequate to the phenomena in question, we must address the view that some contingent events do not require an explanation since genuine brute factuality is a possibility. In the present context, we must confront the suggestion that brute factuality can be attributed to the regularities of nature, that is, the view that the regularities of nature are *mere* regularities that lack any explanation for the patterns they exemplify. The patron saint of this approach is David Hume and the most sophisticated modern articulation of it is given by David Lewis.[53]

In describing the regularities of our world, Lewis's theory takes the fundamental relations to be spatiotemporal: relativistic distance relations that are both space-like and time-like, and occupancy relations between point-sized things and space-time points. Fundamental properties are then *local* qualities: perfectly natural intrinsic properties of points, or of point-sized occupants of points. Everything else supervenes on the spatiotemporal arrangement of local qualities throughout all of history—past, present, and future—hence "Humean supervenience." On this view, natural regularities are simply the theorems of axiomatic deductive systems, and the best system is the one that strikes the optimal balance between simplicity and strength (informativeness). Lewis postulates this "best system" to exist as a brute fact whether we know anything about it or not. As Plantinga points out,[54] we have little conception of what Lewis's "best system" might look like and even less reason to think that there is a uniquely "best" such system as opposed to "a multitude of such systems each unsurpassed by any other." We may add that Lewis's approach, as it stands, is inadequate to deal with quantal nonlocalizability, physical indeterminism, and the undoing of the causal metric of space-time in quantum gravitational theories. Furthermore, quantum-theoretic Bell correlations, while nonlocally and instantaneously coincident, would have to be understood in Lewis's theory in terms of *local* properties manifesting *random values in harmony at space-like separation* without *any* ontological connection or explanation, everything functioning as part of an overarching system of regularities that is in some sense optimal, but which also lacks any explanation for the ongoing order it displays. In short, embracing Lewis's approach requires rejecting the principle of sufficient reason/causality (the principle that every contingent event has an explanation) on a colossal scale.

But following Lewis's programmatic recommendations and rejecting the principle of sufficient reason would destroy knowledge in general and scientific knowledge in particular. Why? In the case of knowledge in general, if there were no sufficient reason why one thing happens rather than another, our current perception of reality and its accompanying memories might be happening for no reason at all, so the world we think we are experiencing might not even exist. How would we know? As far as science is concerned, if it is possible that a physical state of affairs lacks an explanation, then the possibility that there is *no explanation* becomes a competing "explanation" for anything that occurs. Since there is no objective probability and hence no likelihood assignable to something for which there is no explanation, the possibility that there is no explanation becomes an inscrutable competitor to every other proposed explanation, undermining our ability to decide whether there is a scientific explanation for *anything* that happens. Denying that every contingent event has an explanation not only destroys the possibility of science, it opens the door to irremediable skepticism and the destruction of knowledge in general. The principle of sufficient reason is a metaphysical truth that we know a priori; it is a precondition of all knowledge and of the intelligibility of the world.[55]

REVERSING POLARITY: THE DEPENDENCE OF THE PHYSICAL ON THE MENTAL[56]

We have established that an explanation for the contingent regularities of nature is necessary and that no explanation in terms of efficient material causation and physical necessity is possible. What kind of explanation is left? To begin answering this question, let's start with the eminently reasonable assumption that there is a way that the world is, that we can get it right or wrong, and that science is a useful tool in helping us to get it right. In particular, when physical theory backed by experiment demonstrates that the world must satisfy certain formal structural constraints—for example, quantizability, nonlocality as encapsulated in the Bell theorems, nonlocalizability as indicated by the Hegerfeldt-Malament and Reeh-Schlieder theorems, Lorentz symmetries in space-time, internal symmetries like isospin, various conserved quantities as implied by Noether's theorem, and so on—then this *formal* feature of the world may be taken as strong evidence for a certain metaphysical state of affairs. At a minimum, such states of affairs entail that the *structural* constraints empirically observed to hold and represented by a given theory will be preserved (though perhaps in a different representation) by any future theoretical development; thus far structural realism.

Whether this structural realism has further ontological consequences pertaining to the actual furniture of the world (entity realism) is a matter of debate among structural realists. The *epistemic structural realist* believes that there are epistemically inaccessible material objects forever hidden behind the structures of physical theory and that all we can *know* are the structures.[57] The *ontic structural realist* eliminates material objects completely—it is not just that we only *know* structures, but rather that all that *exists* to be known are the structures.[58] Both these versions of structural realism are deficient, though in different ways.

We have seen that quantum theory is incompatible with the existence of material substances, even those of a relationally holistic sort. Given that this is the case, the epistemic structural realist is just *wrong* that there is a world of inaccessible material individuals hidden behind the structures that quantum theory imposes upon the world. The situation would therefore *seem* to default to ontic structural realism. But while the ontic structural realist is correct that there are no *material* objects behind the structures, his position is deficient too because there can be no structures *simpliciter* without an underlying reality that is *enstructured*; we cannot build castles in the air. It would seem, then, that we're in a sort of catch-22 situation. The challenge to making sense of quantum physics is to give an account of what the world is like when it has an objective structure that does *not* depend on material substances. What investigations of the completeness of quantum theory have taught us, therefore, is rather than quantum theory being incomplete, it is material reality (so-called) that is incomplete. The realm that we call the "physical" or "material" or "natural" is *not* self-sufficient but dependent upon a more basic reality that is *not* physical, a reality that remedies its causal incompleteness and explains its insubstantiality, and on which its continued existence depends.

In light of this realization, the rather startling picture that begins to seem plausible is that preserving and explaining the objective structure of appearances in light of quantum theory requires reviving a type of phenomenalism in which our perception of the physical universe is constituted by sense-data conforming to certain structural constraints, but in which there is no substantial material reality causing these sensory perceptions. This leaves us with an ontology of minds (as immaterial substances) experiencing and generating mental events and processes that, when sensory in nature, have a *formal* character limned by the fundamental symmetries and structures revealed in "physical" theory. That these structured sensory perceptions are *not* mostly of our own individual or collective human making points to the falsity of any solipsistic or social constructivist conclusion, but it also implies the need for a transcendent source and ground of our experience. As Robert Adams points out, mere formal structure is ontologically incomplete:

> [A] system of spatiotemporal relationships constituted by sizes, shapes, positions, and changes thereof, is too incomplete, too hollow, as it were, to constitute an ultimately real thing or substance. It is a framework that, by its very nature, needs to be filled in by something less purely formal. It can only be a structure *of* something of some not merely structural sort. Formally, rich as such a structure may be, it lacks too much of the reality of material thinghood. By itself, it participates in the incompleteness of abstractions. . . . [T]he reality of a substance must include something intrinsic and *qualitative* over and above any formal or structural features it may possess.[59]

When we consider the fact that the *structure* of reality in fundamental physical theory is merely phenomenological and that this structure itself is hollow and nonqualitative, whereas our experience is not, the metaphysical objectivity and epistemic intersubjectivity of the enstructured qualitative reality of our experience can be seen to be best explained by an occasionalist idealism of the sort advocated by George Berkeley or Jonathan Edwards. In the metaphysical context of this kind of theistic immaterialism, the *vera causa* that brings coherent closure to the phenomenological reality we inhabit is always and only *agent* causation. The necessity of causal sufficiency is met by divine action, for as Plantinga emphasizes:

> [T]he connection between God's willing that there be light and there being light is necessary in the broadly logical sense: it is necessary in that sense that if God wills that *p*, *p* occurs. Insofar as we have a grasp of necessity (and we do have a grasp of necessity), we also have a grasp of causality when it is divine causality that is at issue. I take it this is a point in favor of occasionalism, and in fact it constitutes a very powerful advantage of occasionalism.[60]

Plantinga is right to emphasize the virtues of occasionalism, but he does not take his argument in the idealist direction that the quantum-theoretic evidence we have considered seems to warrant. Clearly, the philosophical and quantum-theoretic problems for necessitarianism also prohibit a secondary causation account of divine action as the metaphysical basis for natural regularities. Secondary causation requires God to have created material substances to possess and exercise, actively or passively, their own intrinsic causal powers. God acts in the *ordinary* course of nature only as a universal or primary cause that sustains the existence of material substances and their properties as secondary causes. On this view, material substances mediate God's ordinary activity in the world and function as secondarily efficient causes in their own right. Plantinga recognizes that secondary causation inherits many of the philosophical problems associated with necessitarian accounts. Beyond this, however, it also inherits the quantum-theoretic problems that render necessitarianism untenable: the inherent *insubstantiality* of fundamental quantum entities, the *inability* of emergentist accounts of macroscopic objecthood to generate substantial material individuality and identity, and the *operative incompleteness* of this reality in respect of sufficient causation. In the absence of coherent material substances and physical causality, therefore, secondary causation lacks a purchase point in fundamental physical theory. So regardless of whether God *could* have created a world in which there were secondary material causes, it is evident that *he did not do so*. This leaves us with an occasionalist account of natural regularities, which in its "weak" form, as Plantinga is at pains to argue, fares no worse than secondary causation in respect of allowing for libertarian freedom and a resolution of the problem of evil. In fact, if we take advantage of Alfred Freddoso's approach to occasionalism, we can build libertarian freedom into its definition:

> God is the sole efficient cause of every state of affairs in the universe that is not subject to the influence of freely acting creatures.[61]

In other words, God is the only *vera causa* of every state of affairs occurring in "pure" nature, namely, that segment of the universe *not* subject to the causal influence of creatures with libertarian freedom.

In giving an account of the ontological basis for natural regularities under occasionalist idealism, then, the regularities of nature may be formulated as counterfactuals of divine freedom.[62] Rather than understanding God's activity in terms of the divine production of certain behavior in substantial material objects, however, with the perception of the same divinely induced in our material brains, we must instead conceptualize the creaturely experience of mental phenomena as directly communicated to finite immaterial minds by God. So the natural regularities we interpret as "laws of nature" are just specifications of how God would act to produce the phenomena we experience under different complexes of conditions. More precisely, nature's nomological behavior should be understood in the following way: *If collective phenomenological conditions C were realized, all other things being equal—and with a certain quantum-mechanical probability—God would cause us to experience the phenomenological state of affairs S.*

On this view, then, what we take to be material objects are mere phenomenological structures that we are caused to perceive by God and which have no nonmental reality. They exist and are given being in the mind of God, who creates them, and they are perceived by our minds as God "speaks" their reality to us. What we perceive as causal activity in nature is always and only God *communicating* to us—as immaterial substantial minds whose bodies are also phenomenological constructs—the appropriate formally structured qualitative sensory perceptions.

A WORLD WELL LOST

That physics belies physicalism is by now quite clear. Careful consideration of the progress of physics since 1900 reveals that the harder we have looked at the universe's material constitution, the more ephemeral it has gotten, until in the final analysis we are left with a phenomenological reality that does not emanate from a material substratum, for material substances are shown to have no place in fundamental physical theory. The irony for the physicalist is palpable. In seeking an explanation for how the universe works, he turns to science and marshals his resources, restricted as they are to material objects and processes and what can be derived from them. But as he journeys deeper and deeper into the heart of matter, he finds that it dissolves and his whole worldview lacks a metaphysical foundation. Yet the phenomenological universe that constitutes his experience and ours remains, is ever so regular, and is ever so evidently not of human making, for we do not will the experiences we have—they come to us unbidden, sometimes welcome and sometimes not. As we have extracted this metaphysical picture from quantum physics and examined its implications, we have found an explanation of this surprising state of affairs—for it *requires* an explanation—in an occasionalist quantum idealism that has a strong affinity with Berkeley's occasionalist idealism.[63]

When we see what the world is really like in comparison with the world that the physicalist would give us, we understand that the scientific pretensions of physicalism are so far removed from reality as to warrant dismissal as "not even wrong." The physicalist's world of inexplicable brute factuality is a world well lost: rational explanations require purpose, not blind chance, and in the final analysis that is just what fundamental physical theory gives us. The metaphysical requirement that the order and intelligibility of the universe have an explanation in conjunction with the absence of any tenable physical explanation finds its only satisfaction in transcendent causality. Fundamental physical theory does not just reveal the mind of God to us, it reveals to us that *we live in the mind of God*. It is hard to imagine a reality farther removed from physicalism than that.

NOTES

1. There is a weaselly kind of physicalism that tries to adjust the content of the thesis that "all is matter" again and again when a once-favored account of what it means for something to be a material object is rendered untenable by the progress of physical theory. The disingenuous character of this retrenchment strategy is made plain in materialism's confrontation with quantum physics, however, since, as we shall see, there are no sufficient criteria by which to identify and individuate the fundamental constituents of "material" reality in quantum theory, and no sustainable notion (emergent or otherwise) of material substance. Bas Van Fraassen describes the invariant attitude giving rise to the belief that *all is matter* as a form of "false consciousness" ("Science, Materialism, and False Consciousness," in *Warrant in Contemporary Epistemology: Essays in Honor of Plantinga's Theory of Knowledge*, ed. Jonathan L. Kvanvig [Lanham, MD: Rowman & Littlefield, 1996], 170). See also Bruce L. Gordon, "A Quantum-Theoretic Argument against Naturalism," in *The Nature of Nature: Examining the Role of Naturalism in Science*, eds. Bruce L. Gordon and William A. Dembski (Wilmington, DE: ISI Books, 2011), 179–214.

2. Some philosophers and scientists who recognize that physicalism—whether reductive or nonreductive—is a dead end, but who are still committed to evolutionary naturalism, have adopted an immanentistic panpsychist approach to issues in the philosophy of mind and the evidence for teleology in nature. This kind of panpsychism takes mental existence as an irreducible and fundamental component of *everything in*

the universe and of the universe itself. For a survey of the current debate, see Godehard Brüntrup and Ludwig Jaskolla, eds. *Panpsychism: Contemporary Perspectives* (Oxford: Oxford University Press, 2017); see also Thomas Nagel, *Mind and Cosmos: Why the Materialist Neo-Darwinian Conception of Nature is Almost Certainly False* (Oxford: Oxford University Press, 2012).

3. For more on this see Bruce L. Gordon, "A Quantum-Theoretic Argument against Naturalism"; Bruce L. Gordon, "The Necessity of Sufficiency: The Argument from the Incompleteness of Nature," in *Two Dozen (or so) Arguments for God: The Plantinga Project*, eds. Trent Dougherty and Jerry Walls (Oxford: Oxford University Press, 2017, forthcoming); and Bruce L. Gordon, "Divine Action and the World of Science: What Cosmology and Quantum Physics Teach Us about the Role of Providence in Nature," *Journal of Biblical and Theological Studies*, 2, no. 2 (2017): 247–298.

4. Jim Baggott, *The Meaning of Quantum Theory* (Oxford: Oxford University Press, 1992), 1–74; Robert P. Crease and Charles C. Mann, *The Second Creation: Makers of the Revolution in 20th Century Physics* (New York: Macmillan, 1986).

5. Richard P. Feynman, "Probability and Uncertainty—The Quantum-Mechanical View of Nature," *The Character of Physical Law* (Cambridge, MA: MIT Press, 1965), 127–148; Richard P. Feynman, *The Feynman Lectures on Physics*, vol. 3: *Quantum Mechanics* (Reading, PA: Addison-Wesley Publishing Company, 1971).

6. C. J. Davisson, "Are Electrons Waves?" *Journal of the Franklin Institute* 205, no. 5 (1928): 597–623.

7. Alyssa Ney and David Z. Albert, eds. *The Wave Function: Essays on the Metaphysics of Quantum Mechanics* (Oxford: Oxford University Press, 2013).

8. Albert Einstein, Boris Podolsky, and Nathan Rosen, "Can Quantum-Mechanical Description of Physical Reality Be Considered Complete?" *Physical Review* 47 (1935): 777–780.

9. Einstein, Podolsky, and Rosen, "Can Quantum-Mechanical Description," 777.

10. Niels Bohr, "Can Quantum-Mechanical Description of Physical Reality Be Considered Complete?" *Physical Review* 48 (1935): 696–702.

11. For a helpful clarification, see Hans Halvorson and Robert Clifton, "Reconsidering Bohr's Reply to EPR," *Non-locality and Modality*, eds. J. Butterfield and T. Placek (Dordrecht: Kluwer Academic, 2002), 3–18.

12. See John S. Bell, "On the Einstein-Podolsky-Rosen Paradox," *Speakable and Unspeakable in Quantum Mechanics* (Cambridge, UK: Cambridge University Press, 1987 [1964]), 14–21; and John S. Bell, "On the Problem of Hidden Variables in Quantum Mechanics," *Speakable and Unspeakable in Quantum Mechanics*, 1–13.

13. A. Aspect, P. Grangier, and G. Roger, "Experimental Tests of Realistic Theories via Bell's Theorem," *Physical Review Letters* 47 (1981): 460–467; A. Aspect, P. Grangier, and G. Roger, "Experimental Realization of Einstein-Podolsky-Rosen-Bohm *Gedanken-experiment*: A New Violation of Bell's Inequalities," *Physical Review Letters* 48 (1982): 91–94; A. Aspect, J. Dalibard, and G. Roger, "Experimental Tests of Bell's Inequalities Using Time-Varying Analyzers," *Physical Review Letters* 49 (1982): 1804–1807; M. A. Rowe, D. Kielpinski, V. Meyer, C. A. Sackett, W. M. Itano, C. Monroe, and D. J. Wineland, "Experimental Violation of a Bell's Inequality with Efficient Detection." *Nature* 409 (2001): 791–794.

14. John S. Bell, "Bertlmann's Socks and the Nature of Reality," *Speakable and Unspeakable in Quantum Mechanics*, 139–158; Jeffrey Bub, *Interpreting the Quantum World* (Cambridge, UK: Cambridge University Press, 1997); Robert Clifton, ed., *Perspectives on Quantum Reality: Non-Relativistic, Relativistic, and Field-Theoretic* (Dordrecht: Kluwer Academic Publishers, 1996); James T. Cushing and Ernan McMullin, eds., *Philosophical Consequences of Quantum Theory: Reflections on Bell's Theorem* (Notre Dame, IN: University of Notre Dame Press, 1989); Gordon, "A Quantum-Theoretic Argument against Naturalism," 179–214.

15. G. C. Hegerfeldt, "Remark on Causality and Particle Localization." *Physical Review D* 10 (1974): 3320–3321; David Malament, "In Defense of Dogma: Why There Cannot Be a Relativistic Quantum Mechanics of (Localizable) Particles," in *Perspectives on Quantum Reality: Non-Relativistic, Relativistic, and Field-Theoretic*, ed. Robert Clifton (Dordrecht: Kluwer Academic Publishers, 1996), 1–9.

16. Maria Fuwa, Shuntaro Takeda, Marcin Zwierz, Howard Wiseman, and Akira Furusawa, "Experimental Proof of Nonlocal Wavefunction Collapse for a Single Particle Using Homodyne Measurement," *Frontiers in Optics* (Tucson, AZ: *Optical Society of America Technical Digest*, paper FW2C.3, 2014), https://arxiv.org/pdf/1412.7790v1.pdf.

17. Hans Halvorson and Robert Clifton, "No Place for Particles in Relativistic Quantum Theories?" *Philosophy of Science* 69 (2002): 1–28.

18. Erwin Schrödinger, "Die gegenwärtige Situation in der Quantenmechanik," *Naturwissenschaften* 23 (1935): 807–812, 823–828, and 844–849.

19. Guido Bacciagaluppi, "The Role of Decoherence in Quantum Mechanics," *Stanford Encyclopedia of Philosophy*. Edward N. Zalta, ed. (2012), https://plato.stanford.edu/entries/qm-decoherence/

20. Joey Lambert, "The Physics of Superconducting Quantum Interference Devices" (2008) http://www.physics. drexel.edu/~bob/Term_Reports/Joe_Lambert_3.pdf; see also Jim Baggott, *Farewell to Reality: How Modern Physics Has Betrayed the Search for Scientific Truth* (New York: Pegasus Books, 2013), 55.

21. Donald Davidson, "Mental Events," *Essays on Action and Events* (Oxford: Oxford University Press, 1980), 207–224.

22. Jaegwon Kim, *Supervenience and Mind: Selected Philosophical Essays* (Cambridge, UK: Cambridge University Press, 1993), 53–78.

23. Kim, *Supervenience and Mind*, 60.

24. Terence Horgan, "Supervenience and Microphysics," *Pacific Philosophical Quarterly* 63 (1982): 29–43.

25. Horgan, "Supervenience and Microphysics," 36

26. Horgan, "Supervenience and Microphysics," 37.

27. See Arthur Fine, "Hidden Variables, Joint Probability, and the Bell Inequalities," *Physical Review Letters* 48 (1982): 291–295; and Arthur Fine, "Joint Distributions, Quantum Correlations, and Commuting Observables," *Journal of Mathematical Physics* 23 (1982): 1306–1310.

28. See Gordon, "Maxwell-Boltzmann Statistics and the Metaphysics of Modality," 402–407 for a more extensive discussion of related issues.

29. Brian McLaughlin, "Emergence and Supervenience," *Intellectica* 2 (1997): 25–43.

30. Jaegwon Kim, "Making Sense of Emergence," *Philosophical Studies* 95 (1999): 3–36; and Jaegwon Kim, "Being Realistic about Emergence," in *The*

Re-Emergence of Emergence: The Emergentist Hypothesis from Science to Religion, eds. P. Clayton and P. Davies (Oxford: Oxford University Press, 2006), 189–202.

31. Timothy O'Connor and Hong Yu Wong, "The Metaphysics of Emergence," *Nous* 39 (2005): 658–678.

32. McLaughlin, "Emergence and Supervenience," 39.

33. Kim, "Making Sense of Emergence."

34. Barry Loewer, "Review of J. Kim, *Mind in a Physical World*," *Journal of Philosophy* 98, no. 6 (2001): 315–324.

35. Sydney Shoemaker, "Kim on Emergence," *Philosophical Studies* 108 (2002): 53–63.

36. Paul Humphreys, "How Properties Emerge," *Philosophy of Science* 64 (1997): 1–17.

37. Humphreys, "How Properties Emerge."

38. Frederick Kronz and Justin Tiehen, "Emergence and Quantum Mechanics," *Philosophy of Science* 69 (2002): 324–347.

39. Kronz and Tiehen, "Emergence and Quantum Mechanics," 344ff.

40. According to the British emergentists (see Brian McLaughlin, "The Rise and Fall of British Emergentism," in *Emergence or Reduction?* eds. A. Beckermann, J. Kim, and H. Flohr [Berlin: Walter de Gruyter, 1992], 49–93), resultant properties are additive, like force in Newtonian mechanics, whereas emergent properties are not. This *via negativa* is taken as the definition of an emergent property and seems to be motivated by regarding forces as fundamental, then constructing a metaphysical view of emergence by analogy with the way that forces behave.

41. See Paul Teller "Relational Holism and Quantum Mechanics," *British Journal for Philosophy of Science* 37 (1986): 71–81; Paul Teller, "Relativity, Relational Holism, and the Bell Inequalities," in *Philosophical Consequences of Quantum Theory: Reflections on Bell's Theorem*, eds. J. Cushing and E. McMullin (Notre Dame, IN: University of Notre Dame Press, 1989), 208–223.

42. Gordon Fleming, "Nonlocal Properties of Stable Particles," *Physical Review B* 139 (1965): 963–968; Gordon Fleming, "A Manifestly Covariant Description of Arbitrary Dynamical Variables in Relativistic Quantum Mechanics," *Journal of Mathematical Physics* 7 (1966): 1959–1981.

43. Hans Halvorson and Robert Clifton, "No Place for Particles in Relativistic Quantum Theories?"

44. See David Baker, "Against Field Interpretations of Quantum Field Theory," *British Journal for the Philosophy of Science* 60, no. 3 (2009): 585–609.

45. R. Clifton, D. V. Feldman, H. Halvorson, M. L. G. Redhead, and A. Wilce, "Superentangled States," *Physical Review A* 58, no. 1 (1998): 135–145; see also H. Halvorson and R. Clifton, "Generic Bell Correlation between Arbitrary Local Algebras in Quantum Field Theory," *Journal of Mathematical Physics* 41, no. 4 (2000): 1711–1717.

46. Alvin Plantinga, "Law, Cause, and Occasionalism," in *Reason and Faith: Themes from Swinburne*, eds. Michael Bergmann and Jeffrey E. Brower (Oxford: Oxford University Press, 2016), 126–144.

47. For example, Sydney Shoemaker, "Causality and Properties," in *Time and Cause*, ed. Peter van Inwagen (Dordrecht: D. Reidel, 1980), 109–135; Chris Swoyer, "The Nature of Natural Laws," *Australian Journal of Philosophy* 60 (1982): 203–223;

Evan Fales, *Causation and Universals* (London: Routledge, 1990); and Alexander Bird, "The Dispositionalist Conception of Law," *Foundations of Science* 10, no. 4 (2005): 353–370.

48. For example, David Armstrong, *What is a Law of Nature?* (Cambridge, UK: Cambridge University Press, 1983); Fred Dretske, "Laws of Nature," *Philosophy of Science* 44 (1977): 248–268; Michael Tooley, *Causation: A Realist Approach* (Oxford: Clarendon Press, 1987).

49. David Lewis, "New Work for a Theory of Universals," *Australasian Journal of Philosophy* 61 (1983): 166.

50. For example, R. Harré and E. H. Madden, *Causal Powers: A Theory of Natural Necessity* (Oxford: Basil Blackwell, 1975); J. Bigelow and R. Pargetter. *Science and Necessity* (Cambridge, UK: Cambridge University Press, 1990).

51. John S. Bell, *Speakable and Unspeakable in Quantum Mechanics*, 1–21; Michael Redhead, *Incompleteness, Nonlocality and Realism: A Prolegomenon to the Philosophy of Quantum Mechanics* (Oxford: Clarendon Press, 1987), 71–118; Arthur Fine, "Correlations and Physical Locality," in P. Asquith and R. Giere, eds., *PSA 1980*, vol. 2 (East Lansing, MI: Philosophy of Science Association, 1981), 535–562.

52. As an aside, it would be a mistake to think this problem could be solved by appropriating the nonlocal deterministic model associated with the de Broglie-Bohm theory (see Katherine Bedard, "Material Objects in Bohm's Interpretation," *Philosophy of Science* 66 (1999): 221–242; and Michael Dickson, "Are There Material Objects in Bohm's Theory?").

53. David Lewis, *Counterfactuals* (Cambridge, MA: Harvard University Press, 1973); Lewis, "New Work for a Theory of Universals," 343–377; David Lewis, "Humean Supervenience Debugged," *Mind* 103 (1994): 473–490.

54. Plantinga, "Law, Cause, and Occasionalism," 130.

55. For an extended defense of the principle of sufficient reason along these lines, see Alexander Pruss, *The Principle of Sufficient Reason: A Reassessment* (Cambridge, UK: Cambridge University Press, 2006).

56. This section relies on Gordon, "The Necessity of Sufficiency" and Gordon, "Divine Action and the World of Science."

57. John Worrall, "Structural Realism: The Best of Both Worlds?" *Dialectica* 43 (1989): 99–124; Michael Redhead, *From Physics to Metaphysics* (Cambridge, UK: Cambridge University Press, 1995); Tian Yu Cao, *Conceptual Developments of 20th Century Field Theories* (Cambridge, UK: Cambridge University Press, 1997); Tian Yu Cao, "Structural Realism and the Interpretation of Quantum Field Theory," *Synthese* 136 (2003): 3–24; Tian Yu Cao, "Appendix: Ontological Relativity and Fundamentality—Is QFT the Fundamental Theory?" *Synthese* 136 (2003): 25–30; Tian Yu Cao, "Can We Dissolve Physical Entities into Mathematical Structures?" *Synthese* 136 (2003): 57–71.

58. James Ladyman, "What is Structural Realism?" *Studies in the History and Philosophy of Science* 29 (1998): 409–424; Steven French, "Models and Mathematics in Physics: The Role of Group Theory," *From Physics to Philosophy*, ed. J. Butterfield and C. Pagonis (Cambridge, UK: Cambridge University Press, 1999), 187–207.

59. Robert Adams, "Idealism Vindicated," in *Persons: Human and Divine*, eds. Peter van Inwagen and Dean Zimmerman (Oxford: Oxford University Press, 2007), 40.

60. Plantinga, "Law, Cause, and Occasionalism," 137.

61. Alfred Freddoso, "Medieval Aristotelianism and the Case against Secondary Causation in Nature," in *Divine and Human Action: Essays in the Metaphysics of Theism*, ed. Thomas V. Morris (Ithaca, NY: Cornell University Press, 1988), 79–83.

62. Del Ratzsch, "Nomo(theo)logical Necessity," *Faith and Philosophy* 4, no. 4 (1987): 383–402; Plantinga, "Law, Cause, and Occasionalism," 126–144.

63. I arrived at an earlier version of this occasionalist quantum idealism about twenty years ago, but it is encouraging to see a burgeoning interest in and advocacy of Berkeleyan occasionalist idealism by a variety of Christian philosophers and theologians. See, for example, Joshua Farris, S. Mark Hamilton, and James S. Speigel, eds., *Idealism and Christianity, Volume 1: Idealism and Christian Theology* (New York: Bloomsbury Academic, 2016); and Steven B. Cowan and James S. Spiegel, eds., *Idealism and Christianity, Volume 2: Idealism and Christian Philosophy* (New York: Bloomsbury Academic, 2016).

SUGGESTED FURTHER READING

Amstrong, David M. *What is a Law of Nature?* Cambridge, UK: Cambridge University Press, 1983.

Castellani, Elena, ed. *Interpreting Bodies: Classical and Quantum Objects in Modern Physics*. Princeton, NJ: Princeton University Press, 1998.

Davisson, C.J. "Are Electrons Waves?" *Journal of the Franklin Institute* 205, no. 5 (1928): 597–623.

Dretske, Fred. "Laws of Nature." *Philosophy of Science* 44 (1977): 248–268.

Fujita, S. "On the Indistinguishability of Classical Particles." *Foundations of Physics* 21 (1991): 439–457.

Ghirardi, G.C., A. Rimini, and T. Weber. "Unified Dynamics for Microscopic and Macroscopic Systems." *Physical Review D* 34 (1986): 470–491.

Gordon, Bruce L. "Divine Action and the World of Science: What Cosmology and Quantum Physics Teach Us about the Role of Providence in Nature." *Journal of Biblical and Theological Studies* 2, no. 2 (2017): 247–298.

———. "Scientific Explanations Are Not Limited to Natural Causes," In *Problems in Philosophy: An Introduction to the Major Debates on Knowledge, Reality, Values and Government*, edited by Steven B. Cowan. New York: Bloomsbury, forthcoming, 2018.

Gordon, Bruce L., and William A. Dembski, eds. *The Nature of Nature: Examining the Role of Naturalism in Science*. Wilmington, DE: ISI Books, 2011.

Halvorson, Hans. "Reeh-Schlieder Defeats Newton-Wigner: On Alternative Localization Schemes in Relativistic Quantum Field Theory." *Philosophy of Science* 68 (2001): 111–133.

Harré, R., and E.H. Madden. *Causal Powers: A Theory of Natural Necessity*. Hoboken, NJ: Wiley-Blackwell, 1975.

Hegerfeldt, G.C. "Remark on Causality and Particle Localization." *Physical Review D* 10 (1974): 3320–3321.

Kim, Jaegwon. *Supervenience and Mind.* Cambridge, UK: Cambridge University Press, 1993.

Kuhn, Thomas S. *Black-Body Theory and the Quantum Discontinuity, 1894–1912.* Chicago: University of Chicago Press, 1978.

Landsman, N.P. "Between Classical and Quantum." In *Handbook of the Philosophy of Physics*, Part A, edited by Jeremy Butterfield and John Earman. Amsterdam: Elsevier, 2007.

Lo, T. K., and A. Shimony. "Proposed Molecular Test of Local Hidden Variable Theories." *Physical Review A* 23 (1981): 3003–3012.

Maudlin, Tim. "Part and Whole in Quantum Mechanics." In *Interpreting Bodies: Classical and Quantum Objects in Modern Physics*, edited by Elena Castellani. Princeton, NJ: Princeton University Press, 1998.

O'Connor, Timothy, and Hong Yu Wong. "The Metaphysics of Emergence." *Nous* 39 (2005): 658–678.

Pruss, Alexander. *The Principle of Sufficient Reason: A Reassessment.* Cambridge, UK: Cambridge University Press, 2006.

Psillos, Stathis. *Scientific Realism: How Science Tracks Truth.* London: Routledge, 1999.

Putnam, Hilary. "How to Think Quantum-Logically." *Synthese* 29 (1974): 55–61.

Ratzsch, Del. "Nomo(theo)logical Necessity." *Faith and Philosophy* 4, no. 4 (1987): 383–402.

Rea, Michael. *World Without Design: The Ontological Consequences of Naturalism.* Oxford, UK: Oxford University Press, 2002.

Russell, R. J., P. Clayton, K. Wegter-McNelly, and J. Polkinghorne. *Quantum Mechanics: Scientific Perspectives on Divine Action.* South Bend, IN: University of Notre Dame Press, 2001.

Salmon, Wesley. *Causality and Explanation.* New York: Oxford University Press, 1998.

Savellos, Elias, and Ümit Yalçin, eds. *Supervenience: New Essays.* Cambridge, UK: Cambridge University Press, 1995.

Swoyer, Chris. "The Nature of Natural Laws." *Australian Journal of Philosophy* 60 (1982): 203–223.

Tooley, Michael. *Causation: A Realist Approach.* Oxford, UK: Clarendon Press, 1987.

Van Fraassen, Bas C. "The Charybdis of Realism: Epistemological Implications of Bell's Inequality." In *Philosophical Consequences of Quantum Theory: Reflections on Bell's Theorem*, edited by J. Cushing and E. McMullin. South Bend, IN: University of Notre Dame Press, 1989.

Wheeler, John A. "Law without Law." In *Quantum Theory and Measurement*, edited by John A. Wheeler and Wojciech H. Zurek. Princeton, NJ: Princeton University Press, 1983.

Worrall, John. "Structural Realism: The Best of Both Worlds?" *Dialectica* 43 (1989): 99–124.

Chapter 20

Reflections on Christian Physicalism by a Veteran Antiphysicalist

Howard Robinson

When I became a graduate student in 1967, I was not sure whether to work on free will or the mind-body problem. In both cases I was influenced by a desire to defend what I believed to be the Christian understanding of what it is to be a human being, as against what I feared was the nihilistic conception that followed from scientific materialism. In my second term, I fixed on the mind-body problem. I was worried that a materialist account of the human mind would leave the human soul, like Blake's Urizen, drowning in the waters of materialism. (I asked to have Blake's painting on the cover of both my books on physicalism, in 1982 and 2016, but in both cases the publisher said this was not possible!)

When I came to grips with J. J. C. Smart's foundational essay, 'Sensations and brain processes' and, slightly later, with David Armstrong's magisterial *A Materialist Theory of the Mind* I was both shocked and relieved, because the cost of their materialism was to deny that, in being conscious, we had any positive conception of what was in our consciousness; our knowledge of our own lived being was wholly topic neutral. Much as I admired and became good friends with, David Armstrong, this seemed to me to be a very desperate strategy. In the early 70s, thanks mainly to the influence of Rom Harre's classes on Boscovitch, I began to think that the concept of matter was much more problematic than that of consciousness. I also began to work on the nature of universals with a view to relating this to the nature of thought. My consequent belief that human beings must be semantic and not merely syntactic engines was strengthened by John Searle's powerful and lucid Chinese room argument: it was clear that materialist theories were as hopeless about thought as they were about consciousness, and even about materiality itself.

The upshot of all this is that my instinctive reaction to Christian materialism is that it is both an unnecessary sell-out and a stab in the back; if correct, it threatens to render inconsequential everything I had thought worth doing in philosophy. But as it seems obvious to me that materialism is no good, quite apart from religion, what is going on?

Outside of philosophy, but including among many theologians, there is the negative word-magic associated with "dualism." Dualism is associated with the heretical Manichean idea that there two "gods," one of good and the other of evil. It is, less extremely, associated with the Platonic idea that the body is a bad thing, without which in the next life we will be better off. These views are, of course, heretical. Also, an unthought-out, sloganized emphasis on the "incarnational" nature of Christianity is used to shame anyone who is tempted to think Christianity requires dualism.

Among philosophers, I think there are two main concerns. Some Christian philosophers are convinced by the arguments for physicalism, at least to the extent of being impressed by the success of the natural sciences and not wanting to be dependent on a "God of the gaps" claim that inserts religion into the spaces science has not yet conquered. More modestly, there might be the feeling that the truth of the religion should not turn on any particular philosophical doctrine. If physicalism can give an adequate account of human characteristics, then Christianity can accommodate it, as well as it can accommodate any other philosophical anthropology.

The conditional nature of the last sentence is of course important. One reason why a Christian might resist materialism is the belief that materialism involves "the abolition of man," that is, the downgrading or elimination of everything that John McDowell, for example, classifies as "second nature," that is, of all distinctively human forms of psychological life, including all that is normative, so that it eliminates, rather than adequately accommodates, what is special to human nature.

Among protestant analytical philosophers, the tendency toward materialism usually has something to do with the influence of Peter van Inwagen: among Catholics, a certain understanding of Aristotelian hylomorphism is usually to blame.

Van Inwagen's form of materialism seems to rest on his particular style of sparse ontology, and not fundamentally on issues that derive specifically from the philosophy of mind. He affirms that the "only way for a thing to avoid being an abstract object is for it to be a substance."[1] In particular, properties are abstract objects. In fact, van Inwagen treats them like predicates, because he says that they are *true of* the things that *have* them.

> [Certain] properties are, to be sure, mental properties, but that only means that if they are true of or belong to something at a moment, that thing is thinking or feeling at that moment. A parallel definition of "physical property" would be: a property is physical if its being true of something implies that that thing is a physical substance. . . . To call a property physical is not to speak of *its* nature but of the natures of the things it could possibly be true of.[2]

This leads van Inwagen to what one might describe as an ontologically minimal materialism. It is uncontroversial that human beings have or are bodies composed of atoms etc. This, by van Inwagen's definition, makes them physical substances.[3] Properties, being abstract are neither mental nor physical, so, though both mental and physical properties are true of humans, the only substance present is a physical one. I call this position an ontologically minimal materialism because it also rules out standard forms of materialism, such as the identity theory: van Inwagen does not like events or states and thinks that it makes no sense to identify the being true of one property of an object with the being true of another.[4]

It is obvious that most of the metaphysics behind this account is highly contentious, quite independently of how it relates to the mind-body problem. It ignores the fact that the ontological category to which an object belongs, depends on the kind of properties it possesses. Atoms are physical because they possess certain kinds of third-personally available properties, and not ones accessible only from a first-person perspective. It is not that these properties are physical because they belong to substances anteriorly or independently deemed to be physical, rather that something is physical in virtue of possessing the appropriate properties. Something that possesses mental properties, without a reductive understanding of the same (which van Inwagen does not favor) is not, therefore, a straightforwardly physical thing. But this is not the most contentious aspect of his sparse ontology. In Quinean fashion, he seems to deny that there are any truthmakers for the attribution of properties to a substance—the job usually thought to be done by property instances or tropes, which he claims to find unintelligible ideas.[5] So, for van Inwagen, its being true of x that it is F is basic, without there being anything about x that makes this true. I cannot understand how there can fail to be *that about* a thing which makes appropriate the attribution of each of the various predicates true of it. These truthmakers are not abstract objects but spatiotemporally located features of the concrete object.

There is a general point about formal ontology here. There are many philosophers who think that one needs only particulars and universals, and nothing corresponding to property instances or tropes. Armstrong, whose *in re* theory is quite different from van Inwagen's, nevertheless denies that his theory requires no more than universals and particulars. In fact, all theories need a *tertium quid* which expresses the fact that a certain particular and a certain property have come together at a particular time: the mere existence of tallness and Fred is not enough to constitute Fred's being tall. Aristotle, probably, and the Aristotelian tradition certainly, spoke of *individualized* forms as being in this role, and this jargon can be taken to be equivalent to the more modern notion of a property instance. For Plato, the third factor is the *participation* of the individual in the form, and for Armstrong an ontology of states of affairs is introduced to express the idea of the coming together at a particular place and time of the particular and the property. All these ideas involve the presence of the universal in the concrete state of affairs. This is what grounds the true predication of a term to the particular. Even van Inwagen cannot avoid talking of the individual *possessing* the property—again a third factor. That he (and other analytic philosophers) seems to conflate this with a predicate's *being true* of an object no doubt owes something to the Fregean tradition's unfortunate use of "concept to signify a property or universal: concepts *apply* to objects, just as predicates are *true* of them, whereas properties actually *belong to* the objects that *possess* therm.

Without his peculiarly sparse ontology, van Inwagen would not have the resources to describe himself as a materialist, so, in a sense, he is not a materialist according to the usual standards of the mind-body debate, carried on within the normal metaphysical framework.

Catholic flirtation with materialism comes via the conception of the soul as the form of the body. How to understand this expression and whether it is, as Bernard Williams characterizes it, "a polite form of materialism," has been controversial ever since the first commentators on the *De Anima*. It is often presented as a *tertium quid* between materialism and dualism, but usually, in the modern period, with the emphasis that it is not a form of dualism, but a form of "non-reductive naturalism."[6] I have argued elsewhere that Aristotle—and consequently St. Thomas—are firmly dualistic in their theory of the intellect and so of the human soul itself.[7] The problem with hylomorphism in general is that I think that it is complacent to think that it can be preserved as a metaphysical theory independently of its role in scientific explanation, which latter, few if any modern philosophers would want to defend. It was fashionable during the "ordinary language" period of philosophy to say that Aristotle's four causes were not causes in the modern sense, but simply four different forms of explanation, and that the modern sense corresponded to Aristotle's efficient cause. As science concerns only this kind of cause and the others are just different kinds of explanation, science and the general metaphysical scheme cannot come into conflict. This would mean that the distribution of material stuff in the universe could be "closed under physics" and, as long as some explanatory purpose was served by saying *what things are for* (teleological explanation), *what sorts of things they are* (formal cause) and *what things are made of* (material cause) then the Aristotelian categories can retain their validity. I have argued elsewhere that it is important that, for Aristotle and Aquinas, there is no "bottom up" science that definitively describes the distribution of material stuff around the universe; appeal to substantial form is essential for that and that is what gives form metaphysical significance; it actively organizes appropriate matter.[8] And in the case of the human soul, this organizing principle is immaterial. It is possible to accept this last statement—and this is a form of mind-body dualism—but the priority of "top down" explanation that is entailed by a general hylomorphism is incompatible with modern science.[9]

I have said nothing about the vexed question of materialism and the resurrection. Trenton Merricks entertains the idea that intermittent existence is logically possible and that identity across such intermittent existence might simply be stipulated by God. The idea Merricks entertains would seem to imply that God could resurrect two duplicates of you and then decide which one to deem to be you. This does not seem to be a satisfactory solution. It would seem unreasonable to think that complex physical things in general have no identity over and above the appropriate kind of continuity, yet to hold that there could be any content to a divine stipulation. There at least needs to be an "extra fact" available to constitute the identity. The only way I can see how this could be so is if the self is a simple immaterial entity. This is, indeed, Richard Swinburne's view, and he agrees that intermittent existence is possible, but thinks that the sameness is a matter of the bare identity of this simple immaterial substance, not of divine convention. I must admit that I find the notion of intermittent existence problematic and have argued that apparent intermittent existence, as might be said to occur in periods of ordinary unconsciousness, not just at death, can only be reconciled with identity by putting the self fundamentally outside of time.[10] This, of course, also requires that it be nonphysical. Ordinary time is a parameter of the physical world and things that do not operate in just the way standard physical objects behave might stand in a looser relation to it.

These reflections might seem to suggest that I find Christian (or any other kind of) materialism entirely unmotivated. But that is not so. The desire to find a unified conception of the world—a sort of monism—that integrates science seems to me to be deeply natural. The monism that I favor—a form of Berkelian idealism—might seem to do that, but it is not so simple. The problem of embodiment remains, in the form of the difficulty of relating the way certain things are best explained in natural scientific terms with those things that must be explained in terms of reason, norms, and consciousness. John McDowell is not alone in having failed to harmonize what he calls the first and second natures of the universe.

Christian materialism might be motivated, but I remain convinced that, however it is dressed up, all materialisms remain firmly opposed to any spiritual view of the world—whether one construes "spiritual" in an ethical or metaphysical sense. It seems to me that Christian philosophers and theologians should be united in their opposition to all forms of materialism. Christian thinkers should direct their energies to articulating better the metaphysics behind human action and agency, and the nature of human thought and understanding, in ways that free these phenomena from the grasp of a dehumanized or mechanistic science. They should also beware of feeling too obligated to follow the fashions of secular fashion in philosophy and more generally in the intellectual climate. It should also go without saying that this does not involve anything like a direct clash with the hard sciences, as the revival of "first cause" arguments and fine-tuning based versions of design arguments amply show. There is currently a crisis in materialistic thought—as is shown by the growing fashion for panpsychism among even some leading nonreligious philosophers of mind, such as Tom Nagel, Galen Strawson, and David Chalmers. This presents an opportunity for Christian thinkers to put forward positive accounts of human nature that do not simply involve putting oneself outside the mainstream of contemporary thought, as used to be the case. We can prophecy against the age while picking up on leads that current secular thought is handing over for our exploitation!

NOTES

1. Peter van Inwagen, "A Materialist Ontology of the Human Person," in *Persons Human and Divine,* eds. Peter van Inwagen and Dean Zimmerman (Oxford: Oxford University Press, 2007), 208.
2. Van Inwagen, 211–212.
3. Van Inwagen, 209.
4. Van Inwagen, 208–212.
5. Van Inwagen, 202.
6. See, for example, Anthony Kenny, *Aquinas on Mind* (New York: Routledge, 1993) and William Jaworski, *Structure and Metaphysics: How Hylomorphism Solves the Mind-Body Problem* (Oxford: Oxford University Press, 2016).
7. See Howard Robinson, "Aristotelian Dualism," *Oxford Studies in Ancient Philosophy*, 1 (1983): 123–44 and "Form and the Immateriality of the Intellect from Aristotle to Aquinas," in *Aristotle and the Later Tradition*, eds. H. Blumenthal and H. Robinson, *Oxford Studies in Ancient Philosophy, Supplementary Volume* (Oxford: Clarendon Press, 1991), 207–226.
8. Howard Robinson, "Modern hylomorphism and the reality and causal power of structure; a skeptical investigation," *Res Philosophica*, 91 (2014): 203–214.
9. This is strictly true only if one takes science naturalistically. Artifacts, for example, clocks work mechanistically and have parts, such as springs and cogwheels, that only exist because of the contribution they make to the artifact as a whole. If one thought of the universe as a whole as made by God for purposes realized at the macro level and the atoms there as the chosen way of constructing these things, then a sort of reconciliation of science and Aristotelian categories could be made. It is interesting that artifacts are not true—or, at least, paradigm—substances for Aristotle, but are pre-eminent illustrations of the features substances are supposed to possess.
10. Howard Robinson, "The Self and Time," in *Persons, Human and Divine*, 55–83.

Afterword

Gerald O'Collins

This collection challenges not only radical forms of physicalism that deny "conscious awareness, subjective experience, beliefs, and desires," but also more moderate versions that "acknowledge the reality of consciousness, belief, desires, and so on, but identify these with bodily states , processes, or properties that we have as physical, living human animals" (Charles Taliaferro). The whole work shows how philosophy inspired by faith remains healthy and strong in defending appropriately Christian dualism.

Paul Gavrilyuk objects successfully to the insufficiently examined but conventional wisdom about ancient Greek thought endorsing a monolithic soul-body dualism. J. P. Moreland develops his convincing arguments that the appearance and unity of consciousness cannot be explained by strict physicalism. It is the soul, a single "uncomposed" mental substance, that accounts for the unity of consciousness. Felicitous expressions turn up repeatedly in his chapter: for instance, "inseparable parts get their existence and identity from the whole of which they are parts." Moreland rightly insists that "the first-person perspective is not a property persons *have*, it is the thing persons *are*" (emphasis added).

In "Physicalism, Divine Eternality, and Life Everlasting," R. Keith Loftin and R. T. Mullins's treatment of the postresurrection experience of life everlasting encompassing but exceeding our preresurrection experience of life everlasting is helpful. Creation is redeemed, not annihilated and replaced. There is continuity in discontinuity. This chapter argues not only against a timeless life after death but also against a timeless God. Does God experience succession in the divine life and change through the exercise of the divine will? Are souls (understood as immaterial minds) persons, or are they rather what Leftow has called "identity-conferring constituents of persons" (the Thomist view)?

Like other chapters, Angus Menuge's "Physicalism and our Knowledge of God," gives rise to much fruitful thought: for instance, his conclusion that "immaterial souls are more apt to possess divine concepts to know God." As he comments, "this is one way in which we clearly reflect the image of God: we alone among creatures are made so that we can come to know God." Given my remit to write a brief afterword, I am precluded from engaging with substantial discussion on these matters, and for the most part can only record formulations that caught my attention and invited further attention and scrutiny.

Taliaferro's "Physicalism and the Death of Christ" takes up a traditional, more Eastern than Western belief about what happened after the death of Christ: "the harrowing of hell." But, before doing so, he needs to reflect on what happened when the second person of the Trinity became embodied as a finite human being. Taliaferro endorses what he calls "integrative dualism," according to which, "in normal, healthy conditions, the person [the soul] and the body function as a singular unity." In the case of the incarnation, "the person Jesus breathes with his lungs, feels by means of his nervous system and brain, is nourished by food, sees with his eyes, hears with his ears . . . and undergoes events and acts in and as the functional unity of Jesus *qua* embodied person." Thus, such an embodiment involves "taking on fully the many ways in which our mental states are intertwined with our physical states." These remarks are insightful. He does not have much difficulty in dealing with caricatures of dualism that have become common in the aftermath of Gilbert Ryle's dismissal of what Ryle called "the ghost in the machine."

Brandon Rickabaugh's dismantling of objections to bodily resurrection that have been brought against those who hold a mind-body dualism is worth note. He remains somewhat skeptical about the need for bodily identity to ensure that it is the same person who will be gloriously raised from the dead. Yet the "bodily soul view" that he shares with C. Stephen Evans means that the body is "the mode in which we manifest our presence in the world and exercise our agency and relationality." The soul is "both the internal efficient cause and teleological guide for the internal structure *and the development of the body*" (emphasis added). In resurrection the soul will undergo its own transformation, when the whole person, body and soul, is recovered and transformed. Although Rickabaugh never mentions my *bodily, ensouled history*, I like to think that his view is not that far from understanding resurrection as a "reassembly," redemption, and glorification of a such an ensouled, bodily history.

I welcome this valuable work in collaboration that has set itself to expound and develop dualism against various forms of physicalism.

Bibliography

Adams, Marilyn McCord. "The Resurrection of the Body According to Three Medieval Aristotelians: Thomas Aquinas, John Duns Scotus, William Ockham." *Philosophical Topics* 20, no. 2 (1992): 1–33.

———. "Why Bodies as Well as Souls in the Life to Come." In *The Science of Being as Being: Metaphysical Investigations*, edited by Gregory T. Doolan. Washington, DC: The Catholic University of America Press, 2012.

Algra, Keimpe, et al., eds. *The Cambridge History of Hellenistic Philosophy*. Cambridge, UK: Cambridge University Press, 1999.

Althaus, Paul. *The Theology of Martin Luther*. Translated by Robert C. Schultz. Philadelphia: Fortress Press, 1966.

Anselm. *Monologion and Proslogion*. Translated by Thomas Williams. Indianapolis: Hackett, 1995.

Antonietti, Alessandro. "Must Psychologists Be Dualists?" In *Psycho-Physical Dualism Today: An Interdisciplinary Approach*, edited by A. Antonietti, A. Corradini, and E. J. Lowe. Lanham, MD: Rowman & Littlefield, 2008.

Aquinas, Thomas. *Disputed Questions on Spiritual Creatures*. Translated by Mary C. Fitzpatrick and John J. Wellmuth. Milwaukee, WI: Marquette University Press, 1949.

———. *Summa Contra Gentiles: Book Two: Creation*. Translated by James F. Anderson. South Bend, IN: University of Notre Dame Press, 1975.

Aristotle. *Complete Works of Aristotle, Volume 1: The Revised Oxford Translation*, edited by Jonathan Barnes. Princeton, NJ: Princeton University Press, 1984.

Armstrong, David. *What is a Law of Nature?* Cambridge, UK: Cambridge University Press, 1983.

Aspect, A., P. Grangier, and G. Roger. "Experimental Tests of Realistic Theories via Bell's Theorem." *Physical Review Letters* 47 (1981): 460–467.

———. "Experimental Realization of Einstein-Podolsky-Rosen-Bohm *Gedankenexperiment*: A New Violation of Bell's Inequalities." *Physical Review Letters* 48 (1982): 91–94.

Audi, Robert. "Personhood, Embodiment, and Survival Speculations on Life after (Biological) Death." In *Paradise Understood: New Philosophical Essays about Heaven*, edited by T. Ryan Byerly and Eric J. Silverman. Oxford, UK: Oxford University Press, 2017.

Augustine. *Anti-Pelagian Writings*. In *Nicene and Post-Nicene Fathers,* series 1, vol. 5, edited by Philip Schaff. New York: Charles Scribner's Sons, 1902.

———. *Confessions*. Translated by Henry Chadwick. New York: Oxford University Press, 1998.

Ayers, Lewis. "The Soul and the Reading of Scripture: A Note on Henri De Lubac." *Scottish Journal of Theology* 61, no. 2 (2008): 173–190.

Baggott, Jim. *The Meaning of Quantum Theory*. Oxford, UK: Oxford University Press, 1992.

Bailey, Andrew M. "Animalism." *Philosophy Compass* 10 (2015): 867–883.

Baker, Lynne Rudder. "Need A Christian be a Mind/Body Dualist?" *Faith and Philosophy* 12, no. 4 (1995): 489–504.

———. *Persons and Bodies: A Constitution View*. New York: Cambridge University Press, 2000.

———. "Material Persons and the Doctrine of Resurrection." *Faith and Philosophy* 18, no. 2 (2001): 151–167.

———. "Persons and the Metaphysics of Resurrection." *Religious Studies* 43, no. 3 (2007): 333–348.

———. "Christian Materialism in a Scientific Age." *International Journal for Philosophy of Religion* 70, no. 1 (2011): 47–59.

———. *Naturalism and the First-Person Perspective*. New York: Oxford University Press, 2013.

Barnes, Gordon. "Should Property-Dualists Be Substance-Hylomorphists?" *Proceedings of the American Catholic Philosophical Association* 75 (2002): 285–299.

———. "Is Dualism Religiously and Morally Pernicious?" *American Catholic Philosophical Quarterly* 78, no. 1 (2004): 99–106.

Barnett, David. "The Simplicity Intuition and Its Hidden Influence on the Philosophy of Mind." *Noûs* 42 (2008): 308–355.

Barr, James. *Biblical Words for Time*. London: SCM Press, 1962.

Barth, Karl. *Church Dogmatics,* vol. III, no. 2. Translated by Geoffrey W. Bromiley and Thomas F. Torrance. Edinburgh: T. & T. Clark, 1960.

———. *The Resurrection of the Dead*, edited by R. Dale Dawson, translated by H. J. Stenning. Eugene, OR: Wipf & Stock, 2015.

Barthrellos, Demetrios. *Byzantine Christ: Person, Nature, and Will in the Christology of Saint Maximus the Confessor*. Oxford, UK: Oxford University Press, 2005.

Bauckham, Richard. *Bible and Ecology: Rediscovering the Community of Creation*. London: Darton, Longman, & Todd, 2010.

Bayne, Tim. "The Unity of Consciousness and the Split-Brain Syndrome." *The Journal of Philosophy* 105, no. 6 (2008): 277–300.

———. *The Unity of Consciousness*. Oxford, UK: Oxford University Press, 2010.

Bayne, Tim, and David J. Chalmers. "What is the Unity of Consciousness?" In *The Unity of Consciousness: Binding, Integration and Dissociation*, edited by Axel Cleeremans. Oxford, UK: Oxford University Press, 2003.

Beale, G. K. *The Temple and the Church's Mission.* Downers Grove, IL: InterVarsity Press, 2004.

Beauregard, Mario, and Denyse O'Leary. *The Spiritual Brain.* New York: HarperCollins, 2007.

Beeley, Christopher A. *The Unity of Christ: Continuity and Conflict in Patristic Tradition.* New Haven, CT: Yale University Press, 2012.

Behr, J. *Asceticism and Anthropology in Irenaeus and Clement.* New York: Oxford University Press, 2000.

Bennett, M. R., and P. M. S. Hacker. *Philosophical Foundations of Neuroscience.* Malden, MA: Blackwell, 2003.

Berkouwer, G. C. *The Triumph of Grace in the Theology of Karl Barth.* London: Paternoster, 1956.

Bettenson, Henry. *Documents of the Christian Church.* Oxford, UK: Oxford University Press, 1947.

Bielfeldt, Dennis. "Can Western Monotheism Avoid Substance Dualism." *Zygon* 36, no. 1 (2001): 153–177.

Bigelow, John. "Presentism and Properties." *Philosophical Perspectives* 10 (1996): 35–52.

Birch, Bruce E. "Scripture in Ethics: Methodological Issues." In *Dictionary of Scripture and Ethics*, edited by Joel B. Green. Grand Rapids, MI: Baker Academic, 2011.

Bishop, John. *Natural Agency.* Cambridge, UK: Cambridge University Press, 1989.

Blackwell, Ben C. *Christosis: Pauline Soteriology in Light of Deification in Irenaeus and Cyril of Alexandria.* Tübingen, Germany: Mohr Siebeck, 2011.

Blatti, Stephan, and Paul F. Snowdon, eds. *Animalism: New Essays on Persons, Animals, and Identity.* Oxford, UK: Oxford University Press, 2016.

Blocher, Henri. "Yesterday, Today, Forever: Time, Times, Eternity in Biblical Perspective." *Tyndale Bulletin* 52 (2001): 183–202.

Bloom, Paul. *Descartes' Baby: How the Science of Child Development Explains What Makes Us Human.* Cambridge, MA: Basic Books, 2004.

Boethius. *Theological Tractates.* Translated by H. F. Stewart, E. K. R. and, and S. J. Tester. Cambridge, MA: Harvard University Press, 1973.

Bouma, Hessel. "Challenges & Lessons from the Terri Schiavo Case." *Perspectives on Science and Christian Faith* 57, no. 3 (2005): 212–220.

Bourne, Craig. *A Future for Presentism.* Oxford, UK: Oxford University Press, 2006.

Boyd, Jeffrey H. "The Soul as Seen Through Evangelical Eyes, Part I: Mental Health Professionals and 'the soul.'" *Journal of Psychology and Theology* 23, no. 3 (1995): 151–160.

———. "The Soul as Seen Through Evangelical Eyes, Part II: On Use of the Term 'Soul.'" *Journal of Psychology and Theology* 23, no. 3 (1995): 161–170.

———. *Reclaiming the Soul: The Search for Meaning in a Self-Centered Culture.* Cleveland, OH: The Pilgrim Press, 1996.

———. "A History of the Concept of the Soul During the 20th Century." *Journal of Psychology and Theology* 26, no. 1 (1998): 66–82.

Brady, Michael S. *Emotional Insight: The Epistemic Role of Emotional Experience.* Oxford, UK: Oxford University Press, 2013.

Brower, Jeffery E. *Aquinas's Ontology of the Material World: Change, Hylomorphism, and Material Objects*. New York: Oxford University Press, 2014.

Brown, Christopher M. *Thomas Aquinas and the Ship of Theseus*. London: Continuum, 2005.

Brown, David. *Divine Humanity: Kenosis and the Construction of a Christian Theology*. Waco, TX: Baylor University Press, 2011.

Brown, Peter. *The Body and Society: Men, Women and Sexual Renunciation in Early Christianity*. New York: Columbia University Press, 1988.

Brown, Raymond E. *An Introduction to the New Testament*. New York: Doubleday, 1997.

Brown, Warren S., and Brad D. Strawn. *The Physical Nature of Christian Life: Neuroscience, Psychology, and the Church*. New York: Cambridge University Press, 2012.

Brown, Warren S., Nancey Murphy, and H. Newton Malony, eds. *Whatever Happened to the Soul? Scientific and Theological Portraits of Human Nature*. Minneapolis, MN: Fortress Press, 1998.

Brüntrup, Godehard, and Ludwig Jaskolla, eds. *Panpsychism: Contemporary Perspectives*. Oxford, UK: Oxford University Press, 2017.

Bub, Jeffrey. *Interpreting the Quantum World*. Cambridge, UK: Cambridge University Press, 1997.

Burns, J. Patout. "Variations on a Dualist Theme: Augustine on the Body and Soul." In *Interpreting Tradition: The Art of Theological Reflection*, edited by Jane Kopas. Chico, CA: Scholars Press, 1983.

Butterworth, G. W. *Origen: On First Principles*. Gloucester, MA: Peter Smith, 1973.

Bynam, Carolyn Walker. *The Resurrection of the Body in Western Christianity, 200–1336*. New York: Columbia University Press, 1995.

Calhoun, David H. "Human Exceptionalism and the *Imago Dei*: The Tradition of Human Dignity and the Imago Dei." In *Human Dignity in Bioethics: From Worldviews to the Public Square*, edited by Stephen Dilley and Nathan J. Palpant. New York: Routledge, 2013.

Callender, Craig, ed. *The Oxford Handbook of Philosophy of Time*. Oxford, UK: Oxford University Press, 2011.

Cao, Tian Yu. "Structural Realism and the Interpretation of Quantum Field Theory." *Synthese* 136 (2003): 3–24.

Chalmers, David. *The Conscious Mind*. Oxford, UK: Oxford University Press, 1996.

Chan, Michael J., and Brent A. Strawn, eds. *What Kind of God? Collected Essays of Terence E. Fretheim*. Winona Lake, IN: Eisenbrauns, 2015.

Chene, Dennis Des. *Life's Form: Late Aristotelian Conceptions of the Soul*. Ithaca, NY: Cornell University Press, 2000.

Child, J.M. *The Geometrical Lectures of Isaac Barrow*. London: The Open Court Company, 1916.

Chisholm, Roderick. "Mind." In *Handbook of Metaphysics and Ontology*, edited by Hand Burckhardt and Barry Smith. Munich: Philosopia Veril, 1991.

Clayton, Philip. *Mind and Emergence: From Quantum to Consciousness*. Oxford, UK: Oxford University Press, 2004.

Clarke, Samuel. *A Demonstration of the Being and Attributes of God and Other Writings*, edited by Ezio Vailati. Cambridge, UK: Cambridge University Press, 1998.

Coe, John, and Todd Hall. *Psychology in the Spirit: Contours of a Transformational Psychology*. Downers Grover, IL: IVP Academic, 2010.

Cole, Sherwood O. "Don't Disembody Me Just Yet! A Christian Perspective on our Biological Nature." *Journal of Psychology and Christianity* 21, no. 2 (2002): 15–60.

Cooper, John. *Body, Soul and Life Everlasting: Biblical Anthropology and the Monism-Dualism Debate*. Grand Rapids, MI: Eerdmans, 2000.

———. "The Bible and Dualism Once Again: A Reply to Joel B. Green and Nancy Murphy." *Philosophia Christi* 9 (2007): 459–469.

———. "Exaggerated Rumors of Dualism's Demise: A Review Essay on Body, Soul and Human Life." *Philosophia Christi* 11 (2009): 453–464.

———. "The Current Body-Soul Debate: A Case for Holistic Dualism." *Southern Baptist Journal of Theology* 13 (2009): 32–50.

Corcoran, Kevin, ed. *Soul, Body, and Survival*. Ithaca, NY: Cornell University Press, 2001.

———. *Rethinking Human Nature: A Christian Materialist Alternative to the Soul*. Grand Rapids, MI: Baker Books, 2006.

———. "A Critical Appraisal of Francis Beckwith's Defending Life." *Philosophia Christi* 12, no. 2 (2010): 451–457.

Cortez, Marc. "What Does It Mean to Call Karl Barth a 'Christocentric' Theologian?" *Scottish Journal of Theology* 60, no. 2 (2007): 127–143.

———. *Embodied Souls, Ensouled Bodies: An Exercise in Christological Anthropology and Its Significance for the Mind/Body Debate*. London: T&T Clark, 2008.

———. *Christological Anthropology in Historical Perspective: Ancient and Contemporary Approaches to Theological Anthropology*. Grand Rapids, MI: Zondervan, 2016.

———. *ReSourcing Theological Anthropology: A Constructive Account of Humanity in the Light of Christ*. Grand Rapids, MI: Zondervan, 2018.

Craig, William Lane, and Quentin Smith. *Theism, Atheism, and Big Bang Cosmology*. Oxford, UK: Clarendon Press, 1993.

Craig, William Lane, and J. P. Moreland. *Philosophical Foundations for a Christian Worldview*. Downers Grove, IL: InterVarsity Press, 2003.

———, eds. *Naturalism: A Critical Analysis*. New York: Routledge, 2000.

Crease, Robert P., and Charles C. Mann. *The Second Creation: Makers of the Revolution in 20th Century Physics*. New York: Macmillan, 1986.

Crisp, Oliver. *Divinity and Humanity: The Incarnation Reconsidered*. New York: Cambridge University Press, 2007.

———. *God Incarnate: Explorations in Christology*. New York: T & T Clark, 2009.

Crisp, Thomas M., Steven Porter, and Gregg A. Ten Elshof, eds. *Neuroscience and the Soul: The Human Person in Philosophy, Science, and Theology*. Grand Rapids, MI: Eerdmans, 2016.

Cross, Richard. "Philosophy of Mind." In *The Cambridge Companion to Duns Scotus*, edited by Thomas Williams. New York: Cambridge University Press, 2003.

Crouzel, Henri. “Le theme platonicien de ‘véhicule de l’âme’ chez Origène.” *Didaskalia* 7 (1977): 225–237.

Cullman, Oscar. *Immortality of the Soul: Or, Resurrection of the Dead? The Witness of the New Testament*. New York: Macmillan, 1958.

Dainton, Barry. *Time and Space*, 2nd ed. Durham, UK: Acumen, 2010.

Daley, Brian E. “Incorporeality and ‘Divine Sensibility’: The Importance of *De Principiis* 4.4 for Origen’s Theology.” *Studia Patristica* 41 (2006): 139–144.

Dalferth, Ingolf U. *Creatures of Possibility: The Theological Basis of Freedom*. Grand Rapids, MI: Baker Academic, 2016.

Daly, Robert J. *Origen: Treatise on the Passover and Dialogue of Origen with Heraclides and his Fellow Bishops on the Father, the Son, and the Soul*. New York: Paulist Press, 1992.

Davidson, Donald. *Essays on Action and Events*. Oxford, UK: Oxford University Press, 1980.

Davis, Stephen T. *After We Die: Theology, Philosophy, and the Question of Life After Death*. Waco, TX: Baylor University Press, 2015.

Davisson, C. J. “Are Electrons Waves?” *Journal of the Franklin Institute* 205, no. 5 (1928): 597–623.

Dawson, R. Dale. *The Resurrection in Karl Barth*. Farnham, UK: Ashgate, 2007.

Dennett, Daniel C. *The Intentional Stance*. Cambridge, MA: MIT Press, 1990.

———. “Dennett, Daniel C.” In *A Companion to the Philosophy of Mind: Blackwell Companions to Philosophy*, edited by Samuel Guttenplan. Oxford, UK: Basil Blackwell, 1994.

———. *Freedom Evolves*. New York: Viking, 2003.

Descartes, René. *Meditationes de prima philosophia / Meditations on First Philosophy*. Ed. and trans. by George Heffernan. South Bend, IN: University of Notre Dame Press, 1990.

Dilley, Stephen, and Nathan J. Palpant, eds. *Human Dignity in Bioethics: From Worldviews to the Public Square*. New York: Routledge, 2013.

Dillon, John. *The Middle Platonists, 80 B.C. to A.D. 220*. Ithaca, NY: Cornell University Press, 1977.

Dretske, Fred. *Knowledge and the Flow of Information*. Cambridge, MA: MIT Press, 1981.

Drum, Peter. “On the Resurrection of the Body: Discussions with Merricks.” *European Journal for Philosophy of Religion* 3, no. 2 (2011): 451–454.

Duvall, Nancy S. “From Soul to Self and Back Again.” *Journal of Psychology and Theology* 26, no. 1 (1998): 6–15.

Eberl, Jason T. “Aquinas on the Nature of Human Beings.” *Review of Metaphysics* 58, no. 2 (December, 2004): 333–365.

Eccles, John C. *The Human Psyche*. New York: Springer, 1980.

Epsen, Edward. “Eternity is a Present, Time is Its Unwrapping.” *The Heythrop Journal* 51 (2010): 417–429.

Evans, C. Stephen. “Separable Souls: A Defense of ‘Minimal Dualism.’” *Southern Journal of Philosophy* 19, no. 3 (1981): 313–332.

———. “Separable Souls: Dualism, Selfhood, and the Possibility of Life After Death.” *Christian Scholar’s Review* 34 (2005): 333–334.

Evans, Ernest. *Tertullian's Treatise against Praxeas*. London: SPCK, 1948.

Evans, Stephen C., and Brandon Rickabaugh. "What Does it Mean to Be a Bodily Soul?" *Philosophia Christi* 17, no. 2 (2015): 315–330.

Farris, Joshua R. "Bodily-Constituted Persons, Soulish Persons, and the *Imago Dei*: The Problem from a Definite I." *Philosophy and Theology* 28, no. 2 (2016): 455–468.

———. *The Soul of Theological Anthropology: A Cartesian Exploration*. New York: Routledge, 2017.

Farris, Joshua R., and Charles Taliaferro, eds. *The Ashgate Research Companion to Theological Anthropology*. Farnham, UK: Ashgate, 2015.

Fergusson, David. "Barth's Resurrection of the Dead: Further Reflections." *Scottish Journal of Theology* 56, no. 1 (2003): 65–72.

———. *Creation*. Grand Rapids, MI: Eerdmans, 2014.

Feynman, Richard P. *The Feynman Lectures on Physics*, vol. 3*: Quantum Mechanics*. Reading, PA: Addison Wesley, 1971.

Fine, Arthur. "Joint distributions, quantum correlations, and commuting observables." *Journal of Mathematical Physics* 23 (1982): 1306–1310.

Fischer, John Martin, and Mark Ravizza. *Responsibility and Control: A Theory of Moral Responsibility*. Cambridge, UK: Cambridge University Press, 1998.

Flanagan, Owen. *The Science of the Mind*, 2nd ed. Cambridge, MA: MIT Press, 1991.

Fleming, Gordon. "Nonlocal Properties of Stable Particles." *Physical Review B* 139 (1965): 963–968.

———. "A Manifestly Covariant Description of Arbitrary Dynamical Variables in Relativistic Quantum Mechanics." *Journal of Mathematical Physics* 7 (1966): 1959–1981.

Flint, Thomas P. "The Possibilities of Incarnation: Some Radical Molinist Suggestions." *Religious Studies* 37 (2001): 307–320.

Fox, Rory. *Time and Eternity in Mid-Thirteenth-Century Thought*. Oxford, UK: Oxford University Press, 2006.

Ganssle, Gregory E. "Metaphysics, Ethics and Personhood: A Response to Kevin Corcoran." *Faith and Philosophy* 22, no. 23 (2005): 370–376.

Gasparov, Igor. "Emergent Dualism and the Challenge of Vagueness." *Faith and Philosophy* 32, no. 4 (October 2015): 432–438.

Gasser, Georg, ed. *Personal Identity and Resurrection: How Do We Survive our Death?* Farnham, UK: Ashgate, 2010.

Gasser, Georg, and Matthias Stefan, eds. *Personal Identity: Complex or Simple*. Cambridge, UK: Cambridge University Press, 2012.

Gibbard, Allan. "Contingent Identity." *Journal of Philosophical Logic* 4 (1975): 187–221.

Goetz, Stewart, and Charles Taliaferro. "The Prospect of Christian Materialism." *Christian Scholar's Review* 37, no. 3 (2008): 303–321.

———. *A Brief History of the Soul*. Malden, MA: Blackwell Publishing, 2011.

Gordon, Bruce L. "Divine Action and the World of Science: What Cosmology and Quantum Physics Teach Us about the Role of Providence in Nature." *Journal of Biblical and Theological Studies* 2, no. 2 (2017): 247–298.

Gordon, Bruce L., and William A. Dembski. *The Nature of Nature: Examining the Role of Naturalism in Science*. Wilmington, DE: ISI Books, 2011.

Gorman, Michael J. *Abortion and the Early Church: Christian, Jewish & Pagan Attitudes in the Greco-Roman World.* Eugene, OR: Wipf & Stock, 1998.

Green, Joel B. "Eschatology and the Nature of Humans: A Reconsideration of Pertinent Biblical Evidence." *Science and Christian Belief* 14, no. 1 (April, 2002): 48–49.

———, ed. *What About the Soul? Neuroscience and Christian Anthropology.* Nashville, TN: Abingdon Press, 2004.

———. *Body, Soul, and Human Life: The Nature of Humanity in the Bible.* Grand Rapids, MI: Baker Academic, 2008.

Green, Joel B., and Stuart L. Palmer. *In Search of the Soul: Four Views of the Mind-Body Problem.* Downers Grove, IL: InterVarsity Press, 2005.

Grenz, Stanley J. *Theology for the Community of God.* Nashville, TN: Broadman & Holman, 1994.

———. *The Social God and the Relational Self.* Louisville, KY: Westminster, 2001.

Griffin, Carl W., and David L. Paulsen. "Augustine and the Corporeality of God." *Harvard Theological Review* 95, no. 1 (2002): 97–118.

Hall, Todd W. "The Soul or Substantive Self as Experiencer, Actualizer, and Representative in Psychoanalytic Theory." *Journal of Psychology and Theology* 26, no. 1 (1990): 55–65.

Halvorson, Hans, and Robert Clifton. "No Place for Particles in Relativistic Quantum Theories?" *Philosophy of Science* 69 (2002): 1–28.

Hasker, William. *The Emergent Self.* Ithaca, NY: Cornell University Press, 1999.

———. "Materialism and the Resurrection: Are the Prospects Improving?" *European Journal for Philosophy of Religion* 3, no. 1 (2011): 83–103.

Hawley, Katherine. *How Things Persist.* Oxford, UK: Oxford University Press, 2001.

Hebblethwaite, Brian. *Philosophical Theology and Christian Doctrine.* Malden, MA: Blackwell, 2005.

Hegerfeldt, G.C. "Remark on Causality and Particle Localization." *Physical Review D* 10 (1974): 3320–3321.

Hershenov, David B. "Soulless Organisms?" *American Catholic Philosophical Quarterly* 85, no. 3 (2011): 465–482.

Hick, John. "Resurrection Worlds and Bodies." *Mind* 82, no. 327 (July, 1973): 409–412.

———. *Death and Eternal Life.* Louisville, KY: Westminster John Knox, 1994.

Hill, Daniel. "Thomas Aquinas on Human Nature." *British Journal for the History of Philosophy* 15, no. 2 (2007): 383–419.

Hitchcock, Nathan. *Karl Barth and the Resurrection of the Flesh: The Loss of the Body in Participatory Eschatology.* Eugene, OR: Pickwick, 2013.

Hobbel, Arne J. "The *Imago Dei* in the Writings of Origen." *Studia Patristica* 21 (1989): 301–307.

Hoffman, Joshua, and Gary S. Rosenkrantz. *Substance: Its Nature and Existence.* London: Routledge, 1997.

Horgan, Terence. "Supervenience and Microphysics." *Pacific Philosophical Quarterly* 63 (1982): 29–43.

Hudson, Hud. *A Materialist Metaphysics of the Human Person.* Ithaca, NY: Cornell University Press, 2001.

Huffman, Carl. "The Pythagorean Conception of the Soul from Pythagoras to Philolaus." In *Body and Soul in Ancient Philosophy*, edited by Dorothea Frede and Burkhard Reis. Berlin: De Gruyter, 2009.

Hughes, Christopher. "Matter and Actuality in Aquinas." In *Thomas Aquinas: Contemporary Philosophical Perspectives*, edited by Brian Davies. Oxford, UK: Oxford University Press, 2002.

Humphreys, Paul. "How Properties Emerge." *Philosophy of Science* 64 (1997): 1–17.

Hunsinger, George. "Jesus as the Lord of Time According to Karl Barth." *Zeitschrift Für Dialektische Theologie* (2010): 113–127.

Irenaeus. *Against Heresies*. In *The Apostolic Fathers, Justin Martyr, and Irenaeus*, edited by Alexander Roberts, James Donaldson, and A. Cleveland Coxe. *Ante-Nicene Fathers*, vol. 1. Peabody, MA: Hendrickson, 1994.

Jackelen, Antje. *Time and Eternity: The Question of Time in Church, Science, and Theology*. London: Templeton Foundation Press, 2005.

Jackson, Frank. "Epiphenomenal Qualia." *Philosophical Quarterly* 32 (1982): 127–136.

———. "What Mary Didn't Know." *Journal of Philosophy* 83, no. 5 (1986): 291–295.

———. *From Metaphysics to Ethics*. Oxford, UK: Clarendon Press, 1998.

Jaworski, William. *Structure and Metaphysics: How Hylomorphism Solves the Mind-Body Problem.* Oxford, UK: Oxford University Press, 2016.

Jeeves, Malcolm, ed. *From Cells to Souls and Beyond: Changing Portraits of Human Nature.* Grand Rapids, MI: Eerdmans, 2004.

Johnson, Eric L. "Whatever Happened to the Human Soul? A Brief Christian Genealogy of a Psychological Term." *Journal of Psychology and Theology* 26, no. 1 (1998): 16–28.

———. *Foundations for Soul Care: A Christian Psychology Proposal*. Downers Grove, IL: InterVarsity Press, 2007.

Kaiser, Thomas J. "Is DNA the Soul?" *The Aquinas Review* 20 (2015): 90–92.

Kelly, J.N.D. *Early Christian Doctrines.* London: Adam and Charles Black Limited, 1958.

Kelsey, David H. *Eccentric Existence: A Theological Anthropology*. Louisville, KY: Westminster John Knox Press, 2009.

Kenny, Anthony. *Aquinas on Mind.* New York: Routledge, 1993.

Kim, Jaegwon. *Supervenience and Mind: Selected Philosophical Essays.* New York: Cambridge University Press, 1993.

———. *Mind in a Physical World*. Cambridge, MA: MIT Press, 1998.

———. "Making Sense of Emergence." *Philosophical Studies* 95 (1999): 3–36.

———. *Physicalism or Something Near Enough*. Princeton, NJ: Princeton University Press, 2005.

———. *Philosophy of Mind*, 3rd ed. Boulder, CO: Westview Press, 2011.

Kitzler, Petr. "Tertullian's Concept of the Soul and His Corporealistic Ontology." In *Tertullianus Afer: Tertullien et la littérature chrétienne d'Afrique*, *Instrumenta Patristica et Mediaevalia* 70, edited by J. Lagouanere and S. Fialon. Turnhout, Belgium: Brepols, 2015.

Koch, T. "The Challenge of Terri Schiavo: Lessons for Bioethics." *Journal of Medical Ethics* 31 (2005): 376–378.

Koons, Robert C., and George Bealer, eds. *The Waning of Materialism*. New York: Oxford University Press, 2010.

Kvanvig, Jonathan L., ed. *Warrant in Contemporary Epistemology: Essays in Honor of Plantinga's Theory of Knowledge*. Lanham, MD: Rowman & Littlefield, 1996.

Kwan, Kai-Man. *The Rainbow of Experiences, Critical Trust, and God: A Defense of Holistic Empiricism.* New York: Continuum, 2011.

Ladyman, James. "What is Structural Realism?" *Studies in the History and Philosophy of Science* 29 (1998): 409–424.

LaRock, Eric. "Is Consciousness Really a Brain Process?" *International Philosophical Quarterly* 48, no. 2 (2008): 201–229.

———. "From Biological Naturalism to Emergent Substance Dualism." *Philosophia Christi* 15 (2013): 97–118.

Lauber, David. *Barth on the Descent into Hell: God, Atonement, and the Christian Life*. Farnham, UK: Ashgate, 2004.

Lavazza, Andrea, and Howard Robinson, eds. *Contemporary Dualism: A Defense.* New York: Routledge, 2014.

Legrenzi, Paolo, and Carlo Umiltà. *Neuromania: On the Limits of Brain Science.* Trans. Frances Anderson. Oxford, UK: Oxford University Press, 2011.

Lee, Patrick, and Robert P. George. *Body-Self Dualism in Contemporary Ethics and Politics.* Cambridge, UK: Cambridge University Press, 2008.

Le Poidevin, Robin. "The Incarnation: Divine Embodiment and the Divided Mind." *Royal Institute of Philosophy Supplement* 68 (2011): 269–285.

Levering, Matthew. *Jesus and the Demise of Death: Resurrection, Afterlife, and the Fate of the Christian.* Waco, TX: Baylor University Press, 2012.

Levine, Joseph. "Materialism and Qualia: The Explanatory Gap." *Pacific Philosophical Quarterly* 64 (1983): 354–361.

Lewis, Alan E. *Between Cross and Resurrection: A Theology of Holy Saturday.* Grand Rapids, MI: Eerdmans, 2003.

Lewis, David. *Counterfactuals.* Cambridge, MA: Harvard University Press, 1973.

———. *Philosophical Papers,* vol. 1. Oxford, UK: Oxford University Press, 1983.

———. *The Plurality of Worlds.* Oxford, UK: Blackwell, 1986.

Loftin, R. Keith. "On the Metaphysics of Time and Divine Eternality." *Philosophia Christi* 17, no. 1 (2015): 177–187.

Loke, Andrew Ter Ern. *A Kryptic Model of the Incarnation.* Farnham, UK: Ashgate, 2014.

Loose, Jonathan. "Constitution and the Falling Elevator: The Continuing Incompatibility of Materialism and Resurrection Belief." *Philosophia Christi* 14, no. 2 (Winter 2012): 439–449.

Lowe, E. J. *Subjects of Experience.* New York: Cambridge University Press, 1996.

———. *Personal Agency: The Metaphysics of Mind and Action.* Oxford, UK: Oxford University Press, 2008.

———. *More Kinds of Being*, 2nd ed. Malden, MA: Blackwell, 2009.

———. "The Probable Simplicity of Personal Identity." In *Personal Identity: Complex or Simple*, edited by Georg Gasser and Matthias Stefan. Cambridge, UK: Cambridge University Press, 2012.

Lowe, E. J., and Harold W. Noonan. "Substance, Identity, and Time." *Proceedings of the Aristotelian Society* 62 (1988): 61–100.

Machen, J. Gresham. *The Christian View of Man.* New York: Macmillan, 1937.

Madell, Geoffrey. *The Identity of the Self.* Edinburgh: Edinburgh University Press, 1981.

———. *The Essence of the Self.* New York: Routledge, 2015.

Malament, David. "In Defense of Dogma: Why There Cannot Be a Relativistic Quantum Mechanics of (Localizable) Particles." In *Perspectives on Quantum Reality: Non-Relativistic, Relativistic, and Field-Theoretic*, edited by Robert Clifton. Dordrecht, Netherlands: Kluwer Academic Publishers, 1996.

Manley, David, David J. Chalmer, and Ryan Wasserman, eds. *Metametaphysics: New Essays on the Foundations of Ontology.* Oxford, UK: Oxford University Press, 2009.

Marmodoro, Anna, and Jonathan Hill, eds. *The Metaphysics of the Incarnation.* Oxford, UK: Oxford University Press, 2011.

Masai, François. "Les conversions de Saint Augustin et les débuts du spiritualisme en Occident." *Le Moyen Âge* 67 (1961): 1–40.

McCall, Thomas H. *An Invitation to Analytic Theology.* Downers Grove, IL: InterVarsity Press, 2015.

McFarland, Ian A. "Willing Is Not Choosing: Some Anthropological Implications of Dyothelite Christology." *International Journal of Systematic Theology* 9 (2007): 3–23.

McGinn, Colin. *The Mysterious Flame: Conscious Minds in a Material World.* New York: Basic Books, 1999.

McGuckin, John Anthony, ed. *The Westminster Handbook to Patristic Theology.* Louisville, KY: Westminster John Knox Press, 2004.

McKinnon, Neil. "The Endurance/Perdurance Distinction." *Austalasian Journal of Philosophy* 80, no. 3 (September, 2002): 288–306.

McLaughlin, Brian. "Emergence and Supervenience." *Intellectica* 2 (1997): 25–43.

Menuge, Angus J. L. "The Ontological Argument From Reason: Why Compatibilist Accounts of Reasoning Fail." *Philosophia Christi* 13, no. 1 (Summer 2011): 59–74.

———. "Neuroscience, Rationality and Free Will: A Critique of John Searle's Libertarian Naturalism." *Philosophia Christi* 15, no. 1 (Summer 2013): 81–96.

———. "Knowledge of Abstracta: A Challenge to Materialism." *Philosophia Christi* 18, no. 1 (Summer 2016): 7–27.

Merricks, Trenton. "There are no Criteria for Identity Over Time." *Noûs* 32 (1998): 106–124.

———. "The Resurrection of the Body and the Life Everlasting." In *Reason for the Hope Within*, edited by Michael J. Murray. Grand Rapids, MI: Eerdmans, 1999.

———. "Composition as Identity, Mereological Essentialism, and Counterpart Theory." *Australasian Journal of Philosophy* 77 (1999): 192–195.

Middleton, J. Richard. *The Liberating Image.* Grand Rapids, MI: Brazos, 2005.

———. *A New Heaven and a New Earth.* Grand Rapids, MI: Baker Academic, 2014.

Moltmann, Jürgen. *God in Creation: An Ecological Doctrine of Creation.* Edited and translated by Margaret Kohl. London: SCM, 1985.

———. *The Coming of God.* Translated by Margaret Kohl. Minneapolis, MN: Fortress Press, 1996.

Moreland, J. P. "Restoring the Substance of the Soul to Psychology." *Journal of Psychology and Theology* 26, no. 1 (1998): 29–43.

———. "Theories of Individuation: A Reconsideration of Bare Particulars." *Pacific Philosophical Quarterly* 79 (1998): 251–263.

———. "Spiritual Formation and the Nature of the Soul." *Christian Education Journal* 4, no. 2 (2000): 25–43.

———. *Kingdom Triangle: Recover the Christian Mind, Renovate the Soul, Restore the Spirit's Power.* Grand Rapids, MI: Zondervan, 2007.

———. *The Recalcitrant Imago Dei: Human Persons and the Failure of Naturalism.* London: SCM Press, 2009.

———. "Substance Dualism and the Argument from Self-Awareness." *Philosophia Christi* 13, no. 1 (Summer 2011): 21–34.

———. "A Conceptualist Argument for Substance Dualism." *Religious Studies* 49 (March 2013): 35–43.

———. "Exemplification and Constituent Realism: A Clarification and Modest Defense." *Axiomathes* 23 (2013): 247–259.

———. "Tweaking Dallas Willard's Ontology of the Human Person." *Journal of Spiritual Formation & Soul Care* 8, no. 2 (2015): 187–202.

———. "A Critique of and Alternative to Nancey Murphy's Christian Physicalism." *European Journal for Philosophy of Religion* 8, no. 2 (2016): 107–128.

Moreland, James Porter, and Scott B. Rae. *Body & Soul: Human Nature & the Crisis in Ethics.* Downers Grove, IL: InterVarsity Press, 2000.

Morris, Thomas V. *The Logic of God Incarnate.* Ithaca, NY: Cornell University Press, 1986.

Mugg, Joshua, and James T. Turner, Jr. "Why a Bodily Resurrection? The Bodily Resurrection and the Mind/Body Relation." *Journal of Analytic Theology* 5 (2017): 121–144.

Mullins, R. T. "Simply Impossible: A Case Against Divine Simplicity." *Journal of Reformed Theology* 7 (2013): 181–203.

———. "Flint's 'Molinism and the Incarnation' is Too Radical." *Journal of Analytic Theology* 3 (2015): 1–15.

———. "Divine Temporality, the Trinity, and the Charge of Arianism." *Journal of Analytic Theology* 4 (2016): 267–290.

———. *The End of the Timeless God.* Oxford, UK: Oxford University Press, 2016.

Murphy, Francesca Aran, ed. *The Oxford Handbook of Christology.* Oxford, UK: Oxford University Press, 2015.

Murphy, Nancey. *Anglo-American Postmodernity.* Boulder, CO: Westview Press, 1997.

———. *Bodies and Souls, or Spirited Bodies?* New York: Cambridge University Press, 2006.

Murphy, Nancey, and Warren S. Brown. *Did My Neurons Make Me Do It? Philosophical and Neurobiological Perspectives on Moral Responsibility and Free Will.* New York: Oxford University Press, 2007.

Nagel, Thomas. *Mind and Cosmos: Why the Materialist Neo-Darwinian Conception of Nature is Almost Certainly False.* New York: Oxford University Press, 2012.

Newberg, Andrew. *Principles of Neurotheology.* Farnham, UK: Ashgate, 2010.

Ney, Alyssa, and David Z. Albert, eds. *The Wave Function: Essays on the Metaphysics of Quantum Mechanics.* Oxford, UK: Oxford University Press, 2013.

Nichols, Aidan. *The Conversation of Faith and Reason: Modern Catholic Thought from Hermes to Benedict XVI.* Chicago, IL: Liturgy Training Publications, 2011.

Noonan, Harold W. *Personal Identity*, 2nd ed. Abingdon, UK: Routledge, 2003.

Norris, Richard. *The Christological Controversy*. Philadelphia: Fortress Press, 1980.

O'Connell, Robert J. *The Origin of the Soul in St. Augustine's Later Works*. New York: Fordham University Press, 1987.

O'Connor, Timothy, and Hong Yu Wong. "The Metaphysics of Emergence." *Nous* 39 (2005): 658–678.

O'Connor, Timothy, and Jonathan D. Jacobs. "Emergent Individuals and the Resurrection." *European Journal for Philosophy of Religion* 2, no. 2 (2010): 69–88.

Oden, Thomas C. *The Word of Life*. San Francisco, CA: Harper & Row, 1989.

Olson, Eric. *What are We? A Study in Personal Ontology*. Oxford, UK: Oxford University Press, 2007.

Oresme, Nicole. *Le Livre du ciel et du monde*. Edited and translated by Albert D. Menut and Alexander J. Denomy. Madison: University of Wisconsin Press, 1968.

Origen. *Contra Celsum*. Translated by Henry Chadwick. Cambridge, UK: Cambridge University Press, 1965.

Ott, Ludwig. *Fundamentals of Catholic Dogma*. Cork: Mercier Press, 1955.

Pannenberg, Wolfhart. *Systematic Theology*, vol. 3. Grand Rapids, MI: Eerdmans, 1998.

Parfit, Derek. "Personal Identity." In *Personal Identity*. Ed. John Perry. Berkeley: University of California Press, 1975.

Pascal, Blaise. *Pensées*. Translated by A. J. Krailsheimer. New York: Penguin Books, 1966.

Pasnau, Robert. "Olivi on the Metaphysics of Soul." *Medieval Philosophy & Theology* 6, no. 2 (1997): 109–132.

———. *Thomas Aquinas on Human Nature: A Philosophical Study of Summa Theologiae 1a, 75–89*. New York: Cambridge University Press, 2002.

Pasnau, Robert, and Christopher Shields. *The Philosophy of Aquinas*. Boulder, CO: Westview, 2004.

Pelikan, Jaroslav. *The Christian Tradition*, vol. 2. Chicago: University of Chicago Press, 1974.

Penfield, Wilder. *The Mystery of Mind.* Princeton, NJ: Princeton University Press, 1975.

Peters, Ted. "Eschatology: Eternal Now or Cosmic Future?" *Zygon* 36 (2001): 349–356.

Phan, Peter C. "Roman Catholic Theology." In *The Oxford Handbook of Eschatology*, edited by Jerry L. Walls. Oxford, UK: Oxford University Press, 2008, 216.

Pitstick, Lyra. *Christ's Descent into Hell: John Paul II, Joseph Ratzinger, and Hans Urs von Balthasar on the Theology of Holy Saturday*. Grand Rapids, MI: Eerdmans, 2016.

Plantinga, Alvin. *God and Other Minds: A Study of the Rational Justification of Belief in God.* Ithaca, NY: Cornell University Press, 1967.

———. *Warranted Christian Belief.* Oxford, UK: Oxford University Press, 2000.

———. "What is 'Intervention'?" *Theology and Science* 6 (2008): 369–401.

Polkinghorne, John, and Michael Welker, eds. *The End of the World and the Ends of God.* Harrisburg, PA: Trinity Press International, 2000.

Popper, Karl R., and John C. Eccles. *The Self and Its Brain*, 2nd ed. New York: Springer, 1985.

Porter, Steve L. "Sanctification in a New Key: Relieving Evangelical Anxieties Over Spiritual Formation." *Journal of Spiritual Formation & Soul Care* 1, no. 2 (2008): 129–128.

———. "The Gradual Nature of Sanctification." *Themelios* 39, no. 3 (2014): 470–483.

Prestige, G.L. *Fathers and Heretics.* New York: Macmillan, 1940.

Price, Daniel J. *Karl Barth's Anthropology in Light of Modern Thought.* Grand Rapids, MI: William B. Eerdmans, 2002.

Price, Richard. *The Acts of the Council of Constantinople of 553 with Related Texts on the Three Chapters Controversy*, vols. 1 and 2. Liverpool, UK: Liverpool University Press, 2009.

Pruss, Alexander. *The Principle of Sufficient Reason: A Reassessment.* Cambridge, UK: Cambridge University Press, 2006.

Rea, Michael, ed. *Material Constitution: A Reader.* Lanham, MD: Rowman & Littlefield: 1996.

———. *World without Design: The Ontological Consequences of Naturalism.* New York: Oxford University Press, 2002.

Reichenbach, Bruce. *Is Man the Phoenix?* Grand Rapids, MI: Eerdmans, 1983.

Renehan, R. "On the Greek Origins of the Concepts Incorporeality and Immateriality." *Greek, Roman and Byzantine Studies* 21, no. 2 (Summer 1980): 105–138.

Rickabaugh, Brandon. "Eternal Life as Knowledge of God: An Epistemology of Knowledge by Acquaintance and Spiritual Formation." *Journal of Spiritual Formation and Soul Care* 6, no. 2 (2013): 204–228.

Rist, John M. *Augustine.* Cambridge, UK: Cambridge University Press, 1994.

Roberts, Robert C. *Emotions: An Essay in Aid of Moral Psychology.* Cambridge, UK: Cambridge University Press, 2003.

Robinson, Howard. "Aristotelian Dualism." *Oxford Studies in Ancient Philosophy* 1 (1983): 123–144.

———. "A Dualist Perspective on Psychological Development." In *Philosophical Perspectives on Developmental Psychology*, edited by James Russell. Oxford, UK: Basil Blackwell, 1987.

———. "A Dualist Account of Embodiment." In *The Case for Dualism*, edited by John R. Smythies and John Beloff. Charlottesville: University of Virginia Press, 1989.

———. "Modern Hylomorphism and the Reality and Causal Power of Structure; A Skeptical Investigation." *Res Philosophica* 91 (2014): 203–214.

Ross, Susan A. *Anthropology, Engaging Theology: Catholic Perspectives.* Collegeville, MN: Liturgical Press, 2012.

Ryle, Gilbert. *The Concept of Mind*, 3rd ed. London: Penguin Books, 1973.

Sanders, Fred, and Klaus Issler. *Jesus in Trinitarian Perspective.* Nashville, TN: B&H Publishing Group, 2007.

Satel, Sally, and Scott Lilienfel. *Brainwashed: The Seductive Appeal of Mindless Neuroscience.* New York: Basic Books, 2013.

Schaff, Philip. *The Creeds of Christendom,* vol. 3, 3rd ed. New York: Harper & Brothers, 1882.

Schechtman, Marya. *The Constitution of Selves.* Ithaca, NY: Cornell University Press, 2007.

Schopp, Ludwig. *The Immortality of the Soul; The Magnitude of the Soul; On Music; The Advantage of Believing; On Faith in Things Unseen.* Washington, DC: Catholic University of America Press, 1947.

Schurr, George M. "Brunner and Barth on Life after Death." *Journal of Religious Thought* 24, no. 2 (1967): 95–110.

Schwartz, Jeffrey. "A Role for Volition and Attention in the Generation of New Brain Circuitry: Toward a Neurobiology of Mental Force." *Journal of Consciousness Studies*, 6, no. 8–9 (1999): 115–142.

Schwartz, Jeffrey, and Sharon Begley. *The Mind and the Brain: Neuroplasticity and the Power of Mental Force.* New York: HarperCollins, 2002.

Schwarz, Hans. *The Human Being: A Theological Anthropology.* Grand Rapids, MI: Eerdmans, 2013.

Searle, John. *Minds, Brains, and Science.* Cambridge, MA: Harvard University Press, 1984.

Shakespeare. "Romeo and Juliet." In *The New Cambridge Shakespeare*, edited by G. Blakemore Evans. New York: Cambridge University Press, 2003.

Sharpe, Kevin W. "The Incarnation, Soul-Free: Physicalism, Kind Membership, and the Incarnation." *Religious Studies* 53, no. 1 (2017): 117–131.

Shoemaker, David. *Personal Identity and Ethics: A Brief Introduction.* Ontario: Broadview Press, 2009.

———. "Responsibility Without Identity." *The Harvard Review of Philosophy* XVIII (September 2012): 109–132.

Shulman, Robert G. *Brain Imaging: What it Can (and Cannot) Tell Us About Consciousness.* Oxford, UK: Oxford University Press, 2013.

Sider, Theodore. *Four-Dimensionalism.* Oxford, UK: Oxford University Press, 2001.

Singer, Peter. "All Animals Are Equal." *Philosophical Exchange* 1, no. 5 (Summer 1974): 243–257.

Smith, James K. A. *You Are What You Love.* Grand Rapids, MI: Brazos, 2016.

Smith, R. Scott. *In Search of Moral Knowledge: Overcoming the Fact-Value Dichotomy.* Downers Grove, IL: InterVarsity Press, 2014.

———. *Naturalism and our Knowledge of Reality: Testing Religious Truth Claims.* London: Routledge, 2016.

Spira, Andreas, and Ekkehardus Mühlenberg, eds. *Gregorii Nysseni De anima et resurrectione, Opera dogmatica minora*, part III. Leiden, Netherlands: Brill, 2014.

Stanford, Matthew S. *The Biology of Sin: Grace, Hope, and Healing for Those Who Feel Trapped.* Downers Grover, IL: InterVarsity, 2010.

Steinhart, Eric. "The Revision Theory of Resurrection." *Religious Studies* 44 (2008): 66–67.

Stoljar, Daniel. *Physicalism.* New York: Routledge, 2010.

Stroumsa, Gedaliahu. "The Incorporeality of God: Context and Implications of Origen's Position." *Religion* 13 (1983): 345–358.

Stump, Eleonore. *Aquinas*. London: Routledge, 2005.

———. *Wandering in Darkness: Narrative and the Problem of Suffering*. Oxford, UK: Oxford University Press, 2010.

Stump, Eleonore, and Norman Kretzmann, eds. *The Cambridge Companion to Augustine*. Cambridge, UK: Cambridge University Press, 2001.

Suárez, Francisco. *On the Formal Cause of Substance: Metaphysical Disputations XV*. Translated by John Kronen and Jeremiah Reedy. Milwaukee, WI: Marquette University Press, 2000.

Swinburne, Richard. *The Coherence of Theism*. Oxford, UK: Oxford University Press, 1977.

———. *The Evolution of the Soul*. Oxford, UK: Clarendon Press, 1986.

———. "What's So Good About Having a Body?" In *Comparative Theology: Essays for Keith Ward*, edited by Timothy Walter. London: SPCK, 2003.

———. *Mind, Brain & Free Will*. Oxford, UK: Oxford University Press, 2013.

Taliaferro, Charles. *Consciousness and the Mind of God*. New York: Cambridge University Press, 1994.

———. "The Virtues of Embodiment." *Philosophy* 76 (2001): 111–125.

———. "Resurrection." In *The Cambridge Companion to Christian Philosophical Theology*, edited by Charles Taliaferro and Chad Meister. Cambridge, UK: Cambridge University Press, 2010.

Taliaferro, Charles, and Chad Meister. *Contemporary Philosophical Theology*. London: Routledge, 2016.

Taliaferro, Charles, and Alison J. Teply, eds. *Cambridge Platonist Spirituality*. New York: Paulist Press, 2004.

Teller, Paul. "Relational Holism and Quantum Mechanics." *British Journal for Philosophy of Science* 37 (1986): 71–81.

Teske, R. J. "Soul." In *Augustine through the Ages: An Encyclopedia*, edited by Allan D. Fitzerald, et al. Grand Rapids, MI: Eerdmans, 1999.

Thunberg, Lars. *Microcosm and Mediator: The Theological Anthropology of Maximus the Confessor*, 2nd ed. Chicago: Open Court, 1995.

Toner, Patrick. "Hylomorphic Animalism." *Philosophical Studies* 155 (2011): 65–81.

———. "On Hylemorphism and Personal Identity." *European Journal of Philosophy* 19, no. 3 (September, 2011): 454–473.

Torrance, Iain, ed. *Christology After Chalcedon: Severus of Antioch and Sergius the Monophysite*. Norwich, UK: Canterbury Press, 1988.

Torrance, Thomas F. *Incarnation: The Person and Life of Christ*. Downers Grove, IL: InterVarsity Press, 2008.

Turner, James T., Jr. "On the Horns of a Dilemma: Bodily Resurrection or Disembodied Paradise?" *International Journal of Philosophy and Theology* 75, no. 5 (2015): 406–421.

———. "On Two Reasons Christian Theologians Should Reject The Intermediate State." *Journal of Reformed Theology* 11 (2017): 121–139.

———. "Purgatory Puzzles: Moral Perfection and the Parousia." *Journal of Analytic Theology* 5 (2017): 197–219.

Tye, Michael. *Ten Problems of Consciousness: A Representational Theory of the Phenomenal Mind.* Cambridge, MA: Bradford Books, 1995.

Tyndall, John. *Fragments of Science: A Series of Detached Essays, Addresses and Reviews*, vol. 2. New York: D. Appleton and Co., 1896.

Unger, Peter. "The Problem of the Many." *Midwest Studies in Philosophy* 5 (1980): 411–467.

———. *All the Power in the World.* Oxford, UK: Oxford University Press, 2006.

Uttal, William R. *Dualism: The Original Sin of Cognitivism.* Mahwah, NJ: Lawrence Erlbaum Associates, 2004.

Uzgalis, William, ed. *The Correspondence of Samuel Clarke and Anthony Collins, 1707–1708.* Ontario: Broadview Press, 2011.

Vaggione, Richard Paul. *Eunomius: The Extent Works.* New York: Oxford University Press, 1987.

Vallicella, William. "Could a Classical Theist be a Physicalist?" *Faith and Philosophy* 15, no. 2 (1998): 160–180.

van Dyke, Christina. "Human Identity, Immanent Causal Relations, and the Principle of Non-Repeatability: Thomas Aquinas on the Bodily Resurrection." *Religious Studies* 43, no. 4 (December, 2007): 373–394.

———. "Not Properly a Person: The Rational Soul and 'Thomistic Substance Dualism.'" *Faith and Philosophy* 26, no. 2 (2009): 186–204.

van Horn, Luke. "Merricks's Soulless Savior." *Faith and Philosophy* 27, no. 3 (July 2010): 330–341.

van Inwagen, Peter. "The Possibility of Resurrection." *International Journal for Philosophy of Religion* 9, no. 2 (1978): 114–121.

———. *An Essay on Free Will.* Oxford, UK: Oxford University Press, 1983.

———. *Material Beings.* Ithaca, NY: Cornell University Press, 1990.

———. *Metaphysics.* Boulder, CO: Westview Press, 1993.

———. "Dualism and Materialism: Jerusalem and Athens?" *Faith and Philosophy* 12, no. 4 (1995): 475–488.

van Inwagen, Peter, and Dean Zimmerman, eds. *Persons: Human and Divine.* Oxford, UK: Clarendon Press, 2007.

———. "What Do We Refer to When We Say 'I'?" In *The Blackwell Guide to Metaphysics*, edited by Richard M. Gale. Oxford, UK: Blackwell Publishers Ltd, 2002.

Varga, Somogy. *Naturalism, Interpretation, and Mental Disorder.* Oxford, UK: Oxford University Press, 2015.

Voss, Stephen. "Understanding Eternal Life." *Faith and Philosophy* 9, no. 1 (1992): 3–22.

Walls, Jerry. *Heaven: The Logic of Eternal Joy.* New York: Oxford University Press, 2002.

Waltke, Bruce. *Old Testament Theology.* Grand Rapids, MI: Eerdmans, 2007.

Walton, John H. *The Lost World of Genesis One.* Downers Grove, IL: IVP Academic, 2009.

Warfield, B. B. *Two Studies in the History of Doctrine.* New York: The Christian Literature.

Watson, Richard. *Cogito Ergo Sum: The Life of René Descartes*, rev. ed. Boston, MA: David Godine, 2007.

Webster, John, Kathryn Tanner, and Iain Torrance. *The Oxford Handbook of Systematic Theology*. Oxford, UK: Oxford University Press, 2007.

Welker, Michael, ed. *The Depth of the Human Person: A Multidisciplinary Approach*. Grand Rapids, MI: Eerdmans, 2014.

Wellman, Henry. *The Child's Theory of Mind*. Cambridge, MA: MIT Press, 1990.

Wiggins, David. *Identity and Spatio-Temporal Continuity*. Oxford, UK: Blackwell, 1967.

———. "On Being in the Same Place at the Same Time." *The Philosophical Review* 77, no. 1 (1968): 90–95.

Willard, Dallas. "How Concepts Relate the Mind to its Objects: The 'God's Eye View.'" *Philosophia Christi* 1, no. 2 (1999): 5–20.

———. *Renovation of the Heart: Putting on the Character of Christ*. Colorado Springs, CO: NavPress, 2002.

———. *Knowing Christ Today: Why We Can Trust Spiritual Knowledge*. New York: HarperOne, 2009.

Willis, Robert E. *The Ethics of Karl Barth*. Leiden, Netherlands: Brill, 1971.

Wittgenstein, Ludwig. *Philosophical Investigations*, 3rd ed. Translated by G. E. Anscombe. New York: Macmillan, 1953.

Wolfson, Harry. "Immortality and Resurrection in the Philosophy of the Church Fathers." In *Immortality and Resurrection*, edited by Krister Stendahl. New York: Macmillan, 1965.

Worrall, John. "Structural Realism: The Best of Both Worlds?" *Dialectica* 43 (1989): 99–124.

Wright, N. T. *The Resurrection of the Son of God*. Minneapolis, MN: Fortress, 2003.

———. *Surprised by Hope*. New York: HarperOne, 2008.

Yandell, Keith E. "Materialism and Post-Mortem Survival." In *Knowledge and Reality: Essays in Honor of Alvin Plantinga*, edited by Thomas M. Crisp and Matthew Davidson. New York: Springer, 2006.

Yates, Stephen. *Between Death and Resurrection: A Critical Response to Recent Catholic Debate Concerning the Intermediate State*. New York: Bloomsbury, 2017.

Young, Frances. *From Nicaea to Chalcedon: A Guide to the Literature and Background*, 2nd ed. Grand Rapids, MI: Baker Academic, 2000.

Zagzebski, Linda. *Omnisubjectivity: A Defense of a Divine Attribute*. Milwaukee, WI: Marquette University Press, 2013.

Zimmerman, Dean. "Criteria of Identity and the 'Identity Mystics,'" *Erkenntnis* 48 (1998): 281–301.

———. "The Compatibility of Materialism and Survival." *Faith and Philosophy* 16/2 (1999): 194–212.

———. "Bodily Resurrection: The Falling Elevator Model Revisited." In *Personal Identity and Resurrection: How Do We Survive Our Deaths?*, edited by Georg Gasser. Farnham, UK: Ashgate, 2010.

———. "From Experience to Experiencer." In *The Soul Hypothesis*, edited by Mark C. Baker and Stewart Goetz. New York: Continuum, 2011.

Index

About the Contributors

Thomas Atkinson teaches in the Department of Philosophy at the University of Liverpool.

John W. Cooper is professor of philosophical theology at Calvin Theological Seminary.

Marc Cortez is associate professor of theology at Wheaton College.

C. Stephen Evans is University Professor of Philosophy and Humanities at Baylor University.

Joshua R. Farris is assistant professor of theology at Houston Baptist University.

Paul L. Gavrilyuk is Aquinas Chair in Theology and Philosophy at the University of St. Thomas.

Bruce L. Gordon is associate professor of the history and philosophy of science at Houston Baptist University.

Matthew J. Hart is a doctoral candidate at the University of Liverpool.

R. Keith Loftin is associate professor of philosophy and humanities at Scarborough College and Southwestern Baptist Theological Seminary.

Jonathan J. Loose is senior lecturer in philosophy and psychology, and director of learning and teaching at Heythrop College, University of London.

Thomas H. McCall is professor of biblical and systematic theology at Trinity Evangelical Divinity School.

Jason McMartin is associate professor of theology at Rosemead School of Psychology and Talbot School of Theology, Biola University.

Angus Menuge is professor and chair of philosophy at Concordia University.

J. P. Moreland is Distinguished Professor of Philosophy at Talbot School of Theology.

R. T. Mullins is research fellow and director of communications for the Logos Institute for Analytic and Exegetical Theology at the University of St. Andrews.

Gerald O'Collins was, for thirty-three years, professor at the Gregorian University Rome and now teaches at Australian Catholic University and is research fellow at the University of Divinity Melbourne, Australia.

Brandon Rickabaugh is a doctoral candidate at Baylor University.

Howard Robinson is University Professor in Philosophy at Central European University.

R. Scott Smith is professor of philosophy and ethics at Biola University.

Charles Taliaferro is professor of philosophy and chair of the Department of Philosophy at St. Olaf College.

James T. Turner, Jr. teaches Christian studies at Anderson University in South Carolina.

www.ingramcontent.com/pod-product-compliance
Lightning Source LLC
LaVergne TN
LVHW101443200926
848483LV00001B/42
9781498549233